Te Henua Enana
Images and Settlement Patterns in The Marquesas Islands, French Polynesia

Sidsel Nørgaard Millerstrom

Number 67

Contributions of the Archaeological Research Facility

University of California, Berkeley

Archaeological Research Facility
University of California, Berkeley

1st edition 2017
2nd edition 2020

ISBN: 978-0-9982460-5-5 (ebook)
ISBN: 978-0-9982460-6-2 (paper)

Open access available at: http://www.escholarship.org/uc/item/9v38f0zt

This publication is made possible in part by support from the Berkeley Research Impact Initiative sponsored by the University of California, Berkeley, Library.

Cover image of Hatiheu Village, Nuku Hiva, by Sidsel Nørgaard Millerstrom. Cover design: Jerryll Moreno.

Contents

Illustrations

Tables

Foreword

Patrick V. Kirch
Chancellor's Professor Emeritus, University of California, Berkeley

Sidsel Millerstrom is indefatigable as a scholar and researcher of Polynesian rock art. I learned this during several field seasons in the 1990s, working with Sidsel on my archaeological project in the ancient district of Kahikinui, on Maui Island, Hawai'i. On a vast landscape where only a single petroglyph site had been recorded by previous generations of archaeologists, Sidsel discovered more than a dozen new petroglyph panels. She would never be satisfied until she had scoured every potential lava rock face, sometimes going back to the same outcrops repeatedly, knowing that subtle changes in light conditions might suddenly reveal faint glyphs where none had previously been visible. Later, at the well-known petroglyph complex at Nu'u, in adjacent Kaupō district, Sidsel's work revealed that the panels of incised figures extended much farther than anyone had previously realized. Nor was Sidsel merely content just to document these new petroglyph panels, using the noninvasive method she developed of recording them using clear plastic sheets laid over the pecked and incised designs. Sidsel was always seeking to understand the broader archaeological context of these glyphs, testing alternative hypotheses regarding their possible meaning within the ancient Hawaiian culture.

An even greater contribution to Polynesian rock art studies, however, is the present monograph, which in its original form was presented as Sidsel Millerstrom's doctoral dissertation at the University of California, Berkeley. The Marquesas Islands of Eastern Polynesia have long been notable among anthropologists and art historians for their sophisticated plastic arts, ranging from exquisite carvings in bone and wood, to finely woven fans and mats, and especially to the tattooing of human bodies. On all of these media, the traditional Marquesan artisans inscribed intricate geometric, zoomorphic, and anthropomorphic designs reflecting and representing their complex beliefs regarding the world. Largely overlooked, however, was the medium of stone and rock surfaces, to which Marquesan artists also applied their skills. It was this more neglected medium of Marquesan visual representations that Sidsel chose as the focus of her doctoral research.

Beginning in 1984, Sidsel started recording and documenting rock art throughout the Marquesas Islands, initially as a member of a team organized by the former Centre Polynésien des Sciences Humaines, in Pape'ete, Tahiti. She increasingly came to realize, however, that rock art could not effectively be studied in isolation, devoid of the larger archaeological landscape within which it is embedded. This led Sidsel to organize an intensive survey of the Hatiheu Valley, one of the largest and formerly densely settled valleys on the island of Nuku Hiva. Over several field seasons between 1992 and 1997, Sidsel surveyed not just rock art, but the entire settlement pattern of the western portion of Hatiheu Valley. This ambitious project not only involved extensive mapping of stone structural sites, documenting the context of rock art, but also conducting test excavations at selected sites.

The fruits of Sidsel's detailed fieldwork in the Hatiheu Valley are borne out in this monograph. Placing rock art within its settlement pattern context, she shows how the diversity and abundance of glyphs relate to ancient social and political divisions within the valley. Marquesan rock art, along with the other forms of plastic art so well documented by generations of ethnographers and art historians, offered yet another locus for the constant renegotiation of social status and political power in traditional Marquesan society. As Millerstrom so rightly pointed out, while we may never be able to completely comprehend the meaning of every rock art image, by carefully examining the spatial contexts of these images in the larger archaeological landscape, we can gain significant insights into the lives of the ancient Marquesans. Her monograph is a model to which other scholars of rock art, both in Polynesia and elsewhere, should aspire.

Acknowledgments

We were carefully making our way down the mountains of Hatiheu Valley along a steep, narrow, and slippery trail. The heavy rain whipped around us, we were exhausted and hungry. It had been one of those field days when everything got wet, the notepaper crumbled, the pens did not work, and the mosquitoes were fierce. The forest was steaming and the smell of rotting vegetation was overpowering. I glanced at Heidy, my friend and field assistant, walking cautiously ahead of me. Her light pants were caked with a reddish mud from a recent fall and her eyeglasses fogged. Her shoes were soaking wet and so heavy with mud she could barely lift them. I shouted to her: "Heidy, remind me why we are here and what we are doing?" She turned around, flashed one of her beautiful smiles, and answered: "Because we love it!" Heidy was right. Ever since 1984 when I first stepped off Aranui I onto Marquesan soil, with a copy of Linton's Archaeology of the Marquesas Islands securely tucked under my arm, I have loved the islands, the people, the culture, and the many archaeological sites. I know the sites well but never tire of them. There are always new things to discover, more research to do, new information to absorb. Thus, a short archaeological field season turned into a multiyear research project and a long-lasting love affair with the archipelago. This book was possible with the assistance and encouragement of family, friends, and colleagues. It is due to the numerous people and institutions worldwide that I have had the privilege to know and I am grateful to all of you. Subsequent fieldwork was often done in rain but mostly in balmy weather in beautiful lush green valleys.

First and foremost I wish to thank my husband, Ed Millerstrom, without whose loving emotional and financial support this project would have been impossible to complete. Thank you for encouraging me to pursue my passion. My heart-filled thanks extend to our daughters Nikolett and Jessica Gypsy for their assistance and encouragement. Nikolett, with her wonderful humor and common sense helped me keep the household together while she read and commented on my writing. Gypsy was always one of my most inspiring travel companions. Together we explored archaeological sites on many Pacific islands, including the Marquesas. In addition, both Nikolett and Gypsy enthusiastically attended several professional anthropological conferences with me.

My interest in rock art was sparked in 1981 when given the opportunity to work with the late Georgia Lee on her Easter Island Rock Art Project (University of California, Los Angeles). This friendship led to several fieldwork opportunities on Hawaii and Lanai. I am grateful that she took the time to read my dissertation, for the many valuable comments provided, for sharing unpublished research material, and for being a friend and a mentor. Georgia Lee encouraged me to publish this work, and I owe special thanks to her, Frank Morin, Antoinette Padgett, W. Clement Smith, and her staff at the Easter Island Foundation. Their many conversations and comments helped pave the way, and this has been a great benefit.

I am grateful to Christine Hastorf, University of California, Berkeley, for accepting this monograph. Thank you both Christine Hastorf and Nicholas Tripcevich for facilitating the process of getting it published. I could not have done this without the help of Jerryll Moreno, my capable and hardworking editor. I am forever indebted for your insight, your encouragement, and infinite patience. The comments and suggestions from the two anonymous reviewers were especially helpful. Thank you.

The government of French Polynesia receives many thanks for allowing me to do archaeology in the Marquesas. Many thanks go to Maeva Navarro, the previous director of the Département d' Archéologie, Centre Polynésien des Sciences Humaines de Anavaharaiu, currently known as Service de la Culture et du Patrimoine, who invited me to join the Marquesas Rock Art Project. It is the data collected from some of those field seasons that form the basis for this research project. Thank you to Edmundo Edwards, with whom I was fortunate to be able to work; he headed the project and became my working partner from the beginning. His knowledge of Polynesia and his overwhelming generosity and kindness touched everybody.

I am deeply indebted to my former academic advisors for continual critique of this project: Patrick V. Kirch, Margaret W. Conkey, and Joanna G. Williams for reading the original manuscript and providing meaningful remarks and solid academic advice. Throughout my research, they served as a source of inspiration and enlightenment; their many suggestions on behalf of this manuscript are deeply appreciated. Professor Joanna G. Williams of the University of California, Berkeley, Art Department was generous with her time and advice. As always, the towering intellect and academic insight of Professor Patrick V. Kirch informed all of my academic and cultural understandings; his influence is on every page of this book. Thank you for inviting me to join your projects in Kahikinui, Nuu, Maui, and Kalaupapa, Molokai. Your enthusiasm is contagious and always makes it a joy to work with you in the field.

My experience as a graduate student in the Department of Anthropology at University of California, Berkeley, was an enrichment that extends beyond the academic curriculum. After I lost my home and all research material in the Oakland/Berkeley Firestorm in 1991, Professor Margaret W. Conkey gave me free access to her office and allowed me to rummage through her filing cabinets in search of articles to read and copy. Meg possesses the amazing ability to fix personal and academic problems and to place everything in the right perspective. Pat also supplied many new books to rebuild my library. Both Pat and Meg have my greatest respect and admiration.

Thank you to Professors Kent Lightfoot and Ruth Tringham for teaching me to teach and for their wonderful sense of humor. One of my many great experiences as a graduate student was to meet Robert C. Suggs, David W. Steadman, and the late Roger C. Green. Thank you for stimulating conversations, warm smiles, and many bear hugs. Bob Suggs also generously shared unpublished Marquesas manuscripts and maps.

Over the years several people worked with me as field assistants in the Marquesas. Their courage, tenacity, and companionship is much appreciated. These include: Trudy Millerstrom, Heidy Baumgartner-Lesage, Gilles Cordonnier, Seppi Schmidt, Marja Svensson, and Tana Timua. On several occasions Heidy Baumgartner-Lesage took time off from her work to spend the field season with me in the Marquesas. I thank her for her organizational skill, the many wondrous experiences we shared in the islands, and the numerous invitations to live with her and her friends while in Tahiti. I am especially grateful to my Marquesan guide and advisor Tioka Puhetini who patiently provided answers to my many questions and looked out for our safety while in the mountains. I also wish to thank Marimari Kellum, and Julia Nortrap for their friendship, shared knowledge of Polynesia, stimulating conversations, and for numerous stays at their lovely homes.

While working on this research project several Marquesans touched my life in unexpected ways; and I wish to acknowledge some of the numerous people in the islands who assisted me, offered hospitality, and gave me permission to survey on their lands. Thank you Msgr. Chevalier, Bishop of the Marquesas Islands, for his kind permission to excavate on Ototemoui, property of the Catholic Church. Yvonne Katupa, the mayor of the Hatiheu district, who housed and fed me and treated me as a member of her family, receives my sincere gratitude. Her interest in archaeology made it a joy to work in her valley. I wish to acknowledge Lucien Ro'o Kimitete, the late cultural and political leader of the Marquesas Islands and his wife Debora for their friendship and help; he housed us and facilitated our research. Leon Lichte, then the mayor of the island, is the creator of the Community Museum in Vaipaee, the Arboretum in Papuakeikala Valley, the Maritime Museum in Hane, and the Rock Museum in Hokatu. The Rock Museum contains, in addition to various geological specimens, plaster casts of some of the island's many petroglyph panels. In 2003, when the museum was inaugurated, I had the honor of cutting the flower garland, an act that officially opened the museum. Benjamin Teikihuavanaka, the manager of the Arboretum, and Leon Lichtle collected and prepared an impressive array of Marquesan wood species for the comparative collection now housed in the Oceanic Archaeology Laboratory (OAL), Archaeological Research Facilities, Department of Anthropology, University of California, Berkeley. Edward and Diane Stasack who shared all their unpublished rock art material from the Hawaiian Islands, and Gerard O'Regan who kept me informed on New Zealand rock art research, receive my gratitude as well.

Acknowledgements to my OAL and U. C. Berkeley friends and colleagues Pia Anderson, Scarlett Chiu, James Coil, Lisa Holm, Jenny Kahn, Kathy Kawelu, Dana Lepowsky, Peter Mills, Erika Radewagen, Julie Endicott Taomia, and Marshall I. Weisler for reading and commenting on research papers and earlier drafts of my dissertation. I am especially grateful to Pia Anderson for stimulating conversations, sources of inspiration, suggestions, and ideas. Without their encouragement and support, the Ph.D. educational process would have been a lonely road to travel.

The guidance and technical support provided by my friends An Le, Ben Dubois and Jutta Gutenhauser. Jutta taught me time saving features in my word processing software program that could be considered lifesavers. When my computer crashed both An and Ben rescued my data on more than one occasion, and I thank them for retrieving my computer-generated data and the hours spent helping me to scan images and to format the manuscript. Many thanks go to Raul Paoa Ika and Gilles Cordonnier who taught me about illustrations.

I am fortunate to have many friends and family members around the world that encouraged and helped me during this project, and I wish to acknowledge David Addison, Tricia Allen, Henriette Andersen, Wanda Isdahl, Kathy Bolen, Maurice Bonet-Katupa, Andrea Berseman, the late Jean-Louis Candelot and his family, El Casella, Pascal Sellier, Eric Conte, the late Alan Davis-Drake, the late Jeffrey Dhyne, Mark Eddowes, Tana Timua, Fredric, Pernilla, Tess, and Sebastian Fouassier, Dyane Frig, Mary Gleasen, Claire Phillips, Lori D. Hager, John Holson, Solomon Kailihiwa, Marimari Kellum, her father the late Medford Kellum, the late Jan Homsen, Marit and Marius Holmsen, Kirsten Hansen, Carol S. Ivory, Rebecca Gonzalez Lauck, Jimmie Lovell, Kendra Marcus, Netti Martinez, Jan McHague, Diego and the late Oscar Navarro, Per, Sidsel, and Per Christian Nørgaard, Camilla Schnekenburger, Nils Kr. Wiig, Julia Nortrap, Sharyn Jones O'Day, Pierre and Marie-Noëlle Ottino-Garanger, Sherry Pierce Parrish, Barry Rolett, Shareef Salaam, Joan Seaver Kurze, Jean Shelsher, Tanya Smith, Judy Stevenson, Robin Stephenson, Jo Anne Van Tilburg, and John Verano.

This research was in part supported by grants from the following institutions: Department of Anthropology, University of California, Berkeley; the Stahl Endowment and Robert H. Lowie Graduate Scholarships, University of California, Berkeley; the Rock Art Archive, University of California, Los Angeles; the Kon-Tiki Foundation, Oslo, Norway; the Richard Gump Research Stations (# 54); and the Eastern Pacific Foundation. I am thankful to the many people who were instrumental in awarding me these grants.

Sincerely,

Sidsel N. Millerstrom

Chapter 1

Introduction

Archaeology is not what you find, it's what you find out.

— David Hurst Thomas

This book is the culmination of many years of fieldwork and data collection in French Polynesia, the genesis of which was my participation in Georgia Lee's large-scale, long-term rock art project on Rapa Nui (Easter Island) in 1983 and excavation with Linda King in 1984 (Lee 1992). At the same time, research on the Rapa Nui *moai* (monolithic sculptures) was also in progress (Van Tilburg 1994). It soon became obvious that similar research was lacking in the Marquesas Islands and that the distribution and cultural context of petroglyphs, as well as the stone and wood sculptures from the Marquesas, were largely unknown.

My interest in the Marquesas coincided with a project under the direction of Maeva Navarro, then Director of the Departément d'Archeologie du Centre Polynesien des Sciences Humaines, Te Anavarau (CPSH). This government organization is presently known as Service de la Culture et du Patrimoine (hereafter SCP). In 1984, Edmundo Edwards, then Chief Archaeologist with the department headed this long-term project. I was part of the project from the beginning, and the project progressed through several stages during the subsequent five years. Our initial aim was to establish an inventory of petroglyphs and stone sculptures, to estimate the extent of rock art sites in the archipelago, and to assess the carvings' state of preservation. Interpretation naturally followed (e.g., Edwards 1989; Edwards and Millerstrom 1995; Millerstrom 1990, 1997a, 2005, 2006a; Millerstrom and Edwards 1998).The Marquesas Rock Art Project was the first archaeological survey in French Polynesia to focus specifically on rock art. Because of the close connection between rock art and the tiki, carved anthropomorphic sculptures, they were also included in the project. In many respects, we covered previously uncharted territories. Between 1984 and 1989 we amassed a large corpus of survey-generated data from five of the six presently inhabited islands (Figure 1 [all figures are provided in Appendix A]), and it became clear that the carvings were not randomly located. In general, the island-wide fieldwork indicates that, in some valleys, the majority of the rock art was associated with *tohua/tohua koina* (tribal ceremonial places) and *meae/ahu* (sacred sites).

Overall, the Marquesan motif repertoire is limited; it is repetitive and remarkably homogeneous, suggesting that Marquesan archaeological art embodied a collective belief system and that there had been a significant amount of interisland contact (Millerstrom 1990, 1997a). The rock art, however, lacked cultural context, and spatial and temporal diversity remained undocumented. After 1989, and because the majority of rock art sites seemed to be located on Nuku Hiva, the north coast of the island was the initial focus, particularly Hatiheu Valley (Figure 2). I also explored small neighboring valleys, such as Anaho, Haatuatua, Maatea, and Haatavea, and a group of us mapped a large section of Kahuvai, the southwest area of Hatiheu Valley. On Fatu Hiva, I documented one site in the Omoa Valley but did not survey the entire valley. These sites were not included in the Marquesan Rock Art Project.

Research Strategies, Objectives, and Aims

Why is rock art important? Rock art is a global phenomenon, and it is a rich and complex part of the cultural heritage. It is important to indigenous people in various parts of the world because it gives a sense of pride, it expresses ideology, and it is a source of cultural tourism. For example, for some indigenous people such as the Aboriginal Australians, rock art is used to communicate, to record events, to educate, and to connect people to their land and their ancestors. In the Marquesas several villages painted rock art images on walls of schools, restaurants, and administrative buildings. In Hatiheu Valley images are part of decoration on a bridge, motifs are painted on sarongs and tourists regularly visit rock art sites. The Marquesan images became part of tourism and popular culture because those of us that were part of the Marquesan Rock Art Project invited the Marquesans, and especially the school children, to the sites, and we shared the collected information. The next step, of course, is to help make the interpretation part of their cultural history.

Between 1992 and 1997, I surveyed, test excavated, recorded, and mapped a section between Puhioho and Vaiuua rivers in the western part of Hatiheu Valley. The main reason I selected the western section to do my dissertation fieldwork was that numerous rock art sites were associated with architecture. In addition, most of the area was uninhabited at the time. Since I often worked alone, and the mayor of Hatiheu region, Yvonne Katupa, was interested in archaeology, she facilitated much of my fieldwork and looked after my safety.

There were four main objectives within one major research question: how can archaeology be applied to understand interspatial organization of rock images within structures, individual sites, the larger community, and between regions? With that as a key challenge, I investigated: 1) how the images were temporally placed within a cultural context, such as how they are associated with architecture, other cultural features, and the landscape; 2) their links to specific social groups within the society; 3) the changes in image types and their pattern of placement over time; and 4) how the Marquesan rock images were related to other Polynesian islands or island groups.

The research was approached from a holistic standpoint and has drawn on anthropological, ethnohistoric, ethnographic, and linguistic materials. My overall theoretical position is that the images are a form of visual communication that conveys information between individuals, social groups, and communities. This is not a new approach; particularly Wobst (1977) and Conkey (1978), among others, have proposed such viewpoints.

Visual images are not randomly pecked, incised, or painted. Rather, they are part of symbolic systems that carry layers of meanings according to their particular cultural contexts, actors, and intentionalities (e.g., Bahn and Vertut 1988; Conkey 1980; Leroi-Gourhan 1965; Marshack 1972, 1997). Overall, I assume that the images, as the result of social practices, bring into existence or constitute aspects of a range of worldviews, culturally specific attitudes, or ideologies. Relevant questions include whether or not they are highly visible or hidden from view. Did they contain meanings that prompted certain sociopolitical and religious rules that aided in the formation of hierarchical social structures? Did everyone in the population participate in rites where the images were used, or were they reserved for certain members of the population, such as the chiefly class?

After working in the Marquesan Archipelago for several years, I came to view the repetitive symbols as having been in the realm of the chiefly class and their relationship with their gods. These were means to gain and retain social control and served to legitimize and validate the elevated position of the chiefly class. Perhaps the images conceptualized worldviews and helped to create and maintain social inequality.

For Earle (1990:74), the function of an art style and how it develops within material culture is a critical feature of social complexity. Analyzing the iconography of both Hawaiian and Olmec (Mesoamerica) chiefdoms, he showed how the organization of an elite style, interwoven in all sociopolitical structures, empowered leaders and "legitimized systems of inequality and control" (Earle 1990:73). To quote Earle (1990:76), "elite style functioned in the complex chiefdoms of Hawaii to identify a ruling aristocracy and to link it to the gods. The lesson was unmistakable—chiefs were divine and their rule is part of a natural order in the universe."

For instance, the unique crescent-shaped style in elite Hawaiian clothing, such as cloaks and feathered helmets, symbolized social position and served to organize people hierarchically. It is within this theoretical background that I will interpret Marquesan archaeological art.

Issues Related to the Study of Rock Art

That which is collectively referred to as "rock art" has long captured the interest of both social scientists and the general public. However, traditionally the study of prehistoric images, or archaeological art, is in North America—in contrast to the studies of lithic tools and ceramics—academically marginalized in the literature. Despite changes "in our intellectual, social, and political context since the early 1980s" (Conkey 1997:343), rarely are the results of studies of archaeological art accorded the same significance as other prehistoric remains. Far too often, "rock art" is considered apart from mainstream archaeology. Among the main problems is the lack of interpretation and theoretical development, little or no settlement-pattern context, and no material cultural assemblage or temporal association (Millerstrom and Kirch 2002, 2004). However, recent and much welcome academic publications with a focus on ritual practices, sacred landscapes, social transformation, and political economy are, perhaps, part of changing this lack of archaeological excavation and settlement pattern context (David 2010, 2012; Wright 2014; Gilette et al. 2015; McCoy and Codlin 2015).

Rock art is also referred to as petroglyphs and pictographs, as well as archaeological art, pictures, and images. Conkey (1997:358) calls the complex phenomenon of image-making "visual culture." Image-making that involves cognitive processes is among the most difficult part of past human activities to understand (Hawkes 1954). It may, for instance, express a culturally specific belief system, a sacred landscape, political strategies, an ideology, or cosmological structure through the use of arbitrary symbols. Some argue that it is not even possible to consider what people were thinking in the past; their actions are reflected in the material (Binford 1982). Others, however, believe the material record is, in part, the results of a peoples' thought processes (Trigger 1998).

Rock art is a worldwide phenomenon; yet in North America the relationship between anthropology and the study of rock art has been limited (Whitley and Loendorf 1994). Traditionally, the study was in the realm of art historians and rock art enthusiasts or amateurs. The reason, I believe, was mainly due to the lack of academic interest within archaeology that resulted in a body of descriptive work, or short archaeological site reports with little interpretation; depictions of the images were frequently inaccurate or incomplete.

For archaeologists, rock art was problematic because it was considered a conservative force that resisted change (Ashmore and Sharer 1996:187–188). The culture-historic theoretical perspective often conceptualized rock art as a static component of dynamic cultures that remained temporally and spatially unchanged and seen as a "self-contained subsystem," with a "persistent artistic tradition" (Conkey 1981; Conkey et al. 1997). For some, rock art was considered to lie within the realm of paleopsychology and not archaeology (Fritz 1978).

Above all, a primary reason for the peripheral academic status of rock art research after the 1950s is that the images were considered nondatable. In fact, interest in prehistoric aesthetics declined after the advent of radiocarbon dating (Earle 1994). This peripheral status was reinforced by the assumption that hunter-gatherer societies were viewed as static, uncomplicated, and lacking in symbolic elaboration (e.g., Bahn and Vertut 1988; Conkey 1984, 2001; Ouzman 1998; Whitley and Loendorf 1994). Only after the archaeological paradigm expanded in the early 1980s was a wider view promoted about what we might know about human history (e.g., Hodder 1986). For example, the notion that rock art may have multiple alternative interpretations has only been advanced since the early 1990s. Tilley (1991) attempted to demonstrate this in his work at Nämforsen, Sweden, where he viewed the different panels at the Nämforsen rock art site as different pages in a book that could be read as a "text." Tilley looked at these images from three different theoretical and analytical perspectives—structuralist, hermeneutic, and structural Marxist. Do we know more about the meaning of the Nämforsen images after Tilley's analysis? Probably not, but this was not Tilley's intent. His aim was to demonstrate that rock images have several meanings, thus the results were intentionally left inconclusive. Tilley (1991:173) wrote: "I believe that any attempt to establish a totalizing framework, accommodating as many observations as possible with, ideally nothing left out, is doomed to failure. There is always a surplus of meaning."

One basic problem is that we do not know if the images should be read individually or as a group. Nor do we know if all the images were made at the same time (Bahn 1998; Ouzman 1998:32). Harsh critique came from Malmer (1993:113–118), a Swedish archaeologist who named his review of Tilley's book "A Poet's Ambiguity." Malmer claimed that the book centered on Tilley, not on Nämforsen (Malmer 1993:113). Parts of Malmer's critique included Tilley's linguistic misunderstandings in regard to some of the Swedish written sources, problems with statistical probability, and confusion as to the gender of the depicted elk. Because Tilley failed to place the site in a broader regional context, his interpretations were, according to Malmer (1993:117), "imaginative and subjective." Clottes (1993:122) viewed Tilley's work as creative, although the arguments used to build his structuralist case were "a bit thin."

While post-processual critiques raised important issues concerning the narrow adaptationalist stance of processual archaeology (e.g., Hodder 1984; Shanks and Tilley 1987; Tilley 1991), they did not move beyond "the critiques into the more difficult ground of archaeological practice" (Earle 1994:vii). Despite new theoretical approaches to the interpretation of rock art, in addition to the more technical and analytical advances developed in North America in the last decades, the studies of petroglyphs in North America generally remains outside of "mainstream" archaeological inquiry (e.g., Whitley and Loendorf 1994. However, electronic technology in the form of Archaeology of Time Travel is part of making rock art relevant to museum visitors. Take for example the case at Österlens Museum in Simrishamn, Sweden. Digital Heritage Forum, the Faculties of Humanities and Theology, Lund University, Sweden, developed a rock art exhibit *Virtual Rock Arts Experience* (2014) where the visitors can experience a Bronze Age ceremony that may have taken place more than 3400 year ago. Presently Digital Heritage is working on a digital version of events involving a Middle Age church in Dalby, southern Sweden. The project is interdisciplinary and include historical archaeology, theology, architecture, technology, and art history (http://projekt.ht.lu.se/digital-heritage/projects/petroglyphics-virtual-rock-arts-experiences/).

Rock Art vs. Rock Images, or Does it Matter?

Many have discussed the meaning of art and how labeling "art" limits the way we think about it and the questions we ask of it (Conkey 1983; Forge 1991). Interpretive progress comes from asking different sets of questions (Renfrew and Bahn 1991:150). The term *rock art* is convenient and ingrained in our vocabulary. However, for some people, it has a negative connotation. Rock art is a concept and category that has its own specific connotations and biases. How "art" is defined is based on Western value judgments (Forge 1991:39–44; Kaeppler 1979:180–191).

The concept of art in nineteenth- and twentieth-century Europe implies something beautiful, skillfully made, and nonfunctional (Anderson 1990). Westerners often value art because it is labor intensive in terms of our notion of time. In addition, art is often tied to economic issues. Thus David and Wilson (2002:vii) suggest we should hyphenate rock-art to distinguish such practices from the Western artistic output, which is closely tied to a market economy.

While all cultures produce what we would consider to be art, many societies have no concept of it. Even the term *art*, which would coincide with a Western point of view, is absent (Mills 1971). In addition, the concept of art shifts according to fashion and ideology (Layton 1991). For instance, our changing views of hunter-gatherer societies have profoundly affected how we study visual imagery of such groups (Conkey 1984:253–276).

We often assume that other societies share our values and perspectives of imagery, but ethnographic studies illustrate that visual perception is culture specific (e.g., DeBoer 1991; Forge 1970, 1973; Gombrich 1961).

Gombrich (1961:82), for example, wrote: "What a picture means to the viewer is strongly dependent on his [sic] past experience and knowledge. In this respect the visual image is not a mere representation of 'reality' but a 'symbolic system.'" This statement is particularly relevant in relation to image-making in small-scale societies. Consider Forge's study of Abelam flat paintings. The people of this New Guinea group "see" images with multiple outlines. Single outlines are just not "seen" (Forge 1970). Similarly, DeBoer (1991) has written extensively on the decorative style of the modern Shipibo-Conibo group of the Ucayali Basin in South America. He observed, in the government-sponsored schools, Shipibo children of the Amazon doodling on their notebooks in Shipibo-specific designs.

Thus, identifying the limits of art is a thorny issue. I therefore use the term *rock art* with reservation. While some of the Marquesan material arts, such as carvings in bone, wood, and stone, reflect a great deal of aesthetic consideration, the Marquesan petroglyphs embody a belief system that is not, as I demonstrate in

Chapter 4, "art" in the Western sense. I therefore prefer to use the terms *petroglyphs* and *pictographs*. The term *archaeological art* is used here to refer to both petroglyphs and stone sculptures.

Interpreting the Social Landscape Through Settlement Pattern Archaeology

The interest today in social landscapes has its roots in settlement pattern archaeology. As a "landscape" is both constructed and conceptualized, it may embrace sociosymbolic dimensions and ideological, economic, and political perspectives (Ashmore and Knapp 1999). And it may be understood to mean the distribution of any human traces both in and between the rock art sites. As Asmore and Knapp (1991:1) state: a landscape forms the backdrop against which archaeological remains are plotted. For others a landscape is a usefully ambiguous term that is more than an archaeological backdrop in that it involves human behavior and how human actions change and modify the environment (Gosden and Head 1994).

As a general theoretical framework, a settlement pattern approach provides an appropriate methodology for the systematic study of a prehistoric peoples' ways of life and the processes that led to cultural changes. Organization of space, as social scientists have amply demonstrated, is sociopolitically, economically, symbolically, and environmentally constructed (e.g., Allen 2010, 2010; Bourdieu 1973; Chang 1967; Donley 1982; Green et al. 1967; Kirch and Kelly 1975; Willey 1953). Trigger (1967a, 1967b) argued that settlement pattern archaeology could be analyzed at three independent levels: the individual structure, the local settlement, and the distribution of settlements within a region. These levels correspond to the micro, semimicro, and macro levels within British spatial analysis (Clarke 1977). Kirch (1989a:40) delineated four different approaches to the spatial distribution and organization of sites in Polynesian archaeology: 1) spatial distribution of sites as indicators of the ecological adaptation of the population to local environmental conditions and constraints; 2) social dimensions of settlement patterns; 3) reconstruction of economic and political structures, and 4) examining the semiotic value of space in ancient Polynesia.

Settlement pattern analysis as a method of research was perhaps first fully articulated by Chang (1967) and then Clarke (1977). Roger C. Green was inspired to conduct a settlement pattern survey in the Opunohu Valley, Moorea, French Polynesia (Green 1961; Green et al. 1967) and later in Western Samoa (Green and Davidson 1969, 1974). Settlement pattern research has since been a major component of archaeological inquiry in the Pacific that is reflected in archaeological theoretical orientations (e.g., Bellwood 1972; Jennings et al. 1982; Kellum-Ottino 1971; Kirch and Kelly 1975; McCoy 1976; Ottino 1985; Weisler and Kirch 1985).

Traditionally, Polynesian archaeologists interested in settlement perspectives concentrated on monumental stone architecture (e.g., Emory 1928, 1933, 1943; Linton 1925; Stokes 1921). This shift from focus on issues related to ceremonial structures to large regions as units of analysis was, at the time, revolutionary.

Due to cultural resource management work, most Polynesian settlement pattern research has taken place in the Hawaiian Islands (Kirch and Weisler 1994). The earliest settlement pattern survey focused on cultural adaptation to the contrasting ecosystems, such as windward vs. leeward valleys. Settlement variations could also be detected within territorial units or *ahupuaa* (Tuggle and Griffin 1973; Kirch and Kelly 1975; Green 1980; Kirch 1997a, 1997b, 2014). Variations in time and space could be noticed in prehistoric Hawaiian domestic architecture as work in Halawa, Lapakahi, and Makaha demonstrated (Tuggle and Griffin 1973; Kirch and Kelly 1975; Green 1980). During these multiyear, large-scale settlement pattern surveys, petroglyphs were noted and recorded, but cultural contexts were lacking. Few petroglyph sites were excavated. A notable exception is the excavation of a petroglyph site in a lava tube in the Hilina Pali area. Fortunately, a midden deposit had covered some of the figures, and archaeologists were able to determine that the petroglyphs were made after A.D. 1600 (Cleghorn 1980). For recent comprehensive and long-term settlement study at Kahikinui, Maui, see Kirch (2014).

Methodology

This book concentrates on the sociopolitical dimension of settlement patterns and the symbolic value of space in relation to archaeological art in the Marquesas Islands. The empirical focus of the research deals with petroglyphs on individual boulders or on modified stones that are incorporated into architecture. Site designations were assigned according to the identification system laid out by the SCP in 1984. This system was further refined (Conte 1991) and is based on a designated number for the archipelago, island or commune, and district, followed by three letters abbreviating a geographic entity for the site location. A sequential site number follows. Thus, site 331hth 400 translates to the Marquesan Archipelago: (3), Nuku Hiva (3), Hatiheu District (1), Hatiheu Valley (hth), and a specific image boulder number 400. If there is more than one panel, for example if boulder 400 has two panels then the label would read 331hth 400/1 and 331hth 400/2.

All images were documented by tracing on clear polyethylene plastic sheeting with indelible markers. The figures were later reduced to scale and; ink drawings were made and scanned in order to create computer-generated images. A Marquesan archaeological survey form was filled out for each site noting relevant data, such as size, technique, orientation, and cultural and environmental contexts. Because certain images are more visible in indirect sunlight or with artificial light, sites were also visited at night with lanterns whenever possible. Frequently, additional figures were discovered in this manner.

Each architectural structure, including communal breadfruit silos, was given sequential site numbers according to the order in which they were recorded. Associated features were labeled a, b, c, d, and so on after the site number. Each tribal *tohua* (ceremonial complex), a place where all rites and feasts took place, was treated as a separate entity, and their proper names were used. Structures within a tribal ceremonial complex are labeled 1, 2, 3, and so on, and a feature (e.g., pits, uprights, cupule stones) within a structure would be given a letter. Thus a cupule stone on Structure 3 at Tohua Hikokua is labeled Hikokua 3a. In the text, these numbers are placed in parentheses.

In order to spatially place an archaeological site on the landscape and simultaneously take advantage of the contour lines on a map, the Hatiheu part of the Nuku Hiva Island map and the Hatiheu District map have been enlarged to the same scale. The property maps were used to check distance and scale. By using the three existing maps in conjunction with each other, a site map of the western sector of Hatiheu Valley was produced without significantly altering the on-the-ground spatial relationship between architecture and imagery. Due to the steep, rugged terrain and dense vegetation cover, all maps were made with a compass and measuring tape. The speed of this method, I believe, offset the amount of time it would have taken to clear the brush in order to use more sophisticated survey equipment.

Data collected during the course of the survey and excavation were entered into a database management system (FileMaker Pro 14) and were sorted according to location, type, environmental and archaeological associations, and a number of similar attributes. Comparing image types between various Pacific island groups is problematic. In order to minimize this problem and to standardize the classification scheme, I created the Marquesan classification system based on the same system worked out by Georgia Lee (1992) to classify her Rapa Nui database. Lee subsequently used the same system in managing her Hawaiian database (Lee and Stasack 1999). See Appendix B for the Marquesan Classification System.[1]

The Physical Setting

Situated in the eastern Pacific Ocean, the Marquesan Archipelago is some 1,530 km (950 mi.) northeast of Tahiti. Extending approximately 370 km (230 mi.) from the northwest to the southeast, this string of islands is located between 138° 20′ and 140° 30′ west longitude and between 7° 50′ and 10° 35′ south latitude. The Marquesan Archipelago includes eight volcanic islands, a number of islets, and a few surf-beaten rocks.

The archipelago is geographically separated into two distinct groups: in the group to the northwest are the islands of Nuku Hiva, Ua Pou, Ua Huka, the three presently uninhabited islets and banks: Eiao, Hatutu (Hatutaa) and Motu Ono. And in the other group to the southeast are Hiva Oa, Tahuata, Fatu Hiva, and the presently uninhabited islets of Fatu Huku, Moho Tani (Motane), Motu Iti (Hatu Iti), Terihi and Motu Nao (Thomasset). The youngest island, Fatu Hiva, is 1.3 million years old while Eiao, the oldest, is six million years. The total land area is about 1,300 km^2 (Brousse et al. 1978:9–74; Encyclopédie de la Polynésie 5, 1986).

Dramatic ridges rising to 1,200 masl, breathtaking jagged peaks, deep and often narrow valleys, and high cliffs that fall abruptly into the ocean characterize the archipelago. Perennial rivers, waterfalls, and intermittent seasonal streams cut through the dense vegetation on the windward side of the islands. This lush tropical forest is in stark contrast to the arid desert area on the leeward side of the islands. Naturally, the wet-dry dichotomy influenced settlement patterns (Kirch 1984, 1994b, 2000; Kirch and Hunt 1997). Thus, early communities settled in the well-watered windward parts of the islands, while the marginal leeward areas became occupied only when the population increased and the available arable land and resources were diminished.

Conclusion

This chapter contains the theoretical perspective and research strategy. In addition, I have reviewed some of the current problems and limitations associated with rock art research. By reviewing different research approaches, I note the history of settlement pattern research and stress the advantages as a first interpretive step. Despite some setbacks with current dating methods, the future outlook for the development of chronometric dating techniques, and the acceptance of rock art research as part of mainstream archeology looks promising. A major testimony to this is the circumstances surrounding the dated pictographs and petroglyphs in the caves at Wanaham, northern Lifou Island in the Loyalties, New Caledonia (Sand et al. 2006), the discovery of the Cosquer Cave in 1996 (www.bradshawfoundation.com), and the Chauvet Cave in 1994 (Clottes 2003). These discoveries have spurred rock art research and attempts to refine dating techniques. More research on absolute dating of petroglyphs is necessary, but without cultural context chronology will not help with interpretation and theory building.

Chapter 2

Historical Perspective
A Review

Islands and beaches is a metaphor for the different ways in which human beings construct their worlds and for the boundaries that they construct between them. It is a natural metaphor for the oceanic world of the Pacific where islands are everywhere and beaches must be crossed to enter them or leave them, to make them or change them. But the islands and beaches I speak of are less physical than cultural. They are the islands men and women make by the reality they attribute to their categories, their roles, their institutions of "we" and "they."

— Greg Dening, *Islands and Beaches, Discourse on a Silent Land: Marquesas 1774–1880*

While Marquesan ethnohistoric material is extensive, data germane to the study of petroglyphs are fragmented. Early visitors paid more attention to tiki (stone and wood human sculptures), tattoos, and carved images on bone and wood. This may be due to their visibility and location because most tiki are found on ritual sites (Millerstrom 2005a; Millerstrom and Edwards 1998).

Literature will be evaluated in this chapter, particularly the early sources from Western contact in A.D. 1774 to the end of the nineteenth century. Because information on petroglyphs is practically nonexistent, the focus here is on images, such as tattoos and carved objects in other settings, and especially the association and distribution of image types and their contexts. The manuscript written by William Pascoe Crook, the first missionary to the islands in 1797, and the stories told by the beachcomber Edward Robarts are particularly rich. After the 1820s, due to increased European activities, there was a marked change in the Marquesan social fabric. Historic period material consists mostly of observations from whalers, sandalwood traders, French missionaries, and administrators. This period concluded with ethnographic work of von den Steinen in the 1890s.

The historic material contains a confusing array of names for the islands and bays. For instance, in the Northern group, European explorers renamed the islands on four different occasions (Brousse et al. 1978:12, Table 1; Dening 1974:97; Krusenstern 1968 [1813]). In the following discussions, the former place names that are sometimes encountered in the literature are placed in parentheses.

The Protohistoric Period (A.D. 1774–1820)

The Protohistoric, or the early historic period, commenced with the first contact with Europeans, and Captain J. Cook in 1774, and ended in the 1820s after Captain David Porter's two-month stay on Nuku Hiva (Federal Island) in 1813 (Porter 1970 [1822]). The first encounter between Europeans and Te Enana, as the Marquesan people now wish to be called, was brief and brutal.[2] On July 21, 1595, the Spanish commander Alvaro de Mendaña and his Portuguese pilot Pedro Fernández de Quiros came upon Fatu Hiva (Magdalena), an island in the southern group, on their way from Peru to form a colony in the Solomon Islands.[3] On that day the Enana became the first Polynesian island society to be discovered by the Europeans.[4]

Mendaña and his crew anchored in Vaitahu Bay, Tahuata (Santa Christina), on the July 24, 1595 (Dening 1980:9; Markham 1904:15–29). He named the bay Madre de Dios. Two additional islands, Motane (San Pedro) and Hiva Oa (La Dominica), were subsequently discovered. Mendaña named the island group "Las Marquesas de Mendoza," in honor of his patron Don García Hurtado de Mendoza, Marquis de Cañete, Viceroy of Peru (Dening 1980:11).

During their brief visit, Mendaña and his crew raised three wooden crosses, scattered a few corn seeds, held a Mass, and massacred some two hundred islanders (Dening 1980; Markham 1904:15–29). Only Quiros left any record of what they saw during their sojourn in the islands. His account contained the first description of a *meae* (sacred structure) and anthropomorphic sculptures: "Apart from the village there was an oracle surrounded by palisades, with the entrance on the west side. Within there was a house, almost in the middle, in which there were wooden figures badly carved; and here were offerings of food and a pig, which the soldiers took. When the Spanish soldiers attempted to take some of the objects, they were made to understand by the Enana that they had great respect for the place" (Markham 1904:27–28).

As far as the written record is concerned, no Europeans visited the islands during the following 179 years. Western influence first spread after Captain James Cook's three-day visit in 1774. Cook, commander of the *Resolution* and the *Discovery*, also anchored in Vaitahu Bay, Tahuata, April 6, 1774.[5] He renamed it Resolution Bay.

Although he generally was a sagacious observer, Cook had little to say about local customs. Cook's main concern was to check Mendaña's map and to obtain provisions before he sailed to his beloved Tahiti.[6] Nonetheless, engravings by J. Hall after the famed landscape painter William Hodges, the artist on Vaitahu Bay, Tahuata, depicted the Enana with body decorations; canoes were also illustrated. A drawing from Cook's journal of a wooden club and a feather headdress depicted human eyes (Joppien and Smith 1985(2):202–207, Plates 2.105B and C; see also James Cook and the Exploration of the Pacific, 2009, Figure 2.)

In 1791 Joseph Ingraham (1790–1792:45–67) of Boston, Captain of the *Hope*, discovered the northern island group. Although he only observed the island of Nuku Hiva (Federal Island) through his field glasses, he noticed houses on the summit; but trees mostly concealed the houses on the beach. In his description of outrigger canoes, Ingraham (1790–1792:57) observed that "on the upper part at the extremity is a human face rudely carved."

Two months later the French trader Etienne Marchand, on the *Solide*, sighted the northern islands. Believing he was the first to see the island group, he named Nuku Hiva as Isle Baux, after Maison Baux, a merchant house of Marseilles that financed his expedition (Dening 1980:22–23).[7] The most extensive records on early Marquesan culture are derived from the accounts of an English beachcomber, Edward Robarts, and the British missionary William Pascoe Crook. Less is known about another beachcomber, the Frenchman Joseph Baptiste Cabri (aka Jean Cabris, Cadiche, or Joseph Kabrit). Robarts and Cabri, who were bitter enemies, spent a number of years in the islands and served as guides and translators for European visitors. Both were adopted by Marquesan chiefs, became warriors, and married high-status Marquesan women.

Cabri, fifteen years old, was shipwrecked on the north coast of Nuku Hiva in 1796. He immersed himself in the culture and had his body covered with tattoos that identified him with his adopted tribe on Nuku Hiva. Cabri apparently left the islands accidentally with the Krusenstern expedition in 1804; he fell asleep while visiting one of the ships (Krusenstern 1968 [1813]; see also Govor 2010).

Robarts deserted his ship, the *Euphrates*, in 1797, in the Taipivai area (Comptroller's Bay) and lived 8 years in the archipelago (Dening 1974). Robarts was one of 26 members of the "kings club" (the chief's warriors) and was tattooed with an insignia of a rectangular "M" in form "about six inches long and four wide" on his breast (Krusenstern 1968:160 [1813]). This gave him membership to the *tapu* (chiefly, divine, sacred, restricted, or prohibited) class associated with the chief (Dening 1974, 2004).

On June 27, 1797, Captain James Wilson of the *Duff* left William Pascoe Crook and John Harris, two missionaries from the London Missionary Society, in Vaitahu Valley, Tahuata. However, the free-spirited Enana women, led by Tepaihena (Tipanihine), the wife of Chief Teine (Tainai), wanted to examine Harris closely. Accustomed to the European sailors' enthusiasm for the indigenous women's sexual favors, the women were curious as to the abnormal behavior of the missionaries and the *Duff*'s crew. Harris, horrified by the tattooed and inquisitive females, spent a night on the beach, sitting on his trunk, to wait for morning so that he could return to the ship (Calder 1996:146–147; Crook 2007:45–46, 103–104; Dening 1974:4–5; Radiguet 1860; Wilson 1997:103–132 [1797]). Crook, a courageous man of 22 years, spent 19 months in the archipelago and a great deal of the time on Nuku Hiva. After much hardship Crook left on January 8, 1799, with two whalers: ***Euphrates*** and ***Butterworth.*** His chronicle provides the most authoritative statements on early Enana society.[8]

Crook referred to images on wood and stone associated with funeral practices. According to Crook, corpses were commonly deposited on a loft near the habitation place of the deceased or in a sacred enclosure: "Near these lofts, they erect several wooden or stone pillars, not more than two feet high, sometimes carved to resemble a human face. On the top of these pedestals, are laid round flat stones, to hold provisions; and small shrines, with leaves wrapped up, so as to contain Water, placed at the ends of each shrines" (Crook 2007:66 [1797–1799]). Robarts mentioned images in a similar context: "There is a rough image fixt in the ground at the moria [***meae***]. On the head of this log of wood is put a small part of every thing sent. This no one eats, being held sacred. The funeral feast of a prophet is only eat by those belonging to the moria and the surviving prophets" (in Dening 1974:57).

The Russian captain, A. J. Von Krusenstern, who visited Nuku Hiva in 1804, had this to say about images on a sacred site.

> The morai was situated on the top of a pretty high hill, which it cost us some trouble to climb, as the sun was near its height; it was in a thick wood, interwoven with bindweed, and seemed to be nearly impassable: we saw a bier upon a stand, but of the body upon it, nothing but the head was visible. In the outer circle were some statues carved in wood, intended to represent the human figure, and evidently the coarse work of some unskilful artist: near to these statues were some pillars wrapt up in cocoa-leaves and a white cotton stuff. (Krusenstern 1968:127 [1813])

United States Navy captain, David Porter, on the frigate *Essex*, took formal possession of Nuku Hiva Island for the United States in October, 1813.[9] The island became Porter's base while he looked after commerce and political interests in the South Pacific. This was, in part, accomplished by harassing British ships in the Pacific while the United States and Britain were at war. While on Nuku Hiva, Porter engaged in warfare with the Happa against the neighboring Taipi "nation" (Porter 1970 [1822]). On an excursion in Nuku Hiva, Porter (1970:110–111 [1822]) was led to a chief's *meae* where he observed one stone tiki on a platform with several wood tiki (made of breadfruit trees) arranged around it. There were four "splendid war canoes" with a priest, recently killed, seated in the stern of one of the canoes. Offerings, including pig heads and turtles, were placed at the base of two 10.7-m tall obelisks; many humans, some warriors killed in a recent battle, were placed in the canoes or around the *meae.* Porter observed that "the stone god" was so similar to all the anthropomorphic sculptures on the island, including the household gods, the figures on stilts and the handles of the fans, that it may have served as a model (Porter 1970:110 [1822]). Porter's described a house on the *tohua* (tribal ceremonial complex): "Sometimes, indeed, the columns are richly carved in the form of gods, and give the whole an air of grandeur and elegance, which, although in a style differing from that of every other people in the world, does not the less astonish" (Porter 1970:40 [1822]).

Captain David Porter's prolonged stay on Nuku Hiva marked an end to the early historic era. His large crew and his involvement in local warfare adversely affected the native peoples and altered the indigenous social structure. The short but intense sandalwood trade drew to an end about the same time period (Dening 1974). Changes occurred in the material culture, particularly the ***uu*** (war club), a popular collector's item and the most common Marquesan object in museums (Ivory 1993:70). Carvings on war clubs collected after Porter's visit show a noticeable difference compared to the one ***uu*** collected on Cook's voyage.[10] Carol S. Ivory, an art historian who studied 230 clubs from various collections, identified three different styles of ***uu.*** In addition to the ancient and modern types classified by Linton, the third or early type was more plain; the facial features were roughly carved and with few surface decorations (Ivory 1993:63–73, 1994, 1995).[11] Comparing the late historic clubs with the plain ***uu*** collected by Cook, it is clear that the carvings were elaborated with time, especially after Porter's sojourn when woodcarving activity increased due to the demand for souvenirs.

The Historic Period (1820–1920)

Source material from the 1820s and onward comes mainly from French officials (e.g., Delmas 1927; Radiguet 1861; Tautain 1896a, 1896b, 1896c, 1897, 1898), American missionaries (e.g., Thomson 1816–1851[Craig 1989]; Lawson 1861–1867), and French Catholic missionaries from the Congregation of Sacré Coeur de Picpus (e.g., Chaulet 1858–1897; Dordillon 1931; Delmas 1927; Gracia 1843).[12] The majority of missionary accounts remain unpublished.

In 1842, Dupetit-Thouars, then Commander of the *Reine Blanche*, disembarked at Vaitahu, Tahuata, raised the French flag, and took possession of the islands in the name of the King of France (Dening 1980:206).[13] The Marquesans, suitably impressed with French rituals and the firing of the cannons, "accepted the reality of power" (Dening 1980:263; see also Rosenman 1992:145). The northern group was claimed a month later at Taiohae, Nuku Hiva. Catholic conversion soon followed, and the French Catholic missionaries were especially active after the archipelago became French territory.[14] The French missionaries took charge of the islanders' educations and forbid tattooing, singing, dancing, and music related to indigenous ceremonies. The carving tradition, however, continued.

Tautain (1897:675), a French administrator to the islands, 1891–1897, who wrote several articles on the culture in general, remarked "Au mau funéraire, d'un homme, ils auraient inscrit sur une pierre, sur un tronc d'arbre le nom du défunt en figurant les principales pièce du tatouage de sa face; puis plus tard auraient plus ou moins dégrossi une forme humaine" (At the funeral feast for a man, the name of the deceased, the principal part of his facial tattoo would be inscribed on a stone or a tree trunk, then later a rough human form would be added [translated by author]). This strongly indicates that some anthropomorphic faces and figures carved on stones represent individual ancestors.

William Torrey, an American sailor shipwrecked on Tahuata in 1835, made an interesting statement in relation to mortuary practices on Hiva Oa.

> The bodies of kings and those of the royal family are placed, after being wound [wrapped] as before described, in a hut erected on purpose, on or near the tapu ground, in a sitting posture, with the arms raised, the elbow bent to an angle, and supported by sticks of bamboo, driven in the earth. On either side is a man placed, who are taken in war, or stolen from some neighboring tribe, and sacrificed for the purpose, also wound in bark. These are supposed to accompany them in the capacity of servants. (Torrey 1848:123)

Could it be that the numerous anthropomorphic petroglyph figures with arms raised depict important deceased ancestors? Torrey's and Tautain's statements suggest that deceased chiefs, important priests, or other high-status individuals, may have been immortalized as anthropomorphic petroglyphs and pictograph figures in a manner similar to bodies of king and those of the royal family. According to early historic sources, the tiki depicted deified ancestors (Delmas 1927; Tautain 1897:673–678). While the ages of the sculptures are uncertain, it appears that the majority was developed in the late prehistoric period (Suggs 1961). Except for a few cases, tiki were found on the *meae* part of a *tohua*, chiefs' residential units, and fishermen's shrines (Millerstrom and Edwards 1998). Perhaps the symbolic meaning of the ancestor images depicted on stones was transformed over time into carved-stone images in the round.

Ethnography

Karl von den Steinen, a German physician with an interest in the Enana's material culture, directed the first scholarly Marquesan ethnography. After years of research in European museums, he arrived in the archipelago in 1897 (Pagès 1995, 1996). Von den Steinen collected legends (Terrell 1988) and extensively recorded and discussed the material culture that was published in three volumes some thirty years later (von den Steinen 1969(I) [1925], 1969(II) [1928], 1969(III) [1928]). Despite a great amount of information on the decorative art system, petroglyphs were mentioned in general terms only. However, von den Steinen (1969(II):69–172 [1925]) suggests a link between the anthropomorphic faces of the *mata komoe* (anthropomorphic faces) motif and warfare, especially headhunting operations of the free-spirited *kaioi* (sexually oriented adolescent society) (Thomas 1990). Von den Steinen's work at Meae Iipona (also Oipona), Puamau, on Hiva Oa, is of particular interest. The sacred site is both extensive and complex and contains sculptures, petroglyphs, upright stones, platforms, and pavements. According to local informants who claimed they could recite at least 25 generations, von den Steinen (1969(I) 84–86 [1928]) calculated that Meae Iipona was constructed about A.D. 1700–1750. Linton, Heyerdahl, Ottino, and this author conducted research at Iipoka (see Chapter 3).

Demographic Composition

To understand the lack of information on rock art it is useful to see how the population was catastrophically reduced soon after Europeans arrived. At the time of contact, the Marquesan archipelago was well populated. Cook estimated the population for the southern island group to be approximately 50,000, although the population may have exceeded 50,000 for the archipelago (Dening 1974:viii, 1980; McArthur 1968:279–295). Suggs (1961:192) calculated that, at its peak, the population of Nuku Hiva was 30,000, and for the archipelago he proposed a figure of 100,000. However, Rallu (1990) provides a more conservative figure of 43,000 for the island group. As the table shows, in 1926 there were only 90 persons living in Hatiheu Valley (Table 1).

Table 1. Population estimates for the Marquesas Islands

Year	Marquesan Archipelago	Nuku Hiva Island	Hatiheu Valley	References
1774	?	–	–	Captain Cook's arrival
	43,000	–	–	Rallu (1990)
	40,000	–	–	Molle and Conte (2015)
1804	–	18,000–24,000	–	Krusenstern (1968 [1813])
1813	50,000	12,000	–	Porter (1970 [1822]); Gracia (1843:12–13)
1842	20,200	8,000	–	Dupetit-Thouars (1995 [1842])*
1856	11,900	2,700	–	Jouan 1890
1872	6,426	1,560		Eyriaud des Vergnes (1877:37)
1884	4,865	980	195	Clavel (1884:493); von den Steinen (1925:12–13)
1897	–	642	143	Von den Steinen (1925:12–13)
1902	3,562	682	–	Valenziani (1940:179)
1921	1,800	–	–	Handy (1923:5)
1926	2,094	543	90	Adamson (1936:60); Brousse et al. (1978:17)

*First official census

Following Cook's visit, frequent contact with European whalers, sandalwood merchants, and missionaries dramatically reduced the Marquesan population, primarily due to introduced diseases (Kirch and Rallu 2007; McArthur 1968). Intertribal warfare, prolonged drought, famines, and the introduction of firearms and alcohol were certainly contributing factors. The fragile indigenous society became too weak to resist colonization and Christianity.

Hatiheu Bay, Nuku Hiva Island, was not among the main harbors visited by foreign ships. Regardless, the valley population suffered greatly soon after Western contact. It is certain that the prolonged visit of Captain Porter (1970 [1822]) and his crew of 320 men in Taiohae, Nuku Hiva, in 1813 adversely affected the health of the Hatiheu population. Tragically, in September, 1863, the smallpox epidemic that was brought from Peru to Taiohae on the ship *Le Diamant*, resulted in an estimated death of some 150 people in Hatiheu by December of that year ([ORSTOM] Office de la recherche scientifique et technique outre-mer, 1993). The estimated population decline of Hatiheu went from 6,000 to 400 in 40 years (Kellum 1966). For example, in 1888, there were 12 deaths to every birth (Roberts 1929:93). After Western contact, the population decline in the Marquesas was more serious than elsewhere in the Pacific. While it is challenging to estimate the population at initial Western contact, it is clear that the islands supported a substantial population in prehistoric times. In most valleys, a large number of prehistoric and early historic architectural structures extend from the beach to the mountain ridges, suggesting that the indigenous landscape was transformed, modified, and controlled over time.

Social Structure

The Marquesas had a stratified, but highly fluid, society in which status was both prescribed and achieved (Goldman 1970; Kirch 2000; Sahlins 1958; Thomas 1990). In general, there were two classes, the tapu class, which had several subdivisions, and the *mataeinaa* (tribe/people). The *hakaiki* (hereditary chief) descended directly from the gods and was therefore considered tapu. In theory, the chief was the leader of the *mataeinaa*, usually a ramified descent group. However, because the Marquesan *hakaiki* did not, overall, play an active role in tribal ritual matter, his sanctity was sometimes challenged. A person's status within the society depended on his/her genealogical relationship to the *hakaiki* and through his genealogy *mana* (supernatural power) flowed from the gods.

Landless people had low status and were called ***kikino***. The ***kikino*** would frequently attach themselves to landholders. People who owned property, for example, land, trees, and what the trees produced were high-ranking persons within the society. The tapu class then consisted of the *hakaiki* (chief, first-born son), *haatepeiu* (the firstborn daughter of a chief), the *toa* (the chief warrior or war leader), *tuhuna* (skilled artisan), *tuhuna oono* (ceremonial priest), and the *taua* ([the inspirational priests] Crook 2007; Dening 1974:21; Handy 1923; Thomas 1990).[15]

Warfare among tribes was pervasive (Crook 2007; Dupetit-Thouars 1840–1841 (II):358); Thomas 1990). According to Crook, most of the blood was spilled during surprise attacks rather than during staged battles. The ***toa*** may or may not also be the chief (Crook 2007:69); the warriors, the ***mata hoka*** (***mata*** is eye, face, or genealogy; ***hoka*** is courageous), spent a great deal of time preparing for warfare at a special house near or within the ***tohua***; there they were fed by their chief (Crook 2007:69–83–85; Handy 1923:126). Women, in general, did not take an active role in combat. During planned battles, they dressed up as though for a feast (Porter 1970 [1822]) and followed their warrior husbands so they could observe the fighting (Handy 1923:132). From a safe place, the women would shout ***tomoa*** ([cries of encouragement] Dordillon in Handy 1923:133). While the inspirational priests would sometimes stand, uttering a spell on a hilltop overlooking the battle, the ceremonial priest would chant at the ***meae***. Brave and successful warriors or ***toa*** (which also means ironwood, male, brave, and strong) could become high-status persons within their community (Crook 2007 [1797–1799]).

Early voyagers rarely recognized women's sociopolitical position. Yet, "some women in the early contact period were active in political and religious realms in much the same way as men" (Thomas 1990:74). In fact, according to local oral history, at least two tribal ceremonial complexes near Taipi Valley belonged to chiefesses (Millerstrom 1991a).

Certain women, for instance those from a chiefly class, possessed more personal tapu than women in general. Similar to the men from the common class, chiefly women were not allowed near the men's or warriors' houses, nor could they eat with tapu men. However, women could eat with the ***kaioi*** and their male attendants, the ***pekkeyo*** ([also ***pekio***, "second husband"] Crook 2007:52, 55).

The Marquesan tapu system was concerned with social rules and laws. This system may explain why images on boulders occur in some locations and on some architecture and features but not on others. There were general tapu that affected the whole tribe, or specific tapu confined to a family or a single person ordered by the priests. Women were "harshly treated" and were especially saddled with tapu on behavior and food (Crook 2007:48). Normally, turtle and pork were tapu to Polynesian women, yet both women and children ate pork and turtle on certain occasions (Handy 1927:46; Henry 1928:177; Krusenstern 1968:161 [1813]). Porter (1970:116 [1822]) observed that men, women, and children ate together, but from separate dishes. Except during a time of tapu, women consumed pork when they were alone. Even during tapu, high-status women would eat pork in the absence of men or when no one was looking. If they were eating pork, men would look the other way (Porter 1970:116 [1822]). But as Krusenstern (1968:161 [1813]) observed "neither are they forbidden to eat pork, although it is but seldom that they get any." However, Crook observed that women were forbidden to eat pork, turtles, and certain kinds of fish (Crook 2007:48, 69–70). Douglas (1966) quoted in Hastorf (1991:135–136):

> In cultures where the males have fairly clear control over the social relations and moral codes (including power over the sexuality of women and their marriage), such as among the

> Walbiri of central Australia, food and sex taboos tend to be absent. Social relationships are controlled directly by the males, as subordinate subgroups (females) do not have avenues to contest. On the other hand, in cultures where dominance is contested, Douglas notes that ambiguous and contested power between genders is often translated into food taboos that involve restriction of specific foods, on certain activities (speech), and on timing of sexual activities (abstinence during menstruation), thus delimiting areas of control.

It is likely that the Enana women had more autonomy than the written sources show. One of the main reasons for this is that white Western men wrote all the early historic and ethnographic accounts from 1595 until the 1960s, except for the contributions of Willowdean Chatterson Handy (1922, 1925, 1938, 1965, 1973). It is clear from her writing that even Ms. Handy did not concern herself with indigenous women's issues, as these were never discussed in her writing.

Most priests and those who assisted them at the ***meae*** were part of the tapu class. Both men and women with supernatural power, could become an inspirational priest or shaman (Handy 1923:223). ***Taua*** were powerful beings that could influence political relations, sometimes with the use of sorcery (Thomas 1990:74–76). Some inspirational priests/priestesses were considered a living deity or god called ***atua/etua*** (Dening 1974:21). The priests were the keepers of knowledge, and they "are ordain'd, to their office, by predecessors; and do not to inherit it by descent" (Crook 2007:61). Ritual knowledge and oral tradition resided in the ***opu*** (stomach); the intestines were the organs of intelligence, while the liver was the organ of courage (Tautain 1896b:443–452). The common stance of the Marquesan sculptures is bent knees and hands resting on a protruding belly. According to Thornton (1989:147–166), who analyzed linguistic terms related to Polynesian ancestral figures, hands placed on a protruding belly may have been a way to protect these memories. On Easter Island, Lee (1992) noted that hands across the belly indicate protection of ritual knowledge.

The late prehistoric period is characterized as a time when conflicts increased as population rapidly expanded and competition for resources intensified. The sociopolitical dynamics among the hereditary chiefs, warriors, and priests led to prestige rivalry during this period (Kirch 1991a, 2000; Suggs 1961; Thomas 1990).

The Marquesan Decorative System

East Polynesia consists of 38 islands and island groups, including the Australs, Cooks, Rapa Nui, Hawaii, Marquesas, New Zealand, and the Societies. They were the last Pacific islands to be colonized (Kirch 2000) and are both geographically remote and environmentally diverse.

While Oceanic art historians have paid much attention to art collected by the early explorers, they have generally neglected to define an East Polynesian decorative system, and they have not explained how a distinct culture with a unique artistic expression developed in East Polynesia following initial colonization. Conversely, East Polynesia as a specific cultural area has been investigated on several levels, mostly by Oceanic archaeologists (e.g., Kirch 1986; 2000; Lee 1992; Lee and Stasack 1999; Linton 1923; Rolett 1998; Suggs 1961; Van Tilburg 1994; Molle 2013). What then were the social and environmental processes that took place in East Polynesia after the people dispersed from their ancestral Polynesian homeland? In particular, what were the dynamics that led to the development of a cultural specific decorative system in the Marquesas? Lastly, what typifies the Enana's decorative system?

East Polynesia was settled by people from a homeland within the Polynesian Triangle, in the Tonga, Samoa, Futuna, and Uvea regions perhaps as recently as 1,000 years ago (Weisler and Green 2011:223–246). The character of this expansion has been debated for years and is still the subject of some controversy.

Successful colonization on previously uninhabited islands depended upon critical factors, such as the transport of cultigens and domestic animals, and on human reproduction. Successful adaptation strategies would include new fishing technology, exploitation of natural resources, and adaptation to new environments, different from their ancestral homeland. New Zealand, for example, required that the colonists made radical changes in their traditional behavior patterns because of climate. On smaller islands, the environmental conditions were too hostile to sustain a permanent population and were consequently abandoned (Kirch 1986, 2000).

In the Marquesas Archipelago, colonizers had to adjust to local ecological conditions that included steep and rugged high mountains, limited arable land, and highly variable rainfall. Except for a few bays (e.g., Anaho Bay), the Marquesas mostly lacked protective coral reefs, which limited fishing to the deep sea. The common assumption that the steep, high, and inhospitable Marquesan mountains were impassable by foot and thus contributed to isolation between tribes is unfounded and needs to be examined further. This notion originated with Handy (1923:8), and the idea has since been uncritically accepted. It is more likely that the nature of the Enana's social fabric rather than the high mountains led to isolation between tribes, as none of the mountains are inaccessible to humans who want to exploit neighboring valleys on foot.

The early material assemblages reflect stages of experimentation and adjustments to different fishing practices. For instance, the fishhook assemblage that, at the time of settlement, consisted largely of one-piece fishhooks later reflected stylistic innovations in the form of two pieces of shell or bone lashed together (Sinoto 1966). Thus the unique Marquesan environmental conditions resulted in cultural behavior that affected subsistence, settlement patterns, social structure, and material culture, yet East Polynesian still retained a number of features from the ancestral Polynesian society (Kirch 1986, 2000; Rolett 1998).

Some of the ancestral cultural practices discussed below are tattooing, *tapa* (bark cloth work) and anthropomorphic sculptures. In particular, the practice of tattooing developed in the Marquesas to a fine art (P. and M. N. Ottino-Garanger 1998). For instance, in contrast to Samoa, where some 50 names of tattoos have been documented, in the Marquesas, Von den Steinen recorded the names of 174 individual motifs in the late 1890s, about the time the custom was officially forbidden by the French authorities (Gell 1993). In the sixteenth century, Spanish navigator Quiros, among the first Westerners to observe the Marquesan tattoos, wrote in passing that their faces and bodies were "painted" with fish and other patterns in a bluish color (Markham 1904).

Tattooing

Generally the members of the tapu class were densely tattooed or "punctured" as Crook called the practice.[16] The tattoos were believed to represent status variations and a warrior's tribal symbol. Early travelers wrote that tattoos mirrored status and tribal identity (Crook 2007:55–56, 59–60, 86, 92; Porter 1970 [1822]; Roquefeuil 1823:55–56). According to Govor's (2010:145–151) investigation of the Russian expedition, they noticed that tattoos reflected sex, age, social status, and tribal identity. Porter (1970:120 [1822]) observed that every tribe on the island of Nuku Hiva "was tattooed after a different fashion," and "was informed that every line had its meaning, and gave to the bearer certain privileges at their feasts."

Torrey (1848:125–127) related in his personal narrative that each group wore tribal tattoo marks on their faces, breast, or hands, for example. To join the Teheda tribe and go to war against an enemy tribe, the back of Torrey's hand was tattooed "to bear the king's mark." He was thus "considered as the king's own private property and subject entirely to his control and disposition" and could never desert (Torrey 1848:126). Handy (1922:3), however, claimed that tattoos did not appear to be confined to specific classes or ranks.

Handy (1922) believed that she detected distinct stylistic variations in the early tattooing tradition between the southeastern and the northwestern Marquesan Islands. Allen (1992) reevaluated her hypothesis and rejected it. Specific local styles may have existed in the past, but considering that Handy's research took place many years after tattooing was abolished and that she did not take into account all the social changes that took place soon after Western contact, her view is unlikely. Allen pointed to tattoo illustrations observed by early navigators' evidence neglected by Handy, which showed similarities with tattoo styles from the Tuamotus, Austral, and the Society Islands. It has been documented that Polynesians from these islands traveled to the Marquesas as early as 1813 (Dening 1980).

Suggs (1961:150), following von den Steinen (1925(1):69, 172), suggested that there was a link between warfare, the *mata komoe*, and concentric circle motifs, and that the motifs symbolized the sacrificial hunting of *kaioi* youth. Gell (1993) with his insight into the social conditions particular to the Marquesas, opined that tattoos served as a protective skin, a symbolic technique to control sacredness. P. and M. N. Ottino-Garanger (1998), in a generously illustrated volume, reviewed all the literature on tattooing and placed the images in a cultural context. In their analysis, they discuss how tattooing was a rite of integration into the social system. Being tattooed was a social obligation that started early and served several functions that reached beyond high status. Tattoos were marks of recognition, they protected warriors during combat, they concealed, they inspired fear, and they commemorated events. A tattooed person had access to food from certain relatives, and the chief was obliged to feed his warriors.[17]

Specific petroglyph motifs, particularly several variations of the ***mata komoe***, ***atua/etua***, Enana, and the ***ipu*** figures reflect tattoo patterns. Von den Steinen (1969(1) [1925]) believed the complex Marquesan tattoos observed in the historic period derived from the ***mata komoe*** and ***atua/etua*** motifs. Figures 3, 4, and 5 illustrate how some of the tattoo motifs correspond to Marquesan petroglyphs (Millerstrom and Allen 2006). It is clear that in certain architectural contexts, some petroglyph images that are also frequently seen in tattoo motifs reflected tribal recognition, social class, and individual high status. The similarities in historic tattoo motifs and prehistoric petroglyphs imply that some rock image types may have some time depth (Millerstrom 1997a; Millerstrom and Allen 2006).

Mata, *mata komoe*, *ipu*, and *atua/etua* are a few of the numerous tattoo names given to von den Steinen (1969(1), 1969(II) by Enana informants in 1897. These same terms are still used today to describe certain rock art motifs. **Mata*, a proto-Polynesian taxeme, refers to eye and face (Walsh and Biggs 1966). The word has the same meaning on Rapa Nui, Tonga, Samoa, the Cooks, and among the Maori (Walsh and Biggs 1966). ***Maka*** is a Hawaiian cognate. Other glosses for *mata* include "genealogy" (Dordillon 1931; Handy 1923), "clan" (Routledge 1919), "tribe" (Métraux 1940:120), or "status lineage" (Goldman 1970:119). **Ipu* can mean "calabash" and "container," while *atua* or *etua* is translated as "deity" (Walsh and Biggs 1966), "god," or "spirit" (von den Steinen 1969 [1925]). These words are still retained in Polynesian dialects (Walsh and Biggs 1966).

The emphasis on the *mata komoe*, *atua/etua*, and Enana figures may have been a way to memorialize, honor, and venerate ancestors. Concern with genealogy was strong in Polynesia. Genealogy linked people with their ancestors and defined their social position to their chiefs. Genealogies were chanted in rites of adoption, birth of a firstborn, marriages, funerals, and so on (Handy 1923:341–346, 1930b:11). Because the head was considered the seat of ***mana***, ancestral skulls were frequently removed from burials and treated as sacred relics (Handy 1923, 1930b; Suggs 1961:168). Ivory (1990:13–14) suggests that the concepts of *mata* and *ipu* terms linked to images in rock art, tattoo, and material objects are metaphorically connected to "face, eye, skull, gourd, fish, turtle, ray, hair, and belly, and that all of these in turn were associated with the *atua* and the spirit world, and that visually, these could be encapsulated in representations of circular motifs in general, and the face, skull, and eyes, in particular."

Images may have served as visual markers to identify clan members and may have linked families to their ancestral lands. Tautain's (1897:675) statement, that after the death of an important person, a carving was made on stones or trees that represented the deceased's facial tattoos, suggests that some of the images were related to a specific person—probably a chief, priest, or warrior. It also indicates that petroglyph images were socially interrelated with tattooing practices and that they were associated with funeral rites and afterlife. Furthermore, it seems likely that certain motif types belonged to or were associated with specific related tribes or ramages.

The widespread use of the *mata komoe*, *atua/etua*, Enana, and the *ipu* motifs in prehistoric rock images, tattoo motifs, and early historic objects, such as bone and wood carvings in conjunction with the antiquity of the Polynesian terms, indicate that these motif types are connected with concepts that go far back in time (Millerstrom 1997a; Millerstrom and Allen 2006).

Tapa Cloth

Tapa is also known as bark cloth. Few navigators or ethnographers observed the making of Marquesan *tapa*, nor are there many Marquesan *tapa* in museum collections. Written sources on tapa are primarily from around the early 1900s and later and consist mostly of secondhand reports from informants. The most complete information on Marquesan *tapa* derives from Kooijman (1972) who synthesized the fragmented publications especially from the descriptions of Handy (1923), Linton (1923), and von den Steinen (1969 [1925], 1969a [1928], 1969b [1928]).

According to Kooijman (1972:176) Marquesan *tapa* was "both technically and qualitatively inferior compared to the large islands of Tahiti, Hawaii, Samoa, Tonga, and Fiji." This was primarily due to the Marquesan ecological conditions. The lack of arable land and restricted rainfall in many places limited cultivation of ***ute*/*outi*** (paper mulberry), the most important tree for ritual *tapa* making. The ***ute*/*outi*** produced a soft white cloth that was used as ceremonial dress for *meae* assistants, for warriors, for festive occasions, and for gifts and offerings. Often this *tapa* was colored yellow with ***eka*/*ena***, the dye from turmeric root (***Curcuma longa***) and scented with fragrant coconut oil.

Paper mulberry trees (***Broussonetia papyrifera***) were cultivated in small, enclosed protected gardens near the residential units. *Tapa* was also made from the young branches of *aoa*, the sacred banyan tree (***Ficus*** sp.). From this, the Enana produced *hiapo/heapo*, a strong, reddish-colored cloth) and the *hami/maro*, or men's loincloths, that were only used for the firstborn male child, chiefs, warriors, priests, and in marriage ceremonies. Through the combination of the sacred tree that produced a red male loincloth, the *tapa* connected the tapu class with the gods. *Tapa* cloth from the *hiapo/heapo* also played an important part in marriage ceremonies. The cloth made from inner bark of ***mei***, the breadfruit tree (***Artocarpus altilis***), would yield a coarse, light brown, colored cloth. A ***kahu*** (common cape) worn by men and women was made from ***mei***. Sometimes the bark of the *mio*, or Pacific rosewood (***Thespesia populnea***), and a flowering plant, referred to by Jardin (1862:32–59) and Handy (1923:161–162) as *katea* (***Alyxia*** sp.), were used. In Marquesan it is called ***mehe*/*meie*** and was tapu in the past; the priests used the flowers and the branches during ceremonies (Petard 1986:256 [1958]).

Kooijman (1972) attributes the lack of *tapa* specimens in museums to the *tapa* being undecorated. In contrast to the elaborately decorated *tapa* in most Polynesian islands (e.g., Cartmail 1997; Neich and Pendergrast 1997), Marquesan *tapa* was largely undecorated and thus did not catch the interest of collectors. However, in addition to the various colors of *tapa* cloth, with the grooves on the *tapa* beater (usually made of iron wood), the craft person could produce a pattern of watermarks in the shape of diamonds or make the *tapa* look like a woven cloth with "heavy weft and light warp threads" (Linton 1923:414). It is possible that *tapa* was decorated with teeth and perhaps shell. Unpublished material from the mid-1800s by beachcomber/ethnographer Thomas Clifton Lawson describes gifts from the Naiki people of Ua Huka to the Hatiheu-Anaho people of the Taipi tribe that included "blackfish teeth strung two and two together on strips of white tapa" (in Kellum-Ottino 1971:165). Certainly the white *tapa* clothing would draw attention to and accentuate the elaborate bluish tattoo body decoration.

Notable exceptions to plain *tapa* are painted *tapa*-covered human skulls and *tapa*-covered anthropomorphic wooden figures. The eyes are emphasized in both. ***Ipu*** and naturalistic motifs, such as sharks and rays, cover the objects. A roughly formed anthropomorphic figure housed at the Bishop Museum, Honolulu, is composed of a single piece of breadfruit wood and two sticks of *hau* wood (***Hibiscus tiliaceus***), both covered with *tapa* cloth that is decorated with figures in red and black pigment. The *tapa*-covered object depicts large elongated eyes; the lower part of the body is densely covered with the classic ***ipu*** pattern common in tattoos and petroglyphs. A similar image with ***ipu*** and anthropomorphic stick figures is depicted in Trésors des îles Marquises (Panoff 1995:57, Figure 154). Linton (1923:441) believed that these figures were attached to the beams of houses in *meae*. Although the timeframes and functions are unknown, a missionary source suggests that tattoo masters displayed the figures on the *tohua* during large festivals (Panoff 1995:131).

Normally women were the principal *tapa* producers. However, the priests beat the ceremonial loincloths used to dress the tiki (Crook 2007:52). Stewart (1832) described elderly inspirational priests (*tuhuna*) beating loincloth for the gods. The act was so sacred that the priests were only allowed to eat in the evening (Handy 1923:238).

Besides tattooing and *tapa*, the Enana were particularly skillful carvers as evidenced by the objects in wood, stone, ivory, and bone collected by early visitors (Ivory 1990; Millerstrom and Edwards 1998). The decorative motifs, particularly the wood clubs, have been extensively studied by Ivory (1998). However, studies on wood and stone sculptures, housed in many museums around the world, are limited.

Anthropomorphic Sculptures

During the Marquesas Rock Art Project 81 stone tiki were located and documented in 17 valleys on 5 of the presently 6 inhabited islands (Millerstrom and Edwards 1998; Millerstrom 2005b). The majority, 53 images or 65.5 percent, were found on Hiva Oa (Table 2). A total of 43 tiki were documented in situ; the remaining 38 sculptures had been relocated, and their archaeological contexts are uncertain. It is generally assumed that they all represented men. Leaving out sculptured heads from the calculation, there are 58 torsos and fully carved tiki. Of this number, 33 sculptures (57 percent) have no sex depicted, while 17 sculptures (29.3 percent) represent males, and 8 tiki (13.8 percent) represent females. A total of 38.3 percent of the sculptures were carved from sacred red volcanic tuff, but some basaltic tiki were once painted red (Millerstrom and Edwards 1998). Traces of red pigment were found in protected parts of a statue on Meae Iipona, Hiva Oa (Linton 1925:162), and on an image in Punaei Valley, Hiva Oa ([Figure 6] Millerstrom 1985c). The color red was considered sacred in Polynesia.[18] Tiki are associated, for the most part, with ritual places, chief's platforms, or temples located within the *tohua*. While the ages of the sculptures are uncertain, genealogical calculation (von den Steinen 1969(I):84–86 [1928]), excavation at two sites with tiki (Ferdon 1965:117–121; Heyerdahl 1965:123–150), and architecture associated with tiki (Suggs 1961), suggest that the sculptures were first made around A.D. 1600–1700. During his 1997 field season in Vaitahu, Tahuata, Rolett found four sculptured stone tiki heads on the surface of a *meae*. Rolett noted that the tiki probably dated to the late prehistoric or early historic period ca. A.D. 1700–1850 (Donaldson 2004:39–367; Rolett 1997).

Table 2. Frequencies and percentages of tiki according to islands and valleys*

Island	Valley	Tiki (n)	%
Hiva Oa			
	Puamau	30	–
	Atuona	9	–
	Hanaiapa	3	–
	Eiaone	3	–
	Taa Oa	2	–
	Tahauku	2	–
	Mutu Ua	2	–
	Hanapaaoa	1	–
	Punaei	1	–
	Subtotal	**53**	**65,5**
Nuku Hiva			
	Taipivai	12	–
	Hatiheu	4	–
	Taiohae	4	–
	Subtotal	**20**	**24.7**
Ua Huka			
	Hane	4	–
	Vaipaee	1	–
	Hokatu	1	–
	Subtotal	**6**	**7.4**
Tahuata			
	Vaitahu	1	–
	Subtotal	**1**	**1.2**
Ua Pou			
	Hakamoui	1	–
	Subtotal	**1**	**1.2**
Total		**81**	**100**

* The original location of each tiki is not taken into account.

Based on morphology, five main categories are distinguished: 1) statues carved in full round; 2) modified boulders; 3) rectangular blocks with tiki figure in bas-relief; 4) double figures; and 5) other. The last category includes two separate horizontally placed tiki resting on a block and one seated figure. The most common visual characteristics are large circular eyes, wide nose, mouth, flexed knees, and arms placed on a protruding stomach. The head, because it was considered tapu and the seat of *mana* was the most important part of the statue, and it is carved with careful attention to the smallest details. It may include intricate facial features, tattoos, and headbands, while the rest of the body is only roughly outlined. Typically, the size of the head is one-third or more of the body's height. It is usually resting directly on square shoulders lacking a neck. A tiki at Meae Iipona is unusual because of the small size of the head in relation to the body (Figure 7). It is possible that the original head broke off, and it was recarved to fit into a groove in the upper torso. Round, large, and bulging eyes are often encircled with a 1–2-cm-wide raised rim under high arched eyebrows. Sometimes the eyes have a raised or incised curved line that bisects them from the outer part of the eye to the inner corner of each eye. Some tiki have circular indented pits indicating the pupil. The tiki vary in size from 32 to 250 cm above the ground. The majority of the statues, 50.6 percent, measure between 50 and 100 cm.

A broad flat nose with wide nostrils fills out most of the center of the face. The outline of the nose goes upward and connects with the eyebrow. High eyebrows reach down on the outer side of the head and link with the ears, resembling the stem of eyeglasses. A long, wide mouth, sometimes with a protruding tongue or even teeth with great canines, covers the lower part of the face. When tattoos occur, they depict anthropomorphs, dogs, and geometric motifs. Hands with fingers, sometimes 6 fingers, are most often resting on a protruding abdomen. As ritual knowledge, genealogy and oral tradition were believed to be held in the stomach, hands placed on a protruding belly may have been a way to protect these memories (Thornton 1989).

Wide, rounded hips and short, stubby flexed legs rest on a pedestal or a peg base. Some images have the legs only indicated, while other sculptures are cut off below the waist. Ankles may be indicated by raised circular knobs. Only a few images depict toes. The back is often carved with spine and buttocks even when the statues are fitted into a stone platform. Decoration may include headdresses, wreath around the neck, tattoos, short mantles, and hair knots on each side of the head. Some tiki have drilled circular perforation in the earlobes for placement of earplugs. Offerings of human victims, animals, fruit, and vegetables were placed in front of the statue or hung in nearby trees (Stewart 1832:223). Robarts wrote that during the memorial feast, food was always sent to the priests at the *meae*. A small portion of the food was placed on the head of the image. Robarts wrote (quoted in Dening (1974:57): "This no one eats, being held sacred." Garlands strung form the pandanus keys (***Pandanus tectorius***), especially the red type, decorated the neck of tiki during feast (Petard 1886:84). During some ceremonies the tiki was girdled in *tapa* (Handy 1923:238).

The Swedish ethnographer, Hjalmar Stolpe, who studied the relationship between human figures and geometric designs in Polynesian ornaments collected in the early historic period, believed that every geometric form derived from a realistic representation (Stolpe 1890:213). For instance, the human tiki figures on Austral ceremonial paddles, ladles, bowls, and Janus-type or back-to-back figures, gradually transformed to geometrical types in which the prototype was entirely obscured (Stolpe 1927:213–218). Stolpe suggested that the Polynesian images were primarily symbolic before they became decorative.

In general, Polynesian decorative systems are pervasive, especially in Western Polynesia (Green 1979). Both Linton (1923:269) and Porter (1970:110 [1822]) stated that the Marquesan material culture was homogeneous. Burrows (1938:41; Table 1) found that human images were more often depicted on all Marquesan media (e.g., houses, canoes, implements, and ornaments) than on the same media in the other Polynesian islands or island groups. In a similar study, Greiner (1923) numerically demonstrated that the distributions of design elements and motifs used in carving occurred in more Marquesan media than in the islands on Hawaii, Tahiti, Austral, Tonga, Fiji, Samoa, the Cooks, and New Zealand. Unfortunately, Greiner did not have the opportunity to examine Von den Steinen's (1969(I), 1969(II), 1969(III) seminal work on Marquesan material art. His work would have bolstered Greiner's argument.

The similarity of much of the archaeological art may be, in part, because the *tohuna* (craft specialist) apparently could safely travel between islands (Crook 2007; Melville 1964:162 [1846]). However, the homogeneous decorative system demonstrates that the Enana adhered to a common ideology and belief system.

It is useful to examine the imagery within a theoretical framework that DeBoer (1991) has glossed as a pervasive versus partitive decorative organization. The two opposing art systems derive from DeBoer's work among the Shipibo people (in the tropical forest of the central Ucayali Basin in the Peruvian Amazon) and the Chachi (located in the northernmost coast of Ecuador) in South America. A pervasive decorative style exhibits distinct attributes in most media that, in the Shipibo case, was emphasized in pottery, textiles, face and body painting, calabashes, house posts, canoes, canoe paddles, and so forth. A decorative style expressed on the surface of different media identifies and may represent the ideology of a specific culture or group of people. A society agrees on the ideology by unconsciously participating in it. Partitive art systems, however, display no uniform decorative organization in the different media. Instead they have a number of decorative modes, each of which is exhibited in different material objects. DeBoer (1991:157) argued that the partitive decorative organization "experience[s] a much higher rate of change." because the "decorative organization displays weak linkages across various artifactual media, any one medium is freer to change without threatening massive, across-the-board readjustments." DeBoer's theoretical perspective has important implications regarding changes in the Marquesan art system and how these changes reflect social transformation.

Tattoo motifs and carved material objects, such as wooden clubs, heads of stone pounders, stilt footrests, *tapa* masks, head decorations in turtle and sea shells, bone and ivory fan handles, earplugs, hair ornaments, and so on, collected in the early historic period show remarkable similarities in their decorative systems. The similarities seen in Marquesan rock art, tattoo, and material objects suggest that the fundamental principles regarding the symbolic order remained largely unchanged.

Discussion

The Enana's first exposure to Western culture and social rules began with the navigators; this was followed closely by exposure to missionaries and beachcombers. Each visitor arrived with a different agenda and personal bias. Consequently, their descriptions often differ both in respect to their perception and to details concerning the indigenous population, material aspects, and social practices. For instance, Robarts' and Cabri's views on the Enana are biased considering that they were adopted and married into chiefs' families and thus became members of a high-status or tapu class. Crook's account is often considered the best source of information; however, in contrast to Robarts and Cabri, Crook was perhaps largely viewed as an outsider. While a chief adopted Crook according to the Marquesan customs, Crook was never part of the social system. Unlike Robarts and Cabri, he did not contribute food, marry, or partake in warfare. In addition, Crook's missionary activities were probably not respected by the inhabitants in the early part of the contact era.

Major social changes occurred both after Porter's long stay and the serious demographic decline. In 1839, Thomson, a missionary with the London Missionary Society, lamented a culture that was in the process of collapsing (Craig 1980). The French Captain Jules S-C Dumont D'Urville made a comment in 1838 that the people of Nuku Hiva were still unconverted (Rosenman 1992:356). One missionary stated that "the islanders were resistant to the blandishment of Christianity" (Rosenman 1992:143). It took several years and many battles to subdue all the tribes (Oliver 1975:208–209). French interest was never strong and once, in 1859, the administration abandoned the islands altogether (Oliver 1975:208–209).

Notwithstanding the social transformations that took place in the archipelago, the Enana's spiritual organization endured. Even though many rituals were forbidden, meanings were often retained. A letter dated December 20, 1862, to Rev. Damman, Seaman's Chapel, Honolulu, from Thomas Lawson, a missionary, stated:

> A year ago nearly the whole of Nuuhiva [Nuku Hiva] was under the sway of the Roman Catholic missionaries. Now nearly the whole of Nuuhiva has departed the Church and burnt some of the churches down and a few weeks ago some natives fired a musket at one of the Baptist missionaries after stealing a many of thier [*sic*] things the condition of the Protestant church is on the whole favorable but far from what it ought to be there has been one man eaten on Nuuhiva this year it is the first that has been eaten for many years on that Isle and the natives have gone back to their old heathern practices of tatooing and offerings of sacrifices to their [*sic*] old Deities.

A common thread in the spiritual organization can be traced throughout the Enana's history as it is expressed in the decorative system. In particular, the anthropomorphic faces and figures described in funeral context also occurred on clubs, canoes, tattoos, and a number of other material objects. Early visitors observed these motifs both in the southern and the northern island groups, demonstrating that these images occurred throughout the archipelago. Anthropomorphic figures, in various contexts, appear to have been standardized and omnipresent throughout the protohistoric and historic periods. Following DeBoer's arguments, the Enana's ideology and belief system, as it was expressed in the material culture, changed slowly.

Many Europeans perceived the Enana's images as "rude" and "hideous," and their social significance was not seriously considered. Except for human figures on canoes (Ingraham 1790–1792; Wilson 1997 [1797]), the images described were placed on a sacred place or a *meae*; many of them were linked to a funeral *meae*. In addition, figures were also inscribed on stones and tree trunks. While images occurred in various contexts, we cannot assume that the canoe figures were purely decorative and that they did not contain religious or ritual function. Haddon and Hornell (1975:41 [1936, 1937, 1938]), for instance, referred to "the grotesquely carved face used as a figurehead" on canoes as ornaments and decorations. A great deal of danger was involved in each fishing expedition, thus various rituals, some that included portable stone tiki at the fishermen's shrine, preceded each trip. This was both to ensure a good catch and for protection (Handy 1923:165; Meyer 1995:494). The connection between canoes and the spiritual world is interesting. Coffins were often referred to as *vaka*, or "canoes" and were frequently canoe shaped (Handy 1923:111–112; Linton 1923).

Only the descriptions by Crook and Tautain can perhaps be understood to pertain to petroglyphs. Despite the hundred years that separate them, both referred to the images in connection with funeral rites. In regards to tattoo motifs, Crook mentions only geometric designs and "lagoons," or open circles. Facial tattoos, according to early illustrations, consisted of bands, circles, lines, and so on; they rarely depicted human faces (Handy 1922; Ottino-Garanger 1998). Tautain's comments suggested that engravings on trees trunks and stones were made in the historic period. Although no traces of tree trunk carvings have been found, Tautain's statement indicated that both living trees and stones served as mediums. Generally, the early descriptions refer to anthropomorphic sculptures rather than petroglyphs. Perhaps the ancestral tiki sculptures were still part of the Marquesan religious practice in the early Christian era, while rites that involved anthropomorphic stick figures as seen in petroglyphs had been practiced only in the distant past.

The majority of Marquesan petroglyphs are placed on boulders, dressed slabs, and outcrops with no apparent spatial or temporal association. However, several images are incorporated with architecture, even though, in some cases, the architecture is difficult to identify. Later in the manuscript, based on archaeological information, ethnohistoric data, and ethnographic sources, I construct a model of three common architectural complexes: 1) residential household cluster; 2) the communal ceremonial complex; and 3) the various sacred structures. These are all localities where rock art occurs.

Chapter 3

The Marquesan Archaeological Record

Petroglyphs were seen in the islands of Hivaoa, Tahuata, and Uapou, but are by no means common in the Marquesas.

—— Ralph Linton, *Archaeology of the Marquesas Islands*

History of Archaeological Research

In this part I examine the history of archaeological exploration in the Marquesas as it relates to my research, furthermore it includes a discussion of the results of the Marquesas Islands Rock Art Project. Despite a long history of archeological research in the Marquesas Islands (e.g., Linton 1925; Sinoto 1966, 1968; Suggs 1961; Kellum-Ottino 1971; Ottino 1985; Rolett 1998; Allen 2004, 2014; Allen and McAlister 2013; Conte and Molle 2014), rock images have not been systematically documented or analyzed. Prior rock art investigations consist of reports plus a few publications that deal with only certain aspects of the images (e.g., Danielsson 1954; Linton 1925; Métraux 1937; Rolett 1986; Suggs 1961). However, rock art was not the main focus of these research efforts. In contrast, stone and wood anthropomorphic sculptures attracted scholarly attention (Donaldson 2004; Ferdon 1965; Handy 1923; Heyerdahl 1965; Linton 1925; Molle 2013; von den Steinen 1925, 1969(I), [1928], 1969 [1928]).

Thanks to the foresight of Herbert E. Gregory (director of the Bernice Pauahi Bishop Museum, Honolulu, from 1919 onward), an extensive scientific investigation in the Pacific was planned (Krauss 1988:34). The main goal was to gather data on vanishing "classic" Oceanic cultures. A Marquesan expedition was sponsored by the museum and financed by Bayard Dominick of New York and set off for the Marquesas in late 1920. The project members of the Bayard Dominick Expedition were archaeologist Ralph Linton, ethnologist E. S. Craighill Handy, and his wife Willowdean Chatterson Handy (called a volunteer associate). They arrived in the islands on September 21, 1920, and remained until June 21, 1921 (Handy 1923:3). Most of their work was centered on Hiva Oa in the southern group, although Linton mapped architecture and features on all of the six inhabited islands. The Handys focused on cultural aspects including religion, legends (Handy 1923, 1927, 1930a), decorative art (Handy 1922, 1938), string figures (Handy 1925), and music (Handy and Winne 1925). Handy (1965, 1973) subsequently authored two books, one of which is a historic novel. Linton concentrated mainly on ceremonial sites. He mapped and described some 170 structures of which 72 were *tohua*, and 58 *meae*. Some sites were roughly excavated (Linton 1923, 1925). Furthermore, Linton and Wingert (1946) coauthored *Arts of the South Seas*, and Linton (1939) wrote the chapter "Marquesan Culture," in Kardinier's publication *The Individual and his Society, the Psychodynamics of Primitive Social Organization.*

The expedition was a salvage operation intended to obtain information about a "classic" culture that, in many respects, no longer existed. By the time Linton and the Handys arrived, profound social changes had already taken place during the previous century and a half since Western intrusion and occupation. Most of the ancient traditions were no longer practiced, and many of the elders who might have remembered the islands' history had died. Some islanders were reluctant to show ceremonial sites to foreigners, for they believed a great number of them were *tapu*. For instance, although early European settlers reported seeing petroglyphs in the Hatiheu Valley, when Linton visited the area, the residents denied any knowledge of them (Linton

1925:96). Linton declined to speculate on the images' ritual functions. He wrote that "so few petroglyphs were found that it is unsafe to generalize about their occurrence" (Linton 1925:96). During his brief visit to the Hatiheu Valley, Linton only mapped Tohua Nanauhi, a large tribal ceremonial complex located on the east side of the valley. Based on his extensive archaeological work throughout the archipelago, Linton found that most of the images were located on or near ceremonial structures. The most common figures, according to Linton (1925:97), were concentric circles followed by "grotesque faces, with round eyes and oval mouths." He noticed that the eyes of the grotesque faces were similar to the concentric circles and speculated that some of the concentric circles were unfinished faces.

After the Bayard Dominick Expedition, the archipelago was archaeologically ignored until the late 1950s. According to contemporary accepted dogma, attempts to excavate Polynesian sites were thought to be futile since Polynesian occupation was considered a recent occurrence (Dening 1966; Kirch 1982). Moreover, artifact preservation in subtropical climates was believed to be poor (Danielsson 1967; Fowke 1922). In fact, Linton (1925:3) estimated that 90 percent of the Marquesan artifacts were made of wood and vegetable fibers. In addition, Oceanic cultures, for instance the Hawaiian Islands group, had been viewed in the past as uniform (Kirch 1990).

Pioneering stratigraphic excavation was first done by Robert C. Suggs, a member of the American Museum of Natural History Expedition to the archipelago in 1956. Based on artifacts and radiocarbon dates from several excavated sites on Nuku Hiva, Suggs (1961) developed a unilineal evolutionary cultural sequence of five periods. An initial settlement date of approximately 150 B.C. (Shapiro and Suggs 1959; Suggs 1961), the earliest evidence of human colonization in East Polynesia, was based in part on excavations at the Haatuatua (Nhaa 1) sand dune site.

At the same time the Norwegian Expedition, led by the explorer Thor Heyerdahl, was engaged in archaeological research on Nuku Hiva and Hiva Oa. Excavations at Meae Iipona, Hiva Oa (Heyerdahl 1965:123–151), and at Paepae Paeke on Nuku Hiva (Ferdon 1965:117–122) yielded radiocarbon age determinations (uncalibrated) for the architecture that also, they believed, provided relative dates for the stone sculptures. They suggested that the tiki were placed at Meae Iipona approximately A.D. 1400–1500 (Heyerdahl 1965:123–151), while the sculptures at Paepae Paeke were incorporated into the structures around A.D. 1600 (Ferdon 1965:117–122). These dates are significant for placing the images chronologically, since petroglyphs are found at both sites.

Sinoto (1979, 1983), who excavated both at Haatuatua and Hooumi, proposed that the islands were settled approximately A.D. 300. Results of the carbon analyses of Suggs and Sinoto have since troubled archaeologists and led to spirited debates (Kirch 1986; Kirch and Ellison 1994; Spriggs and Anderson 1993). Excavation in Anaho and careful examination of the age determination elsewhere in the islands has led Allen (2004) to rethink the cultural chronology (Table 3).

Table 3. Marquesan cultural chronology

Cultural Period	Suggs (1961)*	Sinoto (1970)	Allen (2004)
Settlement	150 B.C.–A.D. 100	300–600	700–1100
Development	100–1100	600–1300	1100–1300
Expansion	1100–1400	1300–1600	1300–1600
Classic	1400–1790	1600–1800	1600–1970
Historic	1790		

*Unless noted, all dates are A.D.

Fourteen undecorated potsherds, classified as Polynesian Plainware, from three different islands were excavated at Haatuatua and Hooumi, Nuku Hiva, and Hane, Ua Huka (Sinoto 1979; Suggs 1961); two sherds, collected by a young Marquesan during a house construction and thus lacking archaeological context, were in 1985 obtained by Edwards and Millerstrom in the Atuona Valley, Hiva Oa (Kirch et al. 1988). Petrographic analysis on the temper demonstrated that the sherds excavated at Haatuatua were derived from the Western Pacific, specifically the Rewa Delta of Viti Levu in Fiji (Dickinson and Shutler Jr. 1974; Dickinson et al. 1998:119); other sherds indicated local Marquesan origin (Dickinson and Shutler Jr. 1974; Kirch 2000:258; Kirch et al. 1988:101–107, see also Allen et al. 2012).

Controversy surrounding the issues of the earliest Marquesan settlement phase and the islands' chronological cultural sequence led to a multiyear, interdisciplinary project by the University of Hawaii in collaboration with the SCP and the Université Française du Pacifique (Rolett and Conte 1995:195–228). The first fieldwork at Haatuatua took place in 1992. The main goal was to test the area excavated by Suggs (1961) and to examine the large flat region inland from the sand dune. Excavation yielded information on site function, tattooing, technological change, and especially on fishing and tool-making practices. Two key hypotheses of Suggs's, initial human settlement and the extent of human occupation, were evaluated. None of the ten radiocarbon dates supported the hypotheses of early colonization, nor did the data show evidence of occupation in the early periods of Marquesan prehistory as proposed by Sinoto (1970:105-132). Between A.D. 1300 and A.D. 1650, the sand dunes were widespread and intensely occupied. After A.D. 1650 occupation appeared to have been sparse (Rolett and Conte 1995:223–226).

The process of Marquesan initial colonization remains largely unresolved (see Allen 2004:143–196; Anderson and Sinoto 2002:251; Rolett 1993, 1996:531–540, 1998; Sinoto 1996:131–152). Weisler and Green (2011:244) reevaluated the southeastern Polynesian radiocarbon dates and dismiss the dates that support a long cultural sequence (e.g., the Haatuatua 150 B.C. date). They support Allen's (2004) research that places the Marquesas colonization as early as cal A.D. 700 and that settlement on the islands was established a couple of centuries later. In the future, archaeologists will have to grapple with problems such as small and scattered early human traces, as well as sites destroyed by recent human settlement, erosion, or buried through alluvial deposits and natural shoreline erosion. Village settlement in the lower part of the valleys, road construction, and development and modification of infrastructure are part of the destruction of colonization sites as well.

Only Suggs (1961:142–151) examined the stylistic development of rock art, and he established a tentative chronology based on its occurrence at dated sites. All Suggs dates are uncalibrated and his chronology and most of Sinoto dates have been revised by Allen (2004), In this publication I will only use Allen's cultural periods. Based in part on a gallery with figures discovered under several inches of coquina (limestone mixed with crushed shell) in a dry riverbed south of Nhaa 1, area A, at Haatuatua, Suggs (1961:60, Plate 11ab) concluded that the human stick figures, dog, fish, whale, and turtle were present during Development period while the *mata komoe* was a later development.[19] The earliest evidence for human face motifs was excavated in a stratified archaeological site on the north coast of Nuku Hiva. There, at Haataivea Bay, Suggs (1961:66, 145) uncovered a petroglyph slab with a *mata komoe*; it was associated with a fire pit (Site Nhtv1, Stratum II). The site was first occupied Expansion period. Suggs (1961:91, Figure 28d) identified the figure as a "stylized parakeet, inscribed with a basalt flake on a slab of pillow-lava." The figure does look like a parakeet; however, on closer inspection, the figure actually is a human face with a beard, and it is stylistically similar to a *mata komoe* I recorded elsewhere on the north coast of Nuku Hiva.

Overall, the most frequent motifs encountered were squared concentric circles, speared concentric circles (*ipu*), human stick figures, and human faces. In his discussion, Suggs (1961:142–155) compared the rock carving data with similar material from other eastern Polynesian islands and speculated on their link with Melanesia.

In relation to rock art, the human occupation sequence at the Haatuatua dunes is of interest. Inland from the dunes there are remains of extensive ceremonial complexes, including monumental sacred and residential architecture. Many are associated with rock art; this affirms that the interior of the valley was heavily settled at one time (Millerstrom 1988). Because no archaeological excavation has taken place in the interior of Haatuatua Valley, it is uncertain when the valley was occupied. Perhaps settlement shifted from the beach area to the interior of the valley sometime after A.D. 1600. If so, this hypothesis fits well with Allen's general settlement pattern model based on dates from Haatuatua dunes and her dates from the coastal flats at Teavauua, Anaho, (adjacent to Haatuatua). It indicates that, from approximately A.D. 1600 onward, communities were established in the interior (Allen 2004:190). Based on this scenario, the majority of rock art was associated with inland settlement at the height of prehistoric Marquesan cultural development. However, petroglyph sites are also found on many of the Marquesan beaches, although in small numbers.

The Marquesas Rock Art Project

Between 1984 and 1989, Edmundo Edwards and I worked with various volunteers on five of the presently inhabited islands: Nuku Hiva, Hiva Oa, Ua Huka, Ua Pou, and Tahuata.[20] Images have been reported on remote Eiao Island (J. Candelot, pers. comm. 2012), but so far none of these have been documented. Work was carried out in 41 valleys, and over 6,000 individual design elements (petroglyphs), 110 wall paintings (pictographs), and 81 human stone sculptures (tiki) were documented, sketched, and photographed. In addition, numerous associated architectural structures and features were mapped. During the course of the project, efforts were made to involve the Marquesans in our work by sharing information and location of sites discovered. Several local inhabitants enthusiastically assisted in our work, and children from elementary schools visited some of the sites. A few petroglyph locations were already known to the Marquesans, but most were located as a result of our field survey. The vast majority of the sites were previously undocumented.

After 1989, I carried out additional archaeological surveys and several new sites were located and documented by my team. However, none of the islands have been systematically surveyed, and sites continue to be discovered during road and building construction. Many remain undocumented because they have not been reported to archaeologists. Petroglyphs occur on all the islands visited, although site distribution varies from valley to valley. The relatively homogeneous rock art imagery throughout the island group suggests that a great deal of intervalley and interisland contact always existed.

Marquesan rock images are found on boulders, walls of rock shelters, outcrops, beach rocks, and slabs. In geological terms, using the Udden-Wentworth classification, a boulder can be defined as a rock larger than 25.6 cm in diameter (Wentworth 1922). An outcrop is a visible bedrock formation protruding through the soil. Beach rock, locally referred to as *papatea* (white rock), is a slab cut from well-cemented mixtures of gravel, sand, shell, and coral fragments that has formed along shorelines. The Marquesan word ***keetu*** translates to a "stone in red tuff that is easy to work" (Dordillon 1931). Today the Marquesas word refers to a modified block or slab of basalt, beach rock, or red or yellow volcanic tuff.

Our survey was research oriented, and numerous publications resulted from this work (Edwards 1989; Edwards and Millerstrom 1995; Millerstrom 1985a, 1985b, 1988, 1990, 1992, 1997a, 2002, 2003a, 2003b, 2003c; 2005a, 2005b, 2006a, 2006b, 2006c; Millerstrom and Edwards 1998; Millerstrom and Allen 2006; Millerstrom and Coil 2008; Millerstrom and Rogers 2005; Millerstrom and Baumgartner-Lesage 2017). Following is a brief discussion of the rock art research undertaken on each island prior to 1999. It should be noted that not all field information has been analyzed, thus conclusions are tentative.

The Northern Island Group

Nuku Hiva Island

More petroglyph sites were recorded on Nuku Hiva than on any of the other islands visited. This may be due to the more extensive survey conducted on the island. However, compared to the other islands in the archipelago, Nuku Hiva is both the largest island and likely was the most populated island in the past. Most of the rock art sites occurred in the valleys of Hatiheu, Akapa, and Haatuatua, three fertile areas located on the north and northeast side of the island. While some environmentally marginal areas do contain rock art, most of the sites are found in fertile valleys. Taipi Valley is a notable exception. In former times, Taipi was one of the richest and most densely populated valleys in the archipelago. Early Catholic mission records noted the names of 14 *tohua* and 32 *meae* in the valley (Linton 1925:114), yet only 6 individual figures have been documented ([Figure 8] Millerstrom (1990, 1992). However, located on a ridge at Vaihi, along the old trail between Taipi and Taiohae, there is a cluster of boulders with images. A megalithic boulder contains incised historic images and lettering, some superimposed on traditional pecked figures ([Figure 9] Millerstrom and Rogers 2005). At

Paepae Paeke, a site situated on a narrow ridge to the west of the village, there are twelve anthropomorphic sculptures associated with the remains of several platforms and a petroglyph boulder.

Small valleys, such as Anaho on the north coast, have only a few rock art sites; although judging by the numerous house platforms, the valley was well settled at one time. Agricultural terraces can be seen along the rivers on both the east and west sides, and numerous house sites are found on the east side. Allen (2010) recorded approximately 300 stone structures in the valley. Two anthropomorphs and some curvilinear motifs carved on a large basalt boulder, perhaps once part of a fisherman's shrine, are located on the beach to the east valley (Figure 10). Another image boulder is located in a home and its provenience is questionable. In 2002 we discovered a small sculpture on a fisherman's shrine in the center of the beach at Anaho (Figure 11).

Haataivea and Maatea, two isolated valleys between Anaho and Haatuatua, are unique. Two quarries are located on each side of the bay at Haataivea (Suggs 1961). On the east side, a 2–3-m wide, red tuff band is exposed in the mountainside, and a cave is situated above. In the cave, approximately 5 m above the ground, two ***mata komoe*** (anthropomorphic faces) were pecked into the floor (Figure 12a–b). Rectangular outlines, the result of removing blocks of tuff, are visible on the north side of the tuff band. On the surface a roughly outlined tiki is still in place. A basalt adze quarry with a ***papatea*** pavement is situated across the bay.

Maatea, an area by the sea with uplifted limestone or ***papatea***, contained several rock art sites (Figure 13).[21] The images are located on natural raised blocks of ***papatea*** in the vicinity of campsites. Fishermen probably occupied the rockshelters from time to time because pieces of pearl shell fishhooks were observed on the surface around the hearths.

Hatiheu Valley

Hatiheu Valley has an especially rich prehistoric rock carving tradition (Millerstrom 1990, 1997a). Carved outcrops and boulders are distributed throughout the valley, but the majority is clustered on the western side. The rock art typifies the images documented throughout the islands.

Delmas (1927:100) mentioned six ceremonial sites in Hatiheu (Nanauhi, Pahumano, Kamuihei, Maikuku, Hikokua, and Paahaua), but the remains of seven tribal ceremonial complexes are still standing in Hatiheu. For instance, Tahakia, located on the west side of the valley, was not mentioned by Delmas. As these tribal ceremonial centers are part of the study area and linked to my survey and excavations, they will be discussed in Chapter 4. Tohua Pahumano is located near the beach on the east side of the valley. Public gathering places, such as the grammar school, the post office, health clinic, and the town hall, cover the area today. Linton (1925:117–119), although aware of other ceremonial and temple sites in Hatiheu, only mapped Tohua Nanauhi.

Even though the tribal ceremonial complexes vary greatly in size (see Table 8), the size of the ***tohua*** does not seem to be reflected in the amount and complexity of the rock art. For example, no petroglyphs were found on Tohua Nanauhi even though a few occur in the vicinity. However, a great number of petroglyphs are found on Tohua Maikuku, a relatively small tribal complex located in the interior of the valley. The environment is marginal and only a small intermittent stream is located nearby. Approximately 20 structures and several communal breadfruit silos are placed around a 35-×-14-m rectangular dance plaza. Of special interest is an 11-m long narrow alley on the west side with anthropomorphic faces and dogs on several of the boulders (Figure 14); these figures are also the most frequent motifs on the ***tohua***. Located inside the ***tohua*** is a small paved terrace with several upright boulders; the largest upright has a face and several eye motifs (Figure 15). This small paved terrace with several upright boulders, one of which has pecked figures, was probably a sacred or ritual place. On a steep ridge called Te Ivi Maikuku, directly above the tribal ceremonial complex, additional stones with similar motifs were discovered following a grass fire in February, 1995. The majority of the figures depict faces, but dogs are also found.

As of 2014 the carvings in Hatiheu alone totaled more than 3,379 individual motifs representing 478 different outcrops, boulders, and ***keetu*** (Table 4). Geometric motifs dominate with 2,121 figures (62.8 percent), 998 figures (29.5 percent) are anthropomorphs, and 256 figures (7.6 percent) represent zoomorphs.

Table 4. Frequencies and percentages of image types in the Hatiheu Valley as of 2016

Image Type		n	%
Anthropomorph			
Stick Figure		521	
Square-bodied		27	
Double-outlined		25	
Open-bodied		49	
Naturalistic Figure		79	
Profile-squatting Figure		1	
Human and Dog Combination		5	
Human and Bird Combination		1	
Human and Fish Combination		1	
Human-lizard		3	
Exotic		15	
Tiki		4	
Face		267	
	Subtotal	**998**	**29.5**
Zoomorph			
Dog		162	
Lizard		6	
Turtle		32	
Fish		41	
Octopus		3	
Bird		10	
Unidentified		2	
	Subtotal	**256**	**7.6**
Material Objects			
	Subtotal	**4**	**0.1**
Geometric Motifs			
Curvilinear		1,233	
Linear		256	
Cupule		296	
Exotic Figure		336	
	Subtotal	**2,121**	**62.8**
Total		**3,379**	**100**

Ua Huka Island

While working in the Hane Valley, Ua Huka, in 1985, we failed to relocate many of the sites reported by Kellum-Ottino (1971); however, we found several undocumented sites. Other undocumented sites were found in the valleys of Vaipaee, Hokatu, Hatana, Haavei, Hatuana, and Vaikivi (Edwards 1984; Millerstrom 1985c). A *meae* site in Vaikivi, located in the interior, consists of one raised *paepae* on each side of the river and an outcrop with 19 petroglyph panels (Figure 16). Two banyan trees grow on the site. One banyan, dislodging part of the structure with its extensive root system, covered over five petroglyph boulders fitted into the north wall. On the riverbank, a boulder depicts faces with cupules as eyes and two naturalistic pecked figures (Figure 17). The cupule depth ranges from 1.0 cm to 2.5 cm. A rectangular *ma* (fermented breadfruit) pit with figures incised on its clay walls is located in the same valley (Figure 18). Several previously unknown petroglyph sites have since been located on Ua Huka and some have been documented by us.

Ua Pou Island

On Ua Pou, 19 boulders with more than 226 individual images were documented. Although the majority was located in Hakamoui, a valley formerly occupied by chiefs who were never defeated in war (Handy 1923; Linton 1925:133), some images were recorded in Hohoi, Hakahetau, Hakatao, and Hakaohoka (Figure 19). Of special interest is a rectangular pit, situated under a banyan tree on top of a narrow ridge above Hakahau Valley. First recorded by Jean-Louis Candelot of Ua Pou and published in *Les Nouvelles,* April 22, 1987, the approximately 2-m deep pit depicts incised anthropomorphs, turtles, birds, and vertically placed cupules on all four walls (Figure 20). Candelot believed that it functioned as a silo for fermented breadfruit. However, breadfruit fermentation pits with images are neither mentioned in historic records and oral traditions, nor have they, to my knowledge, been examined.

In 2004, a megalithic boulder (6 m in diameter) was found during road construction at Vaipapa, Hakatao. Numerous deeply pecked turtles and a double outlined anthropomorph are depicted. The boulder forms part of a *meae* (Millerstrom 2005a:117–126). In Haakuti a boulder was discovered covered with numerous double outlined anthropomorphs similar to those near Kamuihei, Hatiheu Valley (J. Candelot, pers. comm. 2007).

Eiao Island

Eiao, the northernmost island in the Marquesas, was a center for adze making in the past. Adzes made of superior material from Eiao were traded to other islands in the archipelago (Candelot 1980; Charleux 2011; Linton 1925:106–107; Rolett 1998:189), and as far as Moorea in the Society group, and Mangareva in the Australs

(Weisler 1998:521–532). A few anthropomorphic petroglyphs have been noted on the presently uninhabited Eiao (J. Candelot, pers. comm. 2012). Information regarding the petroglyphs and possible associations are lacking.

The Southern Island Group

Hiva Oa Island

Ten valleys and the Taaovea Ridge on Hiva Oa were surveyed and a total of 73 sites and 592 figures were discovered and documented. Thirty-two sites with 272 figures, including pictographs, were discovered in Eiaone, a narrow valley on the north coast. In contrast, only two sites with two petroglyph boulders and two sculptures were located in Taa Oa, a large amphitheater-like valley on the southwest coast (Figure 21a). Puamau, the valley in the Marquesas that contains the most tiki (Millerstrom and Edwards 1998), has relatively few petroglyphs (Figures 21b), and only nine sites were documented. Several anthropomorphic figures are situated on outcrops at the upper part of the steep and narrow Taaovea Ridge between Eiaone and Puamau valleys. One female figure is located on a smooth rectangular surface (Figure 22). At this point the ridge is barely a meter wide. Several perforations along the edge of the petroglyph boulder appear to have been used to attach ropes, perhaps to assist while climbing up on the cliff-side or for safety while on the ridge. A small platform is located on a flat area below the figures. The narrow and isolated ridge has a spectacular view over the ocean and valleys below. The area could have served as a refuge, a lookout point to watch for enemies, or it could have been the residence of a priest and held some religious significance.

French Polynesia's only known pictographs are situated in rockshelters on the eastern mountain ridge of Eiaone Valley, on the north side of Hiva Oa (Edwards and Millerstrom 1995; Millerstrom 1985b, 1997a). Located on the back wall of 6 rock shelters are 110 anthropomorphs, quadrupeds, marine animals, and geometric figures (Figure 23). While the figures, painted in red pigment, are stylistically unique to the islands, a few associated pecked anthropomorphs are identical to the stick figures found throughout the archipelago. However, three quadrupeds are different from the dogs documented elsewhere in the island group. The choice of a reddish pigment suggests ritual value because red was a sacred color and was associated with high status and *mana* throughout Polynesia (e.g., Burrows 1938; Firth 1967; Handy 1927; Lee 1992; Van Tilburg 1994). Geometric motifs dominate the Eiaone paintings, comprising 45.5 percent (Table 5). Anthropomorphs make up 18.2 percent of the figures (Millerstrom 1997a). A couple of the figures appear to be superimposed with other figures.[22]

The Eiaone pictographs are difficult to see and to photograph. However, slides taken in 1985 that were converted to digital images by Ann Stoll (pers. comm., 2013) and examined through a software program became much more visible. DStretch is an enhancement technique, used in remote sensing, developed by Harman ([2011] www.dstretch.com) and is now used frequently to improve the visibility of rock art images that are difficult to discern.

Table 5. Distribution of pictographs in the Eiaone Valley, Hiva Oa, as of 2016

Image Type	n	%
Geometric Motif	50	45.5
Exotic Form	30	27.3
Anthropomorph	20	18.2
Fish	6	5.5
Quadruped	3	2.7
Sea Craft	1	0.9
Total	**110**	**100.1≈100**

Approximately 80 percent of the architectural remains on the valley floor of Eiaone were mapped (Edwards 1985b). Field surveys indicate that the lower part of the valley floor contained mostly ceremonial structures rather than domestic habitations. Subterranean storage pits for fermented breadfruit, commonly found around domestic sites, are absent in the lower part of the valley. Many of the architectural structures, especially at the head of the valley, have carved images on outcrops and boulders that form part of the house platforms.

A megalithic boulder (9.6 × 3.4 × 2.8 m) in Te Huetu Valley, inland from Tahauku Bay, a small valley east of Atuona (Vevau), depicts figures that have been compared with those from Rapa Nui

(Handy 1943:22–31). Located in a dry part of a streambed, the boulder is part of a *tohua* complex (Linton 1925:149–151).

Chavaillon and Olivier (2007) have documented and mapped numerous other sites on Hiva Oa. In the process they have discovered many previously unknown petroglyph boulders and sculptures. Their publication contains illustrations, maps, and excellent photos; however, their documentation techniques included cleaning the boulders, which is not recommended from a conservation perspective and may adversely affect any future attempts to date the petroglyphs.

Fatu Hiva Island

Inland on Fatu Hiva are two well-known sites. In Pupuauhi, Omoa, several petroglyphs were deeply pecked into the horizontal surface of an approximately 10-×-7-m smooth outcrop (31oomo 1). Marine animals, such as turtles, and a large image, thought to be a whale, are depicted in addition to several *mata komoe* and anthropomorphic stick figures. It is striking that one of the anthropomorphs depicts a female figure stylistically similar to figures on a boulder (331hth 2) in the Kamuihei area, Hatiheu (Figure 24).

François Ollier, then with SCP, recorded a vertical 12-×-3-m red tuff panel in the mountain of Hanativa Valley (Saquet 1990:19, 59). None of the panels appear to be directly linked to architecture. No systematic survey has been conducted on Fatu Hiva, but anthropomorphic figures have been found at the head of Hanavave Valley. Far inland in Omoa, when Meae Pele was restored for the December, 2011, Art Festival, two ***keetu*** with anthropomorphic figures in bas-relief were uncovered (L. Cantois, pers. comm. 2011). According to the same source, a dog petroglyph is located at the base of the sacred mountain Tatu Tea, a headless tiki was found at a site in the center of the valley, and an image of a whale/dolphin was uncovered on his land.

The spatial distribution of the Fatu Hiva sites is interesting, as all of the images are situated at the heads of valleys. While two of the known panels at Omoa and Hanativa are larger in size and the human and animal figures appear more complex, the distribution of image types are similar to the rest of the island group.

Tahuata Island

Five boulders, one of which could be classified as sculpture, were recorded on Tahuata (Millerstrom 1985c, 1992a). Not all of the valleys were visited however. Part of Motopu was briefly surveyed in 1985, but no petroglyphs were discovered. Local residents did not know of any sites. A megalithic boulder, located near the beach in Hapatoni, contains 14 anthropomorphic faces, several concentric circles, and 23 vertically placed cupules (Figure 25). The remaining four boulders are in the valleys of Vaitahu and Hanamiai. All the figures depict anthropomorphic faces and eyes (Figure 26a–d). The emphasis on *mata komoe* figures perhaps corresponds to the late settlement on the island. Excavation at Hanamiai suggested that initial occupation on Tahuata occurred approximately A.D. 1025–1300 (Rolett 1998).

Stones were Possessed with Godly Power

It is impossible to investigate carved images without considering what stones mean to the Polynesians. Some of the pecked and incised stones have been found hidden in the foundations of megalithic architecture. It is unclear what this means. In Hatiheu Valley, these figures depict both anthropomorphic stick figures and *mata komoe* motifs, suggesting that these types of motifs may have been among the earliest form of stone imagery in the valley.

There are several reasons for hiding or covering up the images: 1) the images had ceased to be part of their social practices; 2) they were hiding the figures from the missionaries and local inhabitants who had accepted Christianity; and 3) perhaps the image added ***mana*** or supernatural power to the dwelling. The answer, I believe, is a combination of all these alternatives. In the Polynesian past, all boulders were embodied with *mana* (Burrows 1938; Handy 1927), but some were thought to be more powerful or sacred than others. Boulders, like all things in nature, were believed to grow in the same manner as people and plants. The Enana believed that *keetu* "grows slowly but constantly, a quality peculiar to it" (Linton 1925:165). In fact, carving of stone was so important that certain rituals were observed while quarrying. Workers had to render themselves *tapu* and had to avoid women prior to working with stones or their stone tools would break (Linton 1925:165).

The notion that some carvings were considered special can be inferred from the manner in which the carver selected them for image-making and in the way the carver incorporated natural lines and protrusions in the material. In some instances it appears that the carver had interpreted a particular protrusion in the boulder as an image and modified it to make a conventional bas-relief figure as if to "liberate" the image from the boulder or outcrop. The same technique is noted in Easter Island rock art (Lee 1986:33), as well as in Paleolithic art (Conkey 1981:27). It is certain that some boulders were chosen over others according to specific cultural rules. Frequently a boulder with a smooth surface that appears to be a suitable "canvas" is left undecorated, while a nearby rough and uneven boulder may be covered with images.

That particular boulders were selected over others for their power is not a unique Marquesan cultural phenomenon. Henry (1928:382) wrote that in Tahiti "stone was possessed with godly power." It is also significant that the Polynesian term *papa* (stone) translates as "throne" in Tahitian (Handy 1930b:35). Lee (1992:124), for instance, states that petroglyphs placed on selected Easter Island *moai* (sculptures) are linked to the *mana* of the boulder

Orientation of Figures in the Hatiheu Valley

Table 6. Cardinal orientation of rock art in the Hatiheu Valley

Cardinal Direction		**%**
North		31.6
Northwest		1.7
Northeast		3.1
	Subtotal	**36.4**
South		14.2
Southwest		1.0
Southeast		0.8
	Subtotal	**16.0**
East		14.0
West		15.2
Horizontal		17.3
Reused Boulders		1.1
	Subtotal	**47.6**
Total		**100**

The distribution of the cardinal directions in the Hatiheu Valley demonstrates that of the majority of the figures, 36.4 percent face north, northwest, or northeast, the direction toward the sea (Table 6). The southern directions (16 percent) toward the mountains are less distinct. The strong seaward orientation is culturally significant and is in agreement with the fundamental principles of ancestral Polynesian notion of sacred and secular spatial organization (Firth 1936; Kirch 1996; Ottino 1990). For instance, Firth's ethnographic accounts relate how the indigenous houses of Tikopia, a Polynesian outlier, were spatially structured. The sacred part of the houses faced the beach and the ocean. Men occupied the sacred side of the house. Ancestors were buried beneath the floor, and ritual and sacred paraphernalia were stored in the rafters. The opposite section, the secular part of the house, had female associations. This organization of Tikopia social space was still in place in 1977 (Kirch 1996). Test excavations of two house sites at Tuakamali in Faea (Site TK-20), suggest that the essential sacred/secular distinction dates back to at least A.D. 1400–1800 (Kirch and Yen 1982:126–131). Ottino (1990), examining the symbolic spatial separation of the domestic structure of the Enana, proposed that the sacred (the sleeping area) versus the public part (front platform) of the house foundation is an ancient custom. Because the petroglyphs in other valleys have not been examined as closely as in Hatiheu Valley it is presently unknown if they show the same seaward/mountain, sacred/secular dichotomy.

Discussion and Summary

The investigations of Linton and the Handys, even though recognized as an important contribution to our present knowledge of Marquesan culture, are often regarded as "static" or frozen in time (Dening 1980:278–279; Ivory 1993:66). Moreover, Linton's archaeological survey lacked a diachronic approach (Kirch 1991a:132). However, their research provides an important cultural baseline that is a useful starting point for archaeological investigation. While their work was conducted a century after major social changes took place, they frequently evaluated the data in light of observations made by the early navigators and missionaries. Some of their informants were people who remembered when "the most extensive of ceremonial structures were still in use" (Linton 1925:4). Others were children of Enana who had participated in, or observed, rituals and were cultural masters in legends, genealogy, and chanting (Handy 1923:4). Even though the Marquesan population was drastically reduced when Linton and Handy worked in the islands some of the oral traditions were remembered and had been passed down from generation to generation. For instance, the occupants of the chief's Paepae Poevau, Puamau, Hiva Oa, and the circumstances around several tiki (carved by the Mahoeputiu tribe) placed in the facing walls of a structure, were still remembered by the Enana living in the 1920s (Linton 1925:167). In some respects, Handy's strength lies in his analysis of Marquesan semantics that provided valuable insight into the connections between objects and cultural practices.

Excluding the pictographs in the Eiaone Valley, Hiva Oa, the Marquesan image repertoire represents a homogeneous system. For instance, a female figure on a megalithic boulder at Pupuauhi, Omoa, on Fatu Hiva is stylistically similar to female figures on a megalithic boulder at Kamuihei, Hatiheu, Nuku Hiva. Images on Ua Pou are similar to images on Nuku Hiva. This suggests that the Marquesans closely interacted with each other and that they adhered to the same cultural beliefs.

The Enana pecked most of their images on individual boulders, dressed ***keetu***, and upright stones, most of which have no detectable association. Whereas a few were incised, made in intaglio, or painted, most were pecked. A relatively large number of images were carved in bas-relief. Many of the boulders measure over 4 m in diameter. Frequently, the images are located on rectangular or square-shaped modified slabs incorporated into architectural structures. Most, however, occur in their natural form and are found scattered around house sites, tribal ceremonial complexes, old trails, along riverbanks, and on narrow ridges. A large number of images appear on and in the vicinity of ***tohua*** (Linton 1925:96; Millerstrom 1990b).

Today, houses and ceremonial places are still rebuilt, reused, and modified. Nicely shaped boulders are reused because it is convenient (e.g., for copra drying ovens) and unique-looking boulders are used for decoration in churches and private gardens. In Hatiheu, eight red ***keetu*** were removed in 2003 from a structure (Te Haetaei) in the eastern part of the valley to be part of the altar in the new Catholic Church. Human bones were found behind the stone slabs. However, many Enana still consider specific sculptures and structures ***tapu***, and they requested that we not walk on certain ***paepae***.

In order to compare the Marquesan data set with that of other East Polynesian islands or island groups, the data were classified into seven main groups that, in general, corresponded to Lee's (1992) database from Rapa Nui and the Hawaiian archipelago (Appendix B).

Human figures make up several distinct subclasses with numerous variations in each category (Figure 27). Zoomorphs, such as marine creatures, dogs and birds, compose the rest of the inventory (Figure 28). Material objects and plant forms rarely occur. Island-wide, circular geometric motifs, not human figures, constitute the largest category. Unidentified images make up the rest of the record.

The ubiquitous cupules or cup-shaped impressions on boulders present an interesting phenomenon. Situated both vertically and horizontally, they constitute a unique category. Cupules are most often found with grinding surfaces. In some instances they are associated with rock images and sometimes incorporated into the figures. For example, the eyes of a face may be two cupules. Other boulders contain a large number of cup-shaped impressions indicating some form of accounting or perhaps a gaming board. The connection between the figures and the cupules is unclear. With the exception of two sites where historic motifs are depicted, such as pistols, ponies or horses, sailing ships, and lettering, it appears that petroglyph activity ceased soon after European contact.

Table 7 shows the relationship between the number of individual petroglyphs and the population. Both Nuku Hiva and Hiva Oa are roughly the same size, yet Nuku Hiva has a significantly larger number of individual petroglyphs than Hiva Oa. However, there are more tiki sculptures located on Hiva Oa than Nuku Hiva (Linton 1925; Millerstrom and Edwards 1998). Based on the large number of sculptures on Hiva Oa, Handy (1923) and Linton (1925) speculated that Hiva Oa was the first island in the group to be settled. However, anthropomorphic sculptures appear to be a late prehistoric development, while petroglyphs were probably part of the culture from the onset (Millerstrom 1997a). If the large number of petroglyphs is a criterion for early settlement, then it is likely that Nuku Hiva was the first island in the group to be occupied. Except on Ua Huka and Ua Pou, the land surface area and the estimated prehistoric population correspond to the number of sites. Ua Huka and Ua Pou have the same population, yet it is unclear why the larger islands of Ua Pou contain fewer sites than Ua Huka.

Table 7. Relationships between the number of sites, petroglyphs, land area, and population

Island	Site (n)	Image (n)	Area (km2)	Population in 1842*	Sources**
Nuku Hiva	603+	4,748	330	8,000	Linton (1925); Suggs (1961); Millerstrom (1990); Edwards 1985a)
Hiva Oa	73	592	320	6,000	Millerstrom (1985ab; Edwards 1985ab)
Ua Huka	45	817	77	2,000	Kellum-Ottino (1970); Edwards (1985b); Millerstrom (1985ab)
Ua Pou	17+	226	105	2,000	Linton (1925); Millerstrom (1997a)
Fatu Hiva	4+	47+	80	1,500	Millerstrom (1997a)
Tahuata	9	66	50	700	Linton (1925); Millerstrom (1985b)
Eiao	?	?	52	0	J. Candelot (pers. comm. 2012)
Hatutu	?	–	18	0	none reported
Motane	?	–	15	0	none reported
Fatu Huku	?	–	1.3	0	none reported
Total	**751+**	**6,496+**			

* First official census in the Marquesas
**Geographic and population data from Brousse et al. (1978:15, 17)

Due to the Marquesan Rock Art Project a large corpus of survey-generated data now exist for most of the islands. There are copies of scale drawings, maps, photographs, and reports housed at SCP, Tahiti. Additional copies of some of the material have been deposited at Centre de Documentation des Marquises Paevii, Taiohae, Nuku Hiva, established in 1993.

What follows is a construct of an architectural model based on archaeological, ethnohistoric, and ethnographic sources. Three common architectural complexes are discussed: 1) the *paepae hiamoe*, "residential household cluster"; 2) the *tohua/taha koina*, "tribal communal ceremonial complex"; and 3) *meae/ahu*, the "various sacred and ritual complexes." These are all localities where rock art is found.

A Constructed Architectural Model

Because many of the rock images are linked to architecture it is important to identify the various Marquesan structures. Thus I will discuss the Marquesan house foundations that all archaeologists working in the island will encounter.

Among the various architectural stone remains defined and described in ethnohistoric and ethnographic accounts (e.g., Crook 2007; Handy 1923; Linton 1925; Porter 1970 [1822]), and in more recent publications (Bellwood 1972; Kellum-Ottino 1971; Millerstrom 1997a, 2006a; Ottino 1990; Rolett 1998, 2010; Suggs 1961), three major Marquesan architectural types, both with regard to form and function, have been defined: 1) *paepae hiamoe*, "residential sleeping platform"; 2) *tohua/tohua koina*, "tribal ceremonial complex; and 3) *meae/ahu*, "sacred structure."

In this section the three architectural types are defined and the issues are addressed regarding the occupants of the various platforms within the *tohua* (also referred to *tohua koina* or feasting place). For instance, how are high-status households different than those used by other tribal members? Are the "tabbu houses" as Crook called them, men's houses associated with the residential complexes, or are they part of a *tohua*? Did high-status women occupy a particular place within the *tohua*?

Paepae Hiamoe/Fae Hiamoe (the Residential Sleeping Platform/Sleeping House)

The overall morphology of *paepae hiamoe*, the most common structure, were relatively standardized throughout the archipelago (Crook 2007:41, 78–80; Kellum-Ottino 1971; Krusenstern 1968:159; Linton 1923, 1925; Ottino 1985; Porter 1970:39 [1822]). According to Crook (2007:78–80), all the houses were of similar construction and building material, but differed in size. Every family had a house where they slept, but only the women and children would cook and eat there. However, some men could eat with the women.[23] While size and associated features differ, typically a *paepae* was a single square or rectangular structure. Due to the rugged topography, most platforms were terraced on the hillside. Crook (2007:79) wrote that the raised houses were also to "guard against floods." Generally, the height of the facing wall and the size of the boulders are some features that indicate the status of the occupants. All sides were faced with roughly fitted boulders while the interior was earth and rubble filled. The top surface consisted of two parts; the front part or the veranda was paved with *kiva* (water-worn river stones) or irregular stones set in mud during construction, while the other half, the sleeping area, a raised part, was mostly unpaved. Frequently a stone-lined pit would be situated in the veranda (for discussion of architectural structures with pits see Chapter 5). Sometimes a small part of the sleeping area was paved and served as a storage place for special objects (Lisiansky 1968:72 [1814]).

The roof on the rear of the house extended to the ground, thus simultaneously becoming the roof and the wall. When sleeping, the head faced this "wall," preventing anyone from stepping over the head, the most important part of the body (Handy 1927).

A dividing wall separated the sleeping space and the terrace; it consisted of fitted basalt boulders sometimes intermixed with rectangular *keetu* slabs of red volcanic tuff. In Hatiheu Valley, some of the slabs were of cut beach rock. A house of perishable material (e.g., bamboo poles, pandanus and breadfruit leaves, coconut fronds, and trunks) was constructed over the sleeping section (Handy 1923; Linton 1925; Porter 1970:39–40 [1822]; Rollin 1974; Tautain 1897; von den Steinen 1925;).

Occasionally the open part of the verandah, where many of the household activities took place, was enclosed. An engraving from Dumont d'Urville's voyage to the Marquesas in 1838 depicts a *paepae* enclosed with a wooden fence (Encyclopédie de la Polynésie 5, 1986). Both Melville (1964:99, 110) and Handy (1923:152) referred to similar enclosures. Melville described one fenced *paepae hiamoe*, another *paepae* was part of a *tohua* that appeared to have been a men's or warriors' house (Melville 1964:99). Most likely the fenced part served as a windbreak, perhaps to insure privacy, or in the case of the residential houses, to protect small children.

Crook (2007:52, 81 [1797–1799]) considered the buildings (and the canoes) to be of inferior construction. In Wilson's account it was stated that "the greater part of their houses are miserable hovels" (Wilson 1997:106 [1797]). Individual sleeping houses resembled those on the *tohua* except in "degree of elegance with which they are finished" (Porter 1970:39 [1822]).

In Hanatekua, most of the *paepae hiamoe* were rectangular. House length ranged between 4.6 m and 27.4 m, and the width from 4.5 m to 8 m; the highest *paepae* was 2.4 m (Bellwood 1972:25). Some of the larger *paepae*, approximately 12 m wide, were of composite structure; one *paepae* was built on top of an earlier type (Bellwood 1972:24). The houses in Hane were different than those at Hanatekua. In Hane Valley, the area measurements range from 10 m to 150 m^2 (Kellum-Ottino 1971: Figure 4). The *paepae* ranged in width from less than 2 m to 9.20 m; the maximum height was 1.6 m (Kellum-Ottino 1971:55). While the *paepae* in Hanatekua showed that 20 percent of houses were over 80 m², the corresponding figure was 7 percent for Hane (Bellwood 1972:24–25). In Anaho Allen (2010:354) documented 87 domestic structures. The surface of the *paepae* ranged from 12.3 m² to 174 m², while the mean was 61 m². The majority, or 60 percent, were between 20 m² and 60 m². In the Atikea section of Hatiheu Valley, Addison (2006:114) identified 54 *paepae* with a mean surface area of 72 m².

In addition to the sleeping house, the residential complex consisted of a cookhouse, a house for *fataa* (old men), and a sacred place for burials. The *fataa* was built on posts (Handy 1923:63; Von den Steinen 1928(III):aJ. Figure 7). In some cases, the posts may have been part of a small stone platform. In his *Glossary of Marquesan Native Terms*, Handy (1923:353) defines *fataa* as a "sacred store house." In his text, however, he writes that the *fataa* is a storehouse where food for sacred men were kept and cooked. Apparently old men, because they no longer were associated with women, and thereby always *tapu*, lived on the *fataa*. Here they carved objects, such as bowls, weapons, and ornaments, and ate and also drank *kava*. Kava is a drink consumed in the Pacific Ocean cultures. It is made from the root of the pepper plant (*Piper methysticum*) and in general produce a state of well-being. (Handy 1923:63). Women were not allowed near the *fataa*. Lisiansky (1968:84 [1814]) observed that, in Taiohae, Nuku Hiva, cooking was done in the open air. However, Von den Steinen (1928) photographed a thatched-covered cookhouse built on posts over part of the verandah on the sleeping house.

Tohua/Taha Koina (the Tribal Ceremonial Complex)

A *tohua* or *tohua koina* served as a tribal assembly place where *koina* (feasts) and ceremonies related to the chief took place. The *tohua* complexes are generally easy to recognize, although they differ greatly in size, in the number of stone platforms surrendering an open space, and in attached architectural components. Typically, the *tohua* consisted of a conglomerate of raised-stone *paepae* including sleeping houses, tribunes or spectators seats, terraces, pavements, alignments, and enclosures surrounding a large rectangular, unpaved court (Crook 2007:79–80 [1797–1799], 82–83; Krusenstern 1968:133 [1813]; Handy 1923; Linton 1925; Ottino-Garanger 2006; Porter 1970:105 [1822]). Roquefeuil (1823:48–49) reported that the rectangular section was some 30.5 m long and a quarter of that wide. One *tohua* in Nuku Hiva that Porter (1970:39 [1822]) observed was located high up in the valley at the foot of a mountain, next to a river; it consisted of 16 houses placed around a public square. Special trees shaded the open space (Melville 1964), such as breadfruit, coconut, and *toa* (*Casuarina equisetifolia*). Porter (1970:38–39 [1822]) admired Enana architecture and reflected on the technology and communal labor required to move rocks measuring some 2.44 × 1.22 × 1.22 m from the seaside, and fashion them into "perfect forms." Porter (1970:39 [1822]) noted that, except during periods of feasting, the houses on the *tohua* platforms were unoccupied.

To identify the social groups who occupied the various platforms is challenging. Linton (1925), who recorded numerous structures and features, did not, in general, identify the function of the individual structure. When he did, he depended on informants. Thus, it is unclear who occupied which structures on the *tohua* during ceremonies: the priests/priestesses, the chiefs/chiefesses, women and children, the warriors, or spectators.

In general, one or several socially related groups occupied each Marquesan valley; each group had their own public ceremonial center (Handy 1923; Linton 1925:24). In a few instances, two related tribes shared one *tohua*, and in one case a *tohua* had no chief (Linton 1925). A *tohua* was the domain of the *hakaiki*, hereditary tribal chief, or in some cases a *haatepeiu* (chiefess) and it formed the center of the community where social activities that concerned the tribe took place. The *tohua* in Nuku Hiva were of the enclosed type; they were larger, better constructed, and more complex as compared to the *tohua* in the rest of the islands in the group. Although the chief and his warriors lived in the vicinity of the ceremonial complex, they regularly occupied separate platforms within the *tohua*. During feasts, ritual activities were performed by the *tuhuka oono* (priests) in one end of the *tohua* while musicians, women, children, dancers, drummers, and visiting tribes occupied the other platforms.

Frequently, large communal subterranean silos for the storage of *ma* were located nearby. All public functions related to the *hakaiki* and *haatepeiu* and their families took place on the *tohua*. For instance, the death of an important *hakaiki* or *tuhuka oono*, which created great social stress, was followed by a series of ritual activities, many of which took several years to complete (Handy 1923). *Mau*, commemorative feasts for important persons, took place on the *tohua* (Crook 2007:66–67 [1797–1799]).

In the northern group, the chief's or an elite establishment tended to be unusually large. Some of the architectural features were massive boulders and *keetu*. Allen (2010) made the same observation in Anaho Valley, Nuku Hiva. She suggested that only the elite could afford transportation. In Hatiheu elite structures were constructed of massive stones, large *keetu*, and located in an elevated (up to 3 m high) and prominent position. The chief's residential unit was usually placed along one of the long sides of the dance floor where the chief could overlook the festivities that took place (Handy 1923; Linton 1925). However, the chief's *paepae* did not always overlook the dance space, as was the case at Tohua Tekeia, Hanapaoa Valley, Hiva Oa. Here the chief and the chiefess occupied two *paepae* upstream from the dance floor (Linton 1925:171). According to Krusenstern (1968:159 [1813]) the chief's residence was similar to the other structures except it was better built, larger, and appeared to be two feet higher. At Tohua Tamaeka, Hakamoui Valley, Ua Pou, the chief's *paepae* measured 5 × 8 × 3 m and was situated on a hill. Different platforms of the *tohua* were assigned for visiting tribes where they could build houses in which to live during the festivities. The visiting chief sat in front of the house of the local chief during festivities (Handy 1923; Linton 1925:25).

However, the chief's establishment could also be placed at the short end of the dance floor. For instance, on Nuku Hiva at Tohua Ponaouoho, Hakaui, Tohua Maiki Tokiai, Hooumi, and Tohua Utukua, Taipivai, the chief occupied the *paepae* at the north end of the ceremonial complex (Linton 1925:108, 114, 117, 135). At Tohua Ponaouoho the chief's residence occupied the whole northern end, and it consisted of a sleeping house, a smaller house, and a cookhouse. The wall facing the dance floor measured 1.8 m in height (Linton 1925:109, Figure 3). Many of the chief's *paepae*, whether part of the *tohua* proper or located in other areas of the valley, contained well-carved wooden house posts (Linton 1925:175).

Tohua Pehe Kua was one of several archaeological sites we documented in Puamau, Hiva Oa, during the 1985 field season (Figure 29). A megalithic *paepae* (Structure 1), located at the short west end of the *tohua*, was said to be the residence of Chief Te Hau Moe (V. Heitaa, pers. comm. 1985) and was 22 × 14.5 m; the facing wall was 1.95 m high, while the sidewalls were 2.65 m high. The facing dividing wall was made of rectangular red volcanic slabs; three tiki, and one slab with a face motif were set into the wall. Titoi Ani, daughter of Te Hau Moe and the last chiefess of Puamau Valley, is buried on the platform.[24] Two additional tiki, Mani and Pauto, which are located by the nearby tomb of Chief Te Hau Moe, came from the same *paepae*. Chiefess Titoi Ani occupied *paepae* no. 2. Several other carved stones are associated with the tribal ceremonial complex. A stone slab with a bas-relief human figure was uncovered during the project (Edwards 1985a; Millerstrom 1985c).

Often *tohua* have several large flat boulders set on smaller stones to form a pedestal or referred to as a dance stone. It was on these pedestals that important people displayed their tattoos and ornaments (Linton 1925). Women and children occupied special platform(s) on the *tohua* (Handy 1923:36; Lawson 1861; Linton 1925:25), but there are few references to which platforms they occupied. At Tohua Nanauhi, Hatiheu Valley, Nuku Hiva, a house platform that formed part of the tribune was reserved for women and children (Linton 1925:118). Certain parts of the dance floor were forbidden to women and children (Linton 1925:25).

Warrior's House

It appears that the *fataa*, discussed above, is the same structure that Crook (2007:68, 81 [1979–1799]) referred to as "tabbu house," although it seems that, in 1797, the "tabbu house" was a house for all *tapu* men; it was usually situated near the tribal ceremonial feasting place (Linton 1925:25). Here men gathered during peace, as well as when they were in a *tapu* state, for instance, prior to a journey or warfare. Handy (1923:126–127) writes that the warrior's house was a part of the chief's establishment and was built near his house and the *tohua*. In the southern Marquesan group, warriors often had a separate structure, but on Nuku Hiva the warrior's platforms were frequently part of the *tohua* (Handy 1923:126–127; Linton 1925:118, Figure 6).

This house was sacred, and women and commoners were forbidden to walk by it (Gracia 1843:54–56). White *tapa* banners marked the place (Porter 1970 [1822]). Most men, the *tapu* class, had little to do with the women and lived at the warrior's house near the ceremonial place. Crook noted that "the tabbu house is usually situated upon a raised pavement made of large stones, the bounds of which cannot be passed by women; nor, even, if it belongs to any person of a superior class, by men of the general tabbu class" (Crook 2007:68 [1997–1799]).

Here the men ate their meals, consumed *kava*, and made weapons, ornaments, and drums. The strong young men in the tribe, the warriors, would occupy a specific platform on the *tohua* (Handy 1923; Melville 1964). Sometimes the hereditary chief would also be a chief warrior (Handy 1923). The warrior's house tended to be long, with narrow bed spaces and, like the chief's house, built of massive stones and located at the outer edge of the dance floor (Linton 1925:117, 173). Melville (1964:110) described a long warrior platform measuring more than 61 m in length and 6 m wide. Porter (1970:38 [1822]) reported that some "tabbooed" houses measured 30.5 m in length and were 12.2 m in width.

> The whole front of this latter structure was completely open, and from one end to the other ran a narrow veranda, fenced in on the edge of the pi-pi [*paepae*] with a picket of canes. Its interior presented the appearance of an immense lounging place, the entire floor being strewn with successive layers of mats, lying between parallel trunks of coconut trees, selected for the purpose from the straightest and most symmetrical the vale afforded.

In order to retain his warriors and assure their loyalty, the chief was obliged to house and feed them (Krusenstern 1968:160 [1813]). War paraphernalia and *tapu* objects were kept in the thatch of the sleeping area, the *tapu* place. During warfare, when the men were considered *tapu* and could not be in contact with women, they also slept on the *tohua* (Handy 1923). The chief's *paepae* faced the men's house on the *tohua*; for instance, at Hooumi the warrior's place was at the opposite end of the chief's *paepae*. On other *tohua*, the warrior's house was on the outer edge of the dance floor (Linton 1925).

Spectators' Seats

Suggs (1961:75) applied the term "stadia" to the area for spectators. These seats, part of the tribal ceremonial place, accommodated visitors from other tribes during large feasts. The area was a long, narrow, one-step terrace built of boulders. It was generally located on the long sides of the rectangular *tohua* court (Linton 1925:173). The terrace was either unpaved (e.g., Tohua Maikuku, Nuku Hiva, or fully paved (e.g., Tohua Hikokua (Nuku Hiva)). Crook 2007:82 [1797–1799]) writes that *tohua* in the larger islands could contain 10,000 persons during an important feast. On such occasions, despite the danger of wars and of crossing a large body of water, visitors would flock from across the island and even from neighboring islands.

Meae/Ahu (Sacred Structures)

The *meae* was highly variable and thus difficult to identify (Rolett 2010). Based on construction, a *meae* was similar to residential house platforms (Linton 1925:34).[25] However, the first Westerners to the Marquesas observed that the *meae* was located apart from the village, surrounded by palisades, and placed in the center was a house with wooden tiki (Markham 1904:27). Linton (1925) distinguished between the public *meae* linked to the *tohua* and the separate mortuary *meae*. But there are many different types of sacred ritual places. A *meae* could also be a place for fishermen and be located near the sea, or it could be a *meae* within a household complex (Millerstrom 2009:22–25). Miniature *meae* or perhaps shrines are found on ridges and deep in the valleys on the north coast of the islands of Nuku Hiva and Ua Pou (Millerstrom 2006). Their function is not known.

Local informants and guides told Linton (1925:28) that none of the *tohua* on Nuku Hiva had an attached *meae*. Apparently, on Nuku Hiva, the *meae* was not "a constant feature of the tohua" but there were few secular *tohua* (Linton 1925:29; see also Chapters 4 and 5 herein). Generally, *meae* were more complex in the southern group than in Nuku Hiva (Linton 1925). This was confirmed by Bellwood (1972) in Hanatekua, Hiva Oa, where the *meae* consisted of several terraces, enclosures, and sunken areas. Kellum-Ottino (1971) questioned some of the *meae* identification in the Hane Valley, Ua Huka; at Hanamiai, Tahuata, Rolett (1998:119) was unable to identify the *meae* part of Tohua Aimaha.

Recognizing the challenge of being able to distinguish a *meae* from a high-status residential unit, Rolett (2010:94–102) excavated the Mataiee architectural complex in the Vaitahu Valley, Tahuata. Features that often, but not always, were associated with a *meae*, such as banyan trees, anthropomorphic stone sculptures, and petroglyphs were absent. However, the site was clearly not a high status residential unit. Establishing specific cultural and environmental characteristics between a *paepae hiamoe* and a *meae*, Rolett was able to conclude that the complex was indeed a religious complex.

In early accounts, the *meae* is described as consisting of one or two houses built on a raised, square stone platform. Tall drums were stored inside; placed outside were life-size wood or stone tiki (Radiguet 1978[1860]; Tautain 1897; Wilson 1997:110 [1797]). Wilson (1997:110 [1797]) described an escutcheon with a pyramid-like side "wrought with small reeds" stained in different colors and with the figure of a bird placed on the top.

The following is Lawson's description of the site.

> The Heahu [*ahu* or *meae*] was a large building of stones about four feet from the ground and about 100 yards in length by thirty in [width]. On one side was built up some stones [altars] about six feet high and ten yards square. The whole place was shaded over by large trees and a house was built on one end.... At the back of the house on Heahu assembly ground, there [were] three rows of skulls lashed fast to the rafters. The two upper rows each contained 105 skulls and lower row 103 in all 313. These were the skulls of victims that had been killed in battle at various times and sacrificed to the gods. (Lawson 1861-1867)

Human skulls, as well as stone and wood sculptures, appeared to have been dominant features of the *meae*. The majority of the Marquesan stone sculptures recorded in situ were linked to *meae* (Millerstrom and Edwards 1998). Petroglyphs are often, but not always, found on public *meae*. They were placed on the facing stones, on the surface of the front platform or, in some cases, around the upper part of stone-lined sunken pits. Images were also placed on large boulders incorporated into the *meae* complex (Millerstrom 1997a).

Perhaps one of the main reasons a *meae* is difficult to define is that a chief's *paepae* sometimes became a *meae* after his death. For instance, the *paepae* of Chief Puheputoka, Hakamoui Valley, Ua Pou, built in honor of his mother, became a *meae* after he died sometime in the early part of the 1800s (Linton 1925:134). *Paepae Menaha* is 37 × 18 m; it ranges from 1.8 m to 4.6 m high, according to the contour of the land. The sleeping space is nearly 2.5 m wide and runs the length of the house. It is built with three terraces, each faced with large blocks of *keetu* with bas-relief anthropomorphic figures (one of which is a human-like face), enigmatic motifs, and ornamental adzing in a chevron pattern. Several wood tiki house posts decorated the upper sleeping terrace. This is probably the largest structure still standing in the Marquesas. A *meae* with a tiki, not mentioned by Linton, is placed to the west of the complex (Millerstrom 2006a). Great festivities were associated with the building of the chief's home (Linton 1925:134).

In some cases, tall upright stones were associated with a *meae*. For instance, at the large and complex Meae Iipona, Puamau, Hiva Oa, several uprights were found near the chief's house after restoration in 1990–1991 (Ottino 1996:345–376). A small paved shrine at Tohua Maikuku, Hatiheu, Nuku Hiva, has two upright stones; one measures 1.3 m in height with pecked anthropomorphic figures, faces, circles, and geometric shapes. Smaller upright stones are found at several fishermen's shrines at Ua Pou (Millerstrom 2005a:117–126).

The *meae* were universally associated with special trees or sacred groves. Banyan trees or *aoa* (*Ficus* sp.) were considered to be more sacred than all other trees, and these would grow on or near the platform (Handy 1923:119; Millerstrom 2005a). Temanu (*Calophyllum inophyllum*) is another sacred tree sometimes found near sacred sites. The *meae* was for the gods and thus often *tapu* for the general population. Some *meae* were greatly feared for their *tapu* by everyone but the priests. The meae were always forbidden to women with the exception of inspirational priestesses. During certain ceremonies, no one, not even the chief, could enter (Linton 1923:120). The priestly establishment consisted of an inspirational priest or priestess, and their attendants. Ceremonies were usually accompanied by chants and drumming. While a chief resided over the *tohua*, the *meae* was the domain of the priestly establishment.

Discussion and Summary

Within the tribal ceremonial complex it is clear that the chiefly residences were often more elaborate than the rest of the structures. This is specifically true in regard to the height of the *paepae*, the size of the construction material, and associated architectural components. We have little information about the priest's and the warrior's places. In general, the *meae* part of the *tohua* seems to be placed at the opposite side of the chief's platform, and it appears more complex. The warriors' places were located within the tribal ceremonial place, or they were placed in the vicinity. It was always an extended house platform. Regarding the place of women on the *tohua*, the literature is mostly silent. This is not surprising, considering that almost all observers, from explorers to missionaries to archaeologists, were men.

Robarts (in Dening 1974:89, 100, 111, 155) called the tribal ceremonial place a "play-ground" or "dancing ground." On several occasions he claims to have taken part in festivities with females of high rank and their attendants However, the women Robarts described appear to be of the *tapu* class. Porter (1970:38 [1822]) had this to say about women and the public ceremonial places.

> The women are, on no occasions whatever, allowed to enter their places of feasting, which are houses raised, to the height of six or eight feet on a platform of large stones, neatly hewn and fitted together, with as much skill and exactness, as would be done by our most expert masons: and some of them are one hundred yards [91.5 m] in length and forty yards [36.6 m] in width, surrounded by a square of buildings executed in a style of elegance, which is calculated to inspire us with the most exalted opinion of the ingenuity, taste, and perseverance of a people, who have hitherto remained unnoticed, and unknown to the rest of mankind.

Elsewhere Porter (1970:116 [1822]) wrote that all sacred places and places being *tapu* for women to approach were marked with "a bundle of long sticks, about half the size of the wrist, with the bark stripped off and placed on end."

Four years after Porter's departure in 1817, the French sandalwood trader Camille de Roquefeuil in the ship *Le Bordelais* spent several weeks on the islands; he wrote that in Taiohae, Nuku Hiva, both men and women met at a feasting place (Roquefeuil 1823:48–49). Whether the discrepancies are due to regional or temporal cultural changes, or if they are related to status variation is uncertain. Like Porter, Crook (2007 [1797-1799]), Robarts (in Dening 1974:111), and Cabri (Terrell 1982:110) wrote that females were not allowed to come near the altar or enter sacred places. What Cabri referred to, as a "maraie" may have been the "tapu house" or the men/warrior's house. According to Terrell (1982:110) "these temples are sacred, and each one takes three to four thousand heads to bring it to perfection. Entry is forbidden to women, children and men who have not yet been to war."

Historic sources disagree on the presence or absence of women on the tribal ceremonial complexes. It is certain that the "warrior house," also called the "tapu house," was always off limits to women, even those of the *tapu* class (Crook 2007:68 [1797–1799]); Krusenstern 1868 [1813]; Porter 1970 [1822]). As far as the ritual places, unless a woman was involved in religious ceremonies, such as spiritual priestesses, she could not go near the *meae* (Tautain 1897). However, Tautain states that they are allowed on certain places on the *tohua*;

thus women's access to the *tohua* appears to have been temporally and spatially related. Access depended on the type of rites that were performed and where the men's house and the *meae* were located on the *tohua*; in certain situations it also depended on the woman's personal status. Inspirational priestesses and high-ranking women were often exempt from restrictions put on the men and the women of the tribe or the *mataeinaa* class.

Thus the *tohua* was perhaps divided into sacred and secular spaces with blurred borders between the two, according to the type of ceremonies taking place, and the persons' social standing in the community. Perhaps the women's presence or absence on the *tohua*, as described by the early visitors, also changed according to the seasons and the various ceremonies that were performed during the year.

In the last section a model was constructed of the most common architectural complexes. Identifying the *meae* is challenging. A high-status residence appears to have similar characteristics unless one or more banyan trees mark the site. To complicate the matter, it is known that, in some cases, a chief's residence became a sacred site or *meae* after his death.

While more research on the islands' architecture is certainly needed, this discussion provides a background for my archaeological investigation of the spatial distribution of rock art in the western section of Hatiheu Valley.

CHAPTER 4

The Archaeological Landscape of the Hatiheu Valley, Nuku Hiva

With each new generation the empirical base of archaeology advances, constraining what the next generation of scholars may construct of the past.

—— P. V. Kirch, "On the Road of the Winds," *An Archaeological History of the Pacific Islands Before European Contact*

The field survey in the western part of Hatiheu Valley between 1992 and 1997 included extensive reconnaissance, systematic survey, mapping, recording, and test excavation. Although a number of rock art sites were previously recorded during the Marquesas Islands Rock Art Project, many more undocumented petroglyph boulders were discovered during the field survey.

Hatiheu Valley, on the north coast of Nuku Hiva is a wide amphitheater-like valley hemmed in by irregular ridges; peaks rise up to 800 m (Figure 30). Situated to the south is Te Ava Tapuhiva (490 m); it separates Hatiheu and Taipi. To the southeast lies Anaotako (789 m). Te Ava Maoaoa separates Hatiheu and Anaho to the north. A chain of dramatic pinnacles, the tallest of which is named Te Heu, flank the western side, and jagged peaks run inland from the beach area. Perpendicular spurs run from the mountain chain into the valley, creating small subvalleys. The beachfront is some 1,000 m long and the valley approximately 2 km deep.

In this chapter I examine the landscape of the research area to understand which tribe made the rock art, the area that was most populated where we would expect to find the rock art, and the environment that made it suitable for settlement and agriculture. There are multiple reasons why the western section of Hatiheu was selected for investigation. While petroglyphs are located throughout Hatiheu Valley, they are most numerous in the western section (Millerstrom 1997a). Overall, the majority of the images are not directly associated with structures but, in the western portion of the valley, they are often found in conjunction with both prehistoric and historical remains.

Research Area

The western section of Hatiheu Valley extends between two major watercourses, the Vaiuua and Puhioho. Puhioho, the westernmost, originates from a waterfall at the head of the valley in an uninhabited district named Kahuvai. The Vaiuua River is difficult to define but it appears to start in the upper part of the valley. Both are connected to numerous permanent and intermittent tributary streams or river that flow from underground springs or run from mountain ridges.

Stretching from the beach to the mountain ridge, the western section of the valley is roughly pie shaped, resembling a reversed traditional Polynesian *ahupuaa* (land unit). At the beach, the section is about 200 m wide, and it extends inland approximately 1.5 km to the Kahuvai waterfall at the head of the valley. The central portion is approximately 1,000–1,500 m wide. Parts of an old stone-lined road that connected the Hatiheu and Taipi valleys in the past are, in some areas, still visible. The present road leading to Taipi Valley begins at the beach area of Hatiheu and crosses part of the research area. It is clear that bulldozers destroyed several old structures during road construction and maintenance.

Except for a few houses toward the beach, the western part of the Hatiheu Valley is uninhabited, although some families keep pigs and horses on the land, and copra is regularly collected in some areas. During the 1995–1996 field season, however, the area was uninhabited. The research area was selected on the belief that the main streams delineated past tribal boundaries; traditionally, mountain ridges, rivers and streams, or other natural features demarcated social land boundaries in Polynesia (Crocombe 1964:20; Crook 2007 [1797–1799]; Handy 1923; Henry 1928). This is also the case today (Y. Katupa, pers. comm. 1992). Some contemporary *paheeka* (land boundaries) are marked with an "X" that is incised on the top of round upright stones, on natural outcrops, or the upper part of coconut trees.

My initial goal was to map all visible above-ground structures and features from the beach to the upper part of the valley. This turned out to be an overly ambitious endeavor due to both physical and financial limitations. Generally, archaeological remains are more visible in areas where copra is regularly collected and domestic animals are tethered. Sections of the central and upper areas, however, are practically impenetrable and impossible to map without a crew to cut the underbrush and to survey. I estimated that roughly 80 percent of visible structures and features have been documented and mapped.

Environmental Considerations

Because of the heavy rainfall in Hatiheu, it appears that the valley was always well watered (Adamson 1936:59). In the living memory of the people of Hatiheu, none of the main rivers ever went dry (Y. Katupa, pers. comm. 1992). Four large perennial rivers, emptying out into the ocean, bisect the landscape. The river on the east side of the valley is *Te Ahu Paaoa* (the temple of the dolphin and/or whales), named after the fishermen's shrine located at the mouth of the river; Puaiki and Vaiuua run through the central part of the valley, and Puhioho is located in the western section. Some of the Enana referred to Puhioho River as Kahuvai because it originates at the head of the valley in an area by the same name. Other locals called the stream *Te Ahu Matanui* (the temple of large eye, face, tribe, or genealogy) after a fishermen's shrine by that name, once situated on the beach to the west of the river.

Modern pluviometric data collected from the main islands (Hiva Oa, Nuku Hiva, Ua Huka, Ua Pou, and Fatu Hiva), and Eiao show an average annual rainfall of 70 cm from the leeward side of the islands at Hakahau, Ua Pou, and Vaipaee, Ua Huka, to an average of 148.2 cm for Hatiheu on the windward side of Nuku Hiva. The mean annual rainfall for Hatiheu is approximately 40 percent greater and more evenly distributed through the year than what was recorded for Taiohae on the leeward side of the island. The rainy season generally falls between January and July. In the period between 1962 and 1975, Hatiheu had a higher average annual rainfall compared to the other valleys (Cauchard and Inchauspe 1978:75–105). Similarly, the fertile Taipi Valley, located southeast of Hatiheu, was never seriously affected by a shortage of rainfall (Adamson 1936:59). In 1797, an observer on Wilson's Missionary ship, the *Duff*, stated that "all the valleys about this bay appeared fertile, many of the hills were covered with trees, and the interior parts seemed more habitable than any other of the Marquesas" (Wilson 1997:130 [1797]). Rainfall varied greatly in the archipelago, and more research is needed to examine how the variability affected other parts of the islands. However, the north and northeast part of Nuku Hiva Island appeared to have been well suited for permanent habitation. The climate is subtropical with a temperature ranging from 25° to 29° C (Brousse et al. 1978:57). Devastating cyclones have ravaged the archipelago, and a tsunami hit the islands in 1946, destroying both the Catholic Church and a *tapu* structure placed at the mouth of Puaiki River (Y. Katupa, pers. comm., 1991). The tsunami also exposed the Haatuatua dune site on the northeast coast of Nuku Hiva that Suggs discovered and excavated in the late 1950s. Coral reefs, except in protective areas (e.g., Anaho and Hanaiapa), are mostly lacking in the Marquesas, and none are present in Hatiheu Bay (Chevalier 1978:243–278). Thus, fishing outside the bay always contained an element of uncertainty and danger.

Hatiheu Tribes

The geographic distribution of the tribes and subtribes in Hatiheu is uncertain. Crook (2007:128 [1797–179]) stated that the Puhioho people, who originated from Taipi, inhabited Hatiheu. A hundred years later, in 1897, von den Steinen (1925:16) listed five *ramages* (subtribes) of the Taipi tribe from the same valley. A total of 120–125 people were divided into the following *ramages*: Ati Kea, Puhioho, Ati Puku, Tapatea, and Tuuoho (see also Christian 1910:204; Pechberty 1996:165). *Ati* means tribe, family, people, parents, and descendants (Dordillon 1931:112). Research by Suggs indicates that two major tribes, the Ati Kea and the Puhioho, occupied Anaho and Haitheu in the early nineteenth century. The Ati Heuu branch of the Ati Kea and the Ati Keikahanui branch of the Puhioho lived in Hatiheu (Suggs pers. comm. 2000). Suggs listed three additional subtribes for Hatiheu: the Ati Papua, Ati Puku, and the Tapatea.

According to the local Enana, the Ati Kea group occupied the east part of the valley. Suggs (1961:68) was told by informants that Ati Papua, not listed by von den Steinen, was one of the ramages of the Taipi-Hatiheu and associated with Tohua Hikokua. Apparently the Puhioho tribe lived in the western section near the stream by the same name. Ati Keikahanui, a subgroup, occupied Tohua Keikahanui (also referred to as Tohua Tahakia) and located near the east side of Puhioho stream. Keikahanui was a powerful chief-warrior (Crook 2007:127–151 [1779–1799]). The uncertainty of linking the various tribes with a specific *tohua* perhaps illustrates the shifting alliances of the Marquesan tribes in the pre-contact and postcontact periods. It seems likely then that the Puhioho tribe and several descent groups occupied the western section of Hatiheu where my field research took place.

Spatial and Interspatial Settlement Distribution

Three distinct settlement areas are evident: Zone 1) the flat coastal area or the coastal lowlands, which appear relatively infertile; Zone 2) the central valley floor, a belt of fertile land laying between the coastal flat and the mountains; and Zone 3) the mountainous interior at the head of the valley (Figure 31). Each of these settlement areas will be discussed in terms of the spatial distribution of images, architectural structures, and features.

Zone I: the Coastal Lowlands

Archaeological remains in the first 200 m inland from the beach are sparsely distributed. Generally, remains consist of stone alignments and *paepae* intermixed with modern structures, such as houses, copra-drying sheds, and a concrete staircase from a house destroyed long ago. A few residential complexes were occupied, as late as the 1960s, and in some cases, the names of the occupants are known. Remains of a Chinese trade store, located approximately 120 m inland from the beach, contains pieces of red volcanic tuff and beach rock—stones that were probably removed from an archaeological site, perhaps a high-status residence.[26] The front part of the store was apparently built on stilts. Some of the archaeological architecture still standing in the coastal lowlands seem to span the late precontact and early postcontact eras. In one case, the profession of the past occupant—a sorcerer—is still remembered by local residents (T. Puhetini, pers. comm., 1995). Historic graves were seen next to the house, site and a few house sites contain midden with historic material.

Zone II: The Central Zone

The greatest concentration of archaeological remains is located in the central zone approximately 50 masl. Situated between the old road to Taipi to the east and Puhioho stream to the west, four of Hatiheu's eight ceremonial complexes were built in close proximity to each other. They are Tohua Hikokua, Kamuihei I, Kamuihei II, and Tahakia. Large numbers of images and megalithic *paepae* are clustered around the tribal ceremonial complexes and Meae Te Iipoka. This pattern extends to the south of the Hikokua, Kamuihei, Tahakia, and Te Iipoka areas to Ototemoui ridge, and southwest to a section of Kahuvai, near Puhioho River. The majority of the images are directly integrated or in close proximity to extensive, complex, and very likely high-status megalithic architecture (Figure 32).

An extensive agricultural area is in the western sector, below Kahuvai near Puhioho River. Stone-outlined fields for taro cultivation, now a swamp, cover the valley floor. The valley bottom is presently abandoned and has little drainage. The water level is high, and the stone alignments are barely visible; several house platforms can be seen in this swampy area. Local residents remember that the alignments were more visible some 50 years ago. Stone-outlined taro ponds are seen along the edge of streams in the same area, while agricultural terraces with facing stones are placed on the hillsides.

In the following sections I discuss the rock art documented at each of the tribal ceremonial complexes and some of the adjacent areas. In Table 8 I calculated the overall measurements of the tribal ceremonial complexes as well as the dancefloors. Tohua Tahakia is the largest ceremonial complex in Hatiheu Valley while Kamuihei II is the smallest *tohua*. Pahumano, located by the sea, is today covered over with modern buildings and only a few short alignments are visible. It was challenging to measure Tohua Paahaua and the overall measurements of the complex is uncertain.

Table 8. Sizes of tribal ceremonial complexes in the Hatiheu Valley, Nuku Hiva

Tohua	**Dance Floor (m)**			**Complex (m)**			**References**
	L	W	m²	L	W	m²	
Hikokua	82	24	1,968	125	45	5,625	Suggs (1961:69); Edwards and Millerstrom (1986, 1987)
Kamuihei I	102	18	1,836	168	84	14,112	Edwards and Millerstrom (1986, 1987)
Kamuihei II	50	14	700	84	42	3,528	Edwards and Millerstrom (1986, 1987)
Maikuku	35	14	490	80	42	3,360	Edwards and Millerstrom (1986, 1987)
Nanauhi	107	20	2,140	127	59	7,493	Linton (1925:117–118)
Paahaua	18	45	810	n.a.	n.a.	n.a.	Millerstrom (1996a)
Tahakia	120	19	2,280	155	45	6,975	Millerstrom (1994); Ottino (1998)
Pahumano	?	?	?	?	?	?	Delmas (1927:100)

Tohua Hikokua

Tohua Hikokua is situated approximately 550 m inland from the beach (Figure 33).[27] Built perpendicular to the hillside, it consists of several house platforms and terraced *paepae* surrounding an 82-×-24-m, rectangular flat plaza.

Suggs (1961:68–76) was the first to examine and excavate the Hikokua ceremonial complex during his 1957–1958 archaeological project. An unpublished base map of the *tohua*, kindly made available to me by Suggs, indicates that major changes have taken place in the last 60 years or so. I use Suggs designated letters to indicate the various features and structures.

A *meae* complex that Suggs refers to it as *tuu*, is located at the northern end of the dance plaza. The *tuu* (Structure A) is a walled sacrificial platform and it is the central part of the *tohua*. The walls of the platform consist mainly of alternately placed red and white blocks of volcanic tuff (*keetu*) and beach rock (*papatea*) that form two concentric rectangular enclosures. Three *tiki* are positioned in the south-facing wall (Figures 34a–b). One tiki made of red tuff has deteriorated to the point where it is challenging to see the various features. A communal breadfruit pit (12 × 2 m) is situated on the northeast side, and a communal oven is located at the northwest end of the *meae* complex.

The south end of the *tohua*, the more secular section, is quite disturbed. Some of the structures in the southeast corner, was at one time, incorporated into a copra-drying oven (Tioka Puhetini, pers. comm. 1989). In order to construct the oven, stones were taken from other parts of the complex. This is a common practice and has taken place at several old sites. In 1996 a *koma*-type adze was found on the surface next to a slight depression, 2 m in diameters, on the south side. Without excavating the depression, it is uncertain if it is a breadfruit pit or an oven.

The spectators' *paepae*, or tribune is a long paved terrace, located on the east side. Across from the spectators' section, on the west side, is a well-built raised sleeping house (Structure L) with three terraces that is projecting onto the dance floor. With a height of 2 m, it is the highest platform in the *tohua* and the most visible. Two stone-lined pits are located in the front part of the veranda; the one on the west side is lined with *keetu*. Because of its height, and the prominent placement on the tribal ceremonial plaza, this sleeping house was most likely occupied by a chief. Connected to the chief's house, is the warrior's complex (Structure M), also referred to in the literature as the men's house or the *tapu* house. According to ethnographic and ethnohistoric sources, the men's house was longer than the average sleeping house. *Auti* or *ti* plants (*Cordyline fruticosa*) are now growing on the surface of the chief's sleeping platform. *Auti*, an aboriginal cultivar throughout Polynesia, is an important plant used in both secular and sacred contexts (Petard 1986:101–122 [1958]). The *auti* is only planted with cuttings, therefore those plants were purposefully propagated suggesting the house platform had special significance.

On the south side, near the destroyed structure K, and at the present entrance to the *tohua*, there is a short alignment sunken into the ground.[28] Located on the surface of the alignment is one stick figure (Figure 35a) and a pair of anthropomorphic eyes. The petroglyphs are barely visible today. Suggs found a basalt slab with a *moko* (bas-relief lizard) in the south part of the *tohua* (Figure 35b).[29] In the oven on the west side of the *tuu* Suggs discovered a petroglyph boulder with geometric figures (Figure 35c). A French brandy bottle was also uncovered in the oven testifying that the sites were used in the post-1840 period. According to Suggs (1961:76) broken glass bottles were used as cutting tools in the early part of the post-1840 period. Some surface artifacts collected by Suggs included a musket ball, gunflints, and a pewter coconut grader all typical of the early Historic period. Common in the Classic period were local artifacts such as a number of *koma*-type adze, an unfinished tiki-headed poi pounders, and a grinding pebble. A petroglyph boulder with a face and geometric figures is fitted into the facing wall of the spectators' area (Figure 35d).

Suggs (1961:68–76) excavated 11 trenches and test units at Tohua Hikokua (Site NHe 3). Two trenches excavated through the *tuu* platform exposed two concentric rectangular enclosures that were the remains of an earlier structure (Suggs 1961:71). In the north end of the *tuu*, Suggs found the remains of a banyan tree.[30] None of the charcoal uncovered during Suggs excavation were sufficient for dating, thus the age of the first construction of the *tohua* could not be determined. However, Suggs concluded that the construction of the complex began in the early part of the Expansion period, approximately A.D. 1300 and that the site was still in use into the Historic period.

In 1988 when we mapped and recorded the images at Tohua Hikokua we documented close to 200 cupules placed on the various platforms at the *tohua*. Their various functions are uncertain. Historic graves outlined with stones, some of red tuff, are located on the southeast part of the dance floor. Christian graves are also placed outside the *tohua* proper both to north of the communal breadfruit silo and to the south of the tribal ceremonial complex.

Tohua Hikokua Images

Five image stones and three sculptures were documented on Tohua Hikokua. All the tiki (T2, T3, T4) faced the mountains. They were installed in the outer facing wall during the last construction phase of the *tuu*, probably sometimes after A.D. 1600. The tiki are diverse in shape and size. One tiki (T3) is located to the east and formed as a block, it is 1.0 m high and is made of a reddish tuff. The smallest sculpture (T2) found in situ is located in the center. It is 0.32 m high and carved of red tuff, it is carved at the end of a slab (see Figure 34a and b). Located to the west is another tiki (T4), also in red tuff. It is 1.1 m tall with rounded forms, possibly a female. The tiki is so eroded and fragmented that it is difficult to distinguish some of the facial features. According to informants, one of the two large sculptures represents the goddess Tevanauaua, a principal goddess of Taipi-Hatiheu tribes and a goddess of the subtribe Ati Papua of the Puhioho group (Delmas 1927:100; Suggs 1961:68–69). The goddess Tevanauaua is perhaps represented by T4 because of its height and roundish female form. In the neighboring valley of Taipi, the Puhioho allies have several female tiki with similar rounded forms, which are placed in the facing walls of Paepae Paeke (Linton 1925; Millerstrom and Edwards 1998). Although it is difficult to detect, it appears that T4 is the only one of the three sculptures on Tohua Hikokua that exhibit the classic characteristics, such as large circular eyes, wide noses, long narrow mouths, and hands clasped on a protruding stomach. Local history relates "a captured Hapaa warrior, Tuehu, was immolated on the *tohua* and later raised to the status of a deity by the Hapaa group" (Suggs 1961:68).[31] There are conflicting stories of the identity of the three tiki.

The anthropomorphic stick figure and the anthropomorphic eyes are eroded to the point of being barely visible, suggesting that they are probably part of a paved band, near the destroyed Structure K, that outlined the sides of the *tohua*. These were part of the initial construction phase at the early part of the Expansion period. In 1988 an image stone was found associated with the spectators' *paepae*, lying on the dance floor. The stone fit into the west facade and was replaced in the wall during restoration of the *tohua*, although with the anthropomorphic face and geometric figures facing inward.

A total of 11 petroglyphs and three tiki were documented on Tohua Hikokua. Overall, curvilinear figures (n=7) dominate with a total of 50 percent, while anthropomorphic figures (n=6) represent a total of 42.9 percent (Table 9). Seven of the images are found on the tribune The three tiki, representing 21.4 percent of the total number of figures, are found on the south-facing wall of the platform of the ritual complex, the *tuu* (Table 10).

The distribution of petroglyphs in conjunction with the construction sequence as outlined by Suggs (1961) suggests that anthropomorphic faces and stick figures were part of the initial use of the tribal ceremonial center, while the geometric figures associated with the face were made later. It appears that toward the end of the cultural sequence, the tiki, the bas-relief *moko*, and the abstract geometric motif were added.

Table 9. Tohua Hikokua frequencies and percentages of petroglyphs and tiki

Image Type	n	%
Anthropomorph	–	–
Stick Figure	1	–
Tiki	3	–
Disconnected Body Part	–	–
Face	2	–
Subtotal	**6**	**42.9**
Fauna	–	–
Lizard	1	7.1
Geometric Motif	–	–
Curvilinear	7	50.0
Subtotal	**8**	**57.1**
Total	**14**	**100**

Table 10. Tohua Hikokua frequencies, percentages, and contexts of petroglyphs and tiki

Provenience	BLDR or Tiki (n)*	Image Type	%
Ritual Complex (A)	3	Tiki	21.4
Subtotal	3		**21.4**
Tribune (*aepae* F)	1	Anthropo-morphic Face	7.1
	6	Curvilinear Figures	42.9
Subtotal	**7**		**50.0**
Alignment	1	Anthropo-morphic Eye	7.1
Alignment	1	Anthropo-morphic Stick Figure	7.1
Subtotal	**2**		**14.2**
Dance Plaza	1	Curvilinear Figure	–
Platform (H)	1	Lizard	
Subtotal	**2**		**14.2**
Total	**14**		**99.8≈100**

* BLDR = boulder

Tohua Kamuihei I

Tohua Kamuihei I is centrally located between Tohua Hikokua and Kamuihei II. The complex consists of more than 25 *paepae* and a number of attached terraces, pavements, and alignments surrounding the dance plaza (Figure 36).

Veotapuanui (Structure 1), the name of the ritual complex or the *meae*, is located on an elevated terrace at the east end of the *tohua*. *Paepae* 1 consists of a 6-×-6.5-m paved platform, 1.3 m high. There are no banyan trees marking the ritual place today. Three boulders, two of which contain a total of 11 face images, are placed in front of the platform (Figure 37). An elegant-looking dog (from a Western point of view) is located on a boulder 2 × 2.2 m, approximately 7 m to the east of the *meae* (Figure 38). To the south of the *meae*, Structure 2 is an 18-m long, attached and paved platform with a vertically placed polishing boulder in the facing wall, which suggests the platform was a later addition. Vertically placed cups also occur on Structure 5. Boulders with vertically placed cups are often found in a secondary context. While a communal *ua ma* (breadfruit silo) with a diameter of 7.5 m is located some 30 m to the northeast of the *meae*, a smaller silo is located to the southwest.

Nine of the stone platforms have stone-lined pits; all except one are sleeping platforms. Several of the sleeping platforms are paved with *kiva* (water-worn stones) intermixed with large irregular, flat stones. Part of the sleeping space on the west side of Structure 6 is paved. This type of space served as a place to store sacred objects (Lisiansky 1968:72 [1814]). A rectangular stone-lined pit is located on the front section. Seven grinding stones with one cupule each (one of the stones has a polishing groove) are placed around the dividing wall and the *pakeho*. Similar to Structures L and E, on Tohua Hikokua, the surface of Structure 6 is densely covered with *auti* plants.

The chief's residential complex (Structures 8, 9, 10, 11, 12, and 13) consists of several *paepae*, some of which are terraced. *Paepae* 12, prominently placed on the dance floor, is paved with large boulders; many are *kiva* stones. The *paepae* is 6.4 m long and 6.8 m wide. The facing wall is two courses high and measures 1.7 m.

The front section of the chief's complex faces north, and it is oriented toward the ocean. A *popoi* pounder was found in the rectangular pavement (3.4 × 7.8 m) of Structure 8, located in the back at the edge of what appears to have been a planting area. Considering that the pavement was placed in the back of the chief's residential unit in concert with the *popoi* pounder, this section was perhaps the cookhouse, the most secular part of a residential complex. In the Hane Valley, Kellum-Ottino (1971) only found *popoi* pounders at ceremonial sites.

Structure 7, attached to the chief's residential unit on the east side, is a 42-m long, well-built narrow *paepae hiamoe*. Similar to the chief's unit, the boulders in the facing wall are massive. The maximum height of the facing wall, made up of two courses, is 1 m. An image boulder depicting *ipu* tattoo motifs, was once part of the facing wall and was partly lying on the dance floor. Rather than replace the boulder, the empty slot had been repaired with smaller stones. All the circular figures depict common *ipu* tattoo patterns (Figure 39c–d). Because of the location next to the chief's complex, the carved *ipu* patterns, the length of the structure, and the size of the boulder, this was most likely the warriors' place. Several petroglyphs are on the facing wall of the chief's residential unit (Figure 39c–d).

Located across from the chief's residential unit, probably part of the spectators' place, are low pavements, fully paved *paepae,* with perhaps a miniature temple. To the east, next to the narrow entrance, several boulders have petroglyphs. One worked rectangular block, part of a wall, contains two anthropomorphic stick figures. Two additional cut blocks contain incomplete geometric figures. A cluster of structures and features, some of which appear to be unfinished, border the western side of the dance floor.

Two petroglyph stones with anthropomorphic stick figures, one figure with a forked tail and a pair of eyes are found on portable stones on Structure 10 (Figure 39c–d). There are cut slabs of red tuff and white beach rock, polishing stones, and a rectangular basin in the same area. A facing stone on the connected *paepae* to the west depicts a fish and a dog facing each other. Located on the top of Te Moui, a steep crag directly above Kamuihei I, is a massive but undecorated *paepae;* according to local history it was the home of a chieftess (Tioka Puhetini, pers. comm. 1996).

Tohua Kamuihei I Images

A 0.5-m high tiki (T1) in red tuff and a cut basalt stone with a *mata komoe* were found in 1988 in the center of the dance floor. Both stones are portable and may have been moved from some other location. [32]

The manner in which the image boulder on the warriors' platform once fitted into the facing wall suggests that at least the tattoo motifs were pecked after the boulder fell off the wall. If the stone were in its original position, the anthropomorphic figures would be visible and the tattoo motif would be hidden. It is unknown why the boulder was not replaced in the facing wall. However, if the *tohua* was no longer being kept in order, the tattoo motifs were perhaps an attempt to revive the tribal group when the traditional rituals were no longer in practice.

A total of 38 petroglyphs and one tiki make up the bulk of the petroglyphs with 51.3 percent dipicting anthropomorphs and 16 geometric figures representing 41.0 percent (Table 11). There three zoomorphs representing 7.7 percent. While the anthropomorph with a forked tail is unique for a *tohua*, two boulders in

Table 11. Tohua Kamuihei I frequencies and percentages of petroglyphs and tiki

Image Type	n	%
Anthropomorph	–	–
Stick Figure	5	–
Anthropomorphic Lizard	1	–
Tiki	1	–
Disconnected Body Parts	–	–
Face	13	–
Subtotal	**20**	**51.3**
Fauna	–	–
Dog	2	–
Subtotal	**2**	–
Sea Forms	–	–
Fish	1	–
Subtotal	**1**	**7.7**
Geometric Motif	–	–
Curvilinear	13	–
Cupule	2	–
Exotic Figure	1	–
Subtotal	**16**	**41.0**
Total	**39**	**100**

Table 12. Tohua Kamuihei I frequencies, percentages, and contexts of petroglyphs and tiki

Provenience	BLDR	Image Type	%
Ritual Complex	10	Anthropomorphic Face	–
	1	Anthropomorphic Eye	–
	1	Dog	–
Subtotal	**12**		**30.8**
Spectator's *Paepae*	8	Geometric figures	–
Subtotal	**8**		**20.5**
Warriors' *Paepae*	1	Anthropomorphic Stick Figure	–
	5	Circular Figure	–
Subtotal	**6**		**15.4**
Chief's Complex	2	Anthropomorphic Stick Figure	–
	1	Anthropomorphic Eyes	–
	1	Anthropomorph with Forked Tail	–
	1	Fish	–
	1	Dog	–
Subtotal	**6**		**15.4**
Dance Plaza	1	Anthropomorphic Face	–
	5	Circular Figure	–
	1	Tiki	–
Subtotal	**7**		**18.0**
Total	**39**		**100**

the Kamuihei area show similar figures. The majority of the figures (n=12), or 30.8 percent, are associated with the ritual complex (Table 12). A total of eight figures or 20.5 percent, is linked to the spectators' complex; six figures or 15.4 percent are associated with the warriors' area; 15.4 percent are linked to each the chiefly establishment; seven figures or 18 percent are found on the dance plaza.

Tohua Kamuihei II

Tohua Kamuihei II is located 60 m southwest of Tohua Kamuihei I in the shadow of Ototemoui Ridge. It is the smallest of the four ceremonial complexes in the study area. Six sleeping platforms and one simple platform surround the 50-×-14-m rectangular dance floor. On the northwest corner two large banyan trees grow between a partly paved platform (Structure 5) and a sleeping platform (Structure 6). Von den Steinen, who photographed the *paepae* in 1896, referred to the platforms as being the place for the priest(s) (1928:III. Plate 4). What appear to be graves from the Christian era are located on the dance floor in front of the priest's platforms or the *meae*. The lower section of a *popoi* pounder was found on the dance floor while mapping the tribal ceremonial place. A chief probably occupied the eastern part of the *tohua* complex. On the short end of the *tohua* is a *paepae* with an attached pavement (Structure 1). One portable boulder approximately 30–40 cm wide with an anthropomorphic stick figure was found on the *paepae*.[33] Excavation of a circular pit, located on the dance floor in the corner of structures 5 and 6 yielded lithics, fragments of turtle carapace, and pig and human craniums. The pit was superimposed on a breadfruit pit (Ottino 2003:127–136).

Structure 2 is situated below and at an angle from the dance floor in the northeast corner. It appears detached from the rest of the complex. Perhaps the *tohua* and the *paepae* were constructed at different time periods. Two *pakeho* are placed on the front terrace; the dividing wall consists of a row of red *keetu* slabs. One of the slabs, placed directly above one *pakeho*, depicts a bas-relief human figure. Both circular and linear geometric figures are found on the facing slabs around the second *pakeho*.

Structure 3, perhaps the warrior's platform, has a 1-m long rectangular slab of beach rock placed

Table 13. Tohua Kamuihei II frequencies and percentages of image types

Image Type	n	%
Anthropomorph	–	–
Stick Figure	5	–
Naturalistic Human Figure	1 (bas-relief)	–
Subtotal	**6**	**22.0**
Disconnected Body Parts	–	–
Face	1	–
Subtotal	**1**	**3.7**
Fauna	–	–
Centipede	1	–
Subtotal	**1**	**3.7**
Geometric Motif	–	–
Curvilinear	16	–
Linear	3	–
Subtotal	**19**	**70.4**
Total	**27**	**100**

Table 14. Tohua Kamuihei II contexts, frequencies, and percentages of image types

Provenience	BLDR	Image Type	%
Chief's Complex	1	Anthropomorph	–
	2	Linear Figure	–
	1	Circular Figure	–
	1	Linear Figure	–
	1	Anthropomorphic Stick Figure	–
Subtotal	**6**		**22.2**
Warriors' Platform	1	Centipede	3.7
Priests' Complex	1	Anthropomorphic Face	3.7
Communal Breadfruit Pit	4	Anthropomorphic Stick Figure	14.8
	15	Circular Figure	55.6
Subtotal	**21**		**77.8**
Total	**27**		**100**

on the front paved terrace. A centipede petroglyph is located on the facing wall. Several cupules are associated with the *tohua*. A polishing stone on Structure 1 has one cup on the surface and another one placed at one corner. Structure 4 is paved with a 30-degree-downward sloping angle. The function of this structure is uncertain, but it may have been a place for spectators or warriors.

A communal *ma* pit measuring 12 × 2 m is situated to the southeast of the *tohua*. On the southwest side of the silo a boulder (5.1 × 5.5 × 3.7 m) contains geometric and anthropomorphic stick figures along with modern graffiti.

Tohua Kamuihei II Images

Images on Tohua Kamuihei II are located at the breadfruit storage silo, the chief's complex, and what is perhaps the warriors' platform. Geometric figures (n=19), both circular and linear, make up 70.4 percent of the total, while six anthropomorphic figures and one face (n=7) represent a total of 25.7 percent (Table 13). Most of the images (n=19), a total of 70.4 percent, are linked to the communal breadfruit pit, while a total of six figures or 22.2 percent are associated with the chief's complex (Table 14).

Except for the anthropomorphic stick figure on Structure 1, the various image types, the red tuff, and the well-built platforms with *pakeho* on Structure 2 are consistent with late prehistoric to early historic sites. A portable stone with a stick figure may have been brought onto the site, perhaps when it was constructed. A Polynesian custom in the past dictated that, when a new building was to be made, whether an ordinary residence or a *tohua*, a feast was first prepared; everybody who participated was expected to bring stones (Handy 1923:150, 1927; Henry 1928). In the case of the construction of a new tribal ceremonial place, a stone was removed from an older site and incorporated into the new place. In Tahiti, there was a certain ceremony associated with laying the chief cornerstone of a new national ceremonial center or *marae* (Henry 1928:157). This was to ensure that the new site could absorb some of the *mana* embodied in the stone, and to further tie the old and the new ceremonial site to the land and their lineage.

Tohua Tahakia

Tohua Tahakia, also known as Tohua Keikahanui, is directly to the west of Tohua Kamuihei I and II—across the Hatiheu-Taipi road. To the north and northwest of the *tohua* are extensively irrigated *taro* field systems; to the west is a network of stone-faced terraces in a *mape* (Tahitian chestnut [*Inocarpus fagifer*]) forest. As Table 8 shows, the tribal ceremonial complex Tahakia, with a measurement of 120 × 19 m, has the largest dance plaza of the *tohua* still standing in the valley. The bulk of the images are located on boulders and outcrops incorporated into structures on the long south end and the short east side of the tribal ceremonial complex. It appears that, particularly in the southeast corner, a great number of activities related to petroglyphs took place (Figure 40).

According to local history, Tohua Tahakia was the tribal ceremonial place of the Ati Keikahanui, a subtribe of the Hatiheu Puhioho group (Tioka Puhetini, pers. comm., 1996; Ottino 1998). The historic war chief Keikahanui lived on the *paepae hiamoe* (Structure 1) that occupies the entire east end of the *tohua*, a position on the *tohua* normally reserved for the ritual place or the chief's residential unit. However, local residents insist that Keikahanui was not a chief, but a warrior chief. Apparently Keikahanui's father, the chief of Kamuihei (perhaps chief of Kamuihei I and II), built the *tohua* or perhaps Structure 1 for his son (Tioka Puhetini, pers. comm., 1996). If that is the case, how did Keikahanui manage to occupy a central section of the tribal communal complex? One possible explanation is that the war chief Keikahanui came to power in the late prehistoric early historic period, a time of tribal conflicts in the archipelago (Thomas 1990), and that the location was once a ritual place or was occupied by a former chief. That the tribal ceremonial complex has two names suggests that at least two owners are known to have had control of the tribal ceremonial site.

The Polynesian sociopolitical structure is typically characterized as a chiefdom (Goldman 1970; Kirch 1984, 2000; Sahlins 1958). Based largely on evidence of lexical reconstruction and controlled ethnographic comparison, the aspect of hereditary chieftainship was already part of the ancestral Polynesian social system (Kirch 2000; Kirch and Green 1987). However, a great deal of variation in social organization existed, from the simplest stratified chiefdom particularly found on the atolls and some small islands, to the highly stratified chiefdoms of Hawaii, the Society Islands, Tonga, and Samoa (Kirch 2000). The Marquesan sociopolitical system fell somewhere in between. Goldman (1970), who examined chiefly competition in his comparative and evolutionary study, classified the Marquesas as an "open chiefdom." According to Goldman (1970), in an "open chiefdom," the focus was more on military and political power rather than on religious power. Thus a courageous warrior could gain considerable prestige and wealth within his tribe. In contrast to the Big Man system, the chiefly position was prescribed. Chiefly sanctity also depended on a "network of active alliances and prestige" (Thomas 1990:88). In the Marquesas this was achieved through marriage alliance, adoption, genealogical link, and competitive feasting (Thomas 1990:87–108).

Structure 1 is a well-constructed megalithic *paepae* in excellent condition. Keikahanui's *paepae* is relatively new and not contemporary with the structures on the west end of the *tohua*. Keikahanui's *paepae* seems to have been built with boulders from the southeast section of the complex. While the platform on the west end of the *tohua* is relatively intact, structures on the southeast end have been plundered for boulders. The facing wall of Structure 1 is 16.1 × 11 × 1.85 m. The dividing wall between the front terrace and the sleeping section of the *paepae* contains a 0.4-m high row with 15 *keetu*; the length of the individual dressed tuff slabs varies, but the longest is 1.6 m. A section behind the dividing wall is paved with large stones, a polishing stone, and a 1-m long white beach rock. The sleeping area, normally a dirt floor, is paved with pebbles. Flat smooth stones fitted with pebbles pave the terrace. Four polishing stones are located on the front terrace. The back wall is battered or buttressed, which, according to Linton (1925:6), was a rare occurrence in the islands. It faces the mountain and measures from 0.70 to 1.85 m in height. No stone-lined pits are linked to Structure 1.

One of the upper stones in the northwest corner depicts a *mata komoe* and some geometric figures (Figure 41). This is an odd position for an image stone; the image type is, furthermore, stylistically different from the motifs found elsewhere on Tohua Tahakia. Perhaps the *mata komoe* stone derived from Tohua Kamuihei I, a *tohua* where several face figures occur; it may have been placed as the cornerstone when the structure was built.

Structure 19 is the tallest platform in the complex. Its facing wall measures from 1.5 m to 2.2 m in height and is placed in a prominent position on one of the long sides of the *tohua*. The overall *paepae* measures 11 × 9 m. It was constructed in two terraces, the first of which is 0.9 m high. Part of the sleeping place is paved. Sacred objects were placed in this special area. Two *pakeho* (pits) are situated in the front paved portion of the *paepae*, while a circular pit is situated along the west wall. A 45-cm high, upright stone is set in the southeast corner. Several cup stones are placed on the paved portion. Similar to Structures L and E on Tohua Hikokua, the platform is covered with *auti* plants, indicating that it had special significance in the past. The ceremonial tribal complex is overgrown, difficult to identify, and rarely, if ever, visited by tourists. What appear to be historic graves have been placed on the seaside of the plaza and the complex have been divided by historic boundary walls.[34]

Structure 20, southwest of Paepae Keikahanui, appears to be the place where activities related to warriors took place. The sleeping platform measures more than 20 m in length and is 11 m wide, longer than the average sleeping house, and it is located near the war chief Keikahanui's *paepae*. The facing wall is 1.2 m high. Four grinding stones and a 1-m deep *pakeho* are found on the *paepae*. Situated on the dance floor, 0.55 m directly north of Structure 20 is a large 1.8-×-0.8-m tall polishing stone. Some 20 cup-shaped depressions (some overlapping), 4 polishing grooves, and a circular polishing area are found on the top surface. Prehistoric Marquesan male and female activities have not been examined in details. However, it is generally assumed that only men made stone tools and worked with tools associated with, for example, house structures and canoe building (Firth 1967:213–225; Handy 1923:149; Linton 1925:165; Métraux 1940:137). Two anthropomorphic stick figures are on the north side, and one is on the south side of the grinding and polishing surface. Three outcrops on Structure 20 are covered with images. One outcrop with two turtles superimposed on a dog and an unfinished anthropomorph shows that the panel was used on several occasions (Figure 42). It is possible that the dog petroglyph, similar to the one found on Tohua Kamuihei I, was a tribal symbol of affiliation but at a later period turtles became prevalent at this ceremonial complex. An exceptionally large, fully pecked human-like figure with three fingers on each hand had been lightly pecked on the left side.

The west end of the complex is the oldest and least disturbed part of the *tohua*. While no banyan tree marks the *meae*, a cluster of structures (16 and 17) located on the southwest side stands out in regard to its complexity and high-status architectural components, suggesting ritual function. Structure 16 is unique. It consists of an approximately 1.2-m high, walled square (2.8 × 2.4 m), outlined with 15 whole and fractured pieces of red *keetu*. One *keetu* depicts a 43 cm long bas-relief anthropomorph.[35] The slab had been freshly broken and revealed a bright and even-colored, but fragile, red tuff (Munsell 2.5YR 4/8).

Connected to the walled structure on the east side, is a well-constructed megalithic sleeping platform (17) measuring 10 × 10 m; it has a massive, 2-m tall, facing wall. A petroglyph boulder, 1.20 × 0.5 × 0.5 m (331hth 117), is located in the upper second course of the wall facing the dance floor. Panel 1 contains vertically placed lightly pecked figures; a 2-cm deep cupule with radiating lines is incorporated into the panel (Figure 43). Panel 2 is horizontal and depicts an anthropomorphic face connected to circular geometric motifs. Some of the figures are difficult to see except at night with artificial light. The *paepae hiamoe* has a *pakeho* measuring 1.0 × 1.5 m. Five polishing stones are located on the front terrace. Three red *keetu* are part of the dividing wall. A large piece of branch coral was tucked into the upper part of the east wall. Pieces of coral are frequently found at special sites; coral pieces were found on some of the structures during restoration of Meae Eia in Hapatoni, Tahuata (Pierre Ottino, pers. comm. 1999). While picking bird eggs on a sea stack northeast of Haataivea Bay, I once noted several large branch coral pieces in a small rock shelter. The practice of placing branch coral on special structures is also seen in Hawaii. At Kipapa and Nakaohu, two *ahupuaa* in the district of Kahikinui, Maui, offerings of branch coral pieces are almost always found on ritual structures (Kirch 1997b:23). Corals pieces that are part of architectural component on temple structures in Moorea were dated with 230Th/U demonstrating that all the structures were built over a relatively short time period (Sharp et al. 2010).

The spectators' area is located both on the north and the south long sides of the dance floor. One petroglyph boulder (331hth 119) is incorporated into the back wall of the spectators' terrace (Feature 18). Only abstract geometric figures, mostly curvilinear, are depicted. Activities related to breadfruit preparation and storage took place at the northwest end; one small storage pit is located on the dance floor (Feature 9), and there are possibly three *ma* preparation pits on Structure 10, in addition to three unidentified circular pits located on and near Platforms 11, 12 and 13. A circular shallow depression (24), possibly a communal breadfruit silo, and a rectangular stone-lined pit (Feature 25) are located near Structure 19 on the south side of Tohua Tahakia.

Stone-stacked fences crossing the tribal communal complex and several historic graves show that parts of the complex were occupied and used after the valley was converted to Christianity. Incorporated into the foundation of Structure 2, is a roughly built sleeping *paepae* on the northeast side. A 1.10-×-1.15-m polishing stone with 14 vertically-placed cupules was reused and placed in the first course in the southwest corner of the facing wall. It is split in the middle, indicating that it was damaged during transport or construction. A stone enclosure with a historic grave that extends onto the dance floor is attached on the south side. Outlined with several pieces of red tuff, the grave today rests amid a dense ginger stand. Low (0.5–0.6 m) stone-stacked fences cross the dance floor in five different places. One wall extends from Structure 8, encloses a breadfruit pit (Feature 9), and continues over Structure 14, a 1.35-m high terrace. A similar but smaller wall encloses a pit on the east side of Structure 19.

Table 15. Tohua Tahakia frequencies and percentages of image types

Image Type	n	%
Anthropomorph	–	–
Stick Figure	23	–
Naturalistic	2	–
Subtotal	**25**	**29.7**
Disconnected Body Part	–	–
Face	2	–
Subtotal	**2**	**2.4**
Fauna	–	–
Dog	7	–
Turtle	5	–
Subtotal	**12**	**14.3**
Geometric Motif	–	–
Curvilinear	33	–
Linear	11	–
Cupule	1	–
Subtotal	**45**	**53.6**
Total	**84**	**100**

Tohua Tahakia Images

With the exception of Tohua Maikuku, images on Tohua Tahakia are more numerous, show a greater variety of types, and depict more complex panels than on any of the other tribal ceremonial places in the valley. A total of 84 petroglyphs were documented at Tohua Tahakia. Slightly more than half of the figures (n=45) or 53.6 percent, are geometric, with curvilinear motifs (n=33) dominant (Table 15). Anthropomorphs, both figures and disconnected body parts, comprise a total of 27 figures or 32.1 percent of the images. The remaining 12 images are zoomorphs a total of 14.3 percent.

A total of 53 petroglyphs or 63.1 percent of the images are associated with the warrior's complex, while 8 image represents 9.5 percent of the petroglyphs that is linked directly to war chief Kekanhanui (Table 16). Taken together (9.5 percent + 63.1 percent), these represents 72.6 percent of the total repertoire. The tribal complex appears to have been occupied for an extended period, but how far into the past its history reaches is yet unknown. Images, such as the bas-relief figures and megalithic architecture, are believed to be a late prehistoric–early historic development.

Keikahanui and his family apparently owned land with several residential units in the western part of the valley near Puhioho River. Keikahanui's mother, for example, is supposed to have occupied a residential complex northwest of Tohua Tahakia.[36] War Chief Keikahanui, because of his power and influence, may have usurped the former chief of Tohua Tahakia. Certainly the presence of four tribal ceremonial places in close proximity caused tension among the social groups. Most likely, War Chief Keikahanui, because of his large family and his important position, was able to gather sufficient support to overturn a weaker chief. Some of the unusually large and the superimposed images may have been one way to display a new social order and the power of a fearless war leader.

Hikokua-Kamuihei-Te Iipoka-Tahakia Area

A total of 802 individual petroglyphs are located on and in the neighborhoods of the Hikokua, Kamuihei I and II, Tahakia tribal ceremonial sites, and Meae Te Iipoka. Most figures are in the proximity of Tohua Kamuihei I and II, and Meae Te Iipoka complex to the west. A great number of the images are not associated with any visible structures. Situated between mountain peaks, Te Heu and Te Moui, the Meae Te Iipoka complex was

Table 16. Tohua Tahakia frequencies, percentages, and contexts of image types

Provenience	BLDR	Image Type	%
Ritual Complex	1	Anthropomorph	–
	4	Anthropomorph	–
	1	Anthropomorphic Face	–
	6	Geometric, Curvi-linear	–
Subtotal	**12**		**14.3**
Spectators' Place	7	Geometric, Curvi-linear	–
	2	Geometric, Linear	–
Subtotal	**9**		**10.7**
Warriors' Complex	2	Anthropomorph	–
	1	Geometric, Curvi-linear	–
	9	Anthropomorph	–
	5	Turtles	–
	1	Dog	–
	9	Geometric, Curvi-linear	–
	5	Geometric, Linear	–
	3	Anthropomorph	–
	4	Dog	–
	1	Geometric, Curvi-linear	–
	1	Geometric, Linear	–
	4	Anthropomorph	–
	2	Dog	–
	4	Geometric, Curvi-linear	–
	2	Geometric, Linear	–
Subtotal	**53**		**63.1**
War Chief Kekahanui's *Paepae*	1	Anthropomorphic Face	–
	6	Geometric, Curvi-linear	–
	1	Geometric, Linear	1.2
Subtotal	**8**		**9.5**
Dance Plaza	2	Anthropomorph	–
Subtotal	**2**		**2.4**
Total	**84**		**100**

an important sacred place in the past. A gigantic banyan tree some 40 m in circumference, grows over part of Structure 1. A deep, stone-lined pit is located in the same platform. Associated with the *meae*, 2 m to the north of the banyan tree, there is a large petroglyph boulder (331hth 5) with human figures and abstract geometric motifs (Figure 44).

Uphill from Meae Te Iipoka, just below Ototemoui ridge, several petroglyph boulders are distributed along both the east and west side of the stream. Two megalithic boulders (331hth 1 and 2) located at the base of the Ototemoui ridge are impressive. Images of *mata komoe*, human stick figures, fish, a hammerhead, and turtles are placed on several of the faces of 331hth 1. A bulge on the rock has been modified to depict a bas-relief turtle. There are numerous figures, especially on the west panel (Figure 45). Twelve of the 32 turtle motifs recorded in Hatiheu are found here.

In Polynesia, turtles were considered sacred. Historically turtles were surrounded by *tapu* and their consumption was restricted to persons of high rank (Crook 2007:70, 106 [1797–1799]; Henry 1928:380; Jardin 1862:74). Crook (2007:93 [1797–1799]) noted that on Tahuata "the priests alone have a right to feed upon Turtle, it being always when caught devoted to the Atua [god], and sacrificed." Kirch (1994a:225–298) was of the opinion that ceremonial cooking and consumption of turtles, a regular part of Futuna's (Western Polynesia) dry-season ritual, was also a key aspect of ritual practices in eastern Polynesia. Turtles were frequently substituted for human offerings in religious ceremonies. (Crook 2007:106 [1797–1799]; Jardin 1862:74). In Tahiti, the turtle was considered the shadow of the gods of the ocean (Henry 1928:384). Emory (1934, 1947:59–93, 1975:217) described turtle feasts in the Tuamotu Islands and found a large number of turtle bones, especially skulls, on most of the ceremonial *marae*. A wooden turtle effigy was collected from what Emory (1947:32–34) describes as a burial cave at Makatea Island, and coral turtle images were found on the island of Fagatau. As turtles have the ability to travel on land, as well as in the sea, Valeri (1985:23) suggested that they serve as a "metaphor for the relation between the dead and the living." Following a similar line of reasoning, Rolett (1986:78–87), who interpreted the turtle images within the framework of ethnohistoric and ethnographic sources, argued that turtles, because of their ability to travel between the physical (land) and the spiritual world (sea), served as a conduit between priests and the other world.

While image Boulder 331hth 1 was one of the few petroglyph stones known to the people of Hatiheu, image Boulder 331hth 2 was discovered by our team in 1984. These two boulders are located within a few meters of each other. Boulder 331hth 2 is of great interest because it depicts both female and male figures (Figure 46). The female figures are stylistically similar to a female figure in Omoa, Fatu Hiva, and a figure on a boulder on Ua Pou. Fish, exotic anthropomorphs, and geometric figures are also found on the same boulder.

A small *meae* complex (Structures 207 and 208) is found southeast of the Kamuihei-Te Iipoka-Tahakia area (Figure 47). Some claim that the *moko* or lizard boulder in bas-relief, said to be moved from Tohua Hikokua to Taiohae more than 60 years ago (Suggs 1961:145), originally came from this complex. Situated on a hill, the complex consists of a *paepae hiamoe* and a long, narrow, fully paved platform. The sleeping platform (207) measures 8.5 × 9.5 m. The east-facing wall consists of one course with five stones; it is 1 m high and is massive. The sleeping section, which faced the mountain to the south, was the weakest part of the *paepae* and was constructed with smaller stones; subsequently the south wall collapsed. Two polishing stones are part of the pavement, and two upright stones lie along the east and south walls. One of the uprights came from Te Heu, one of the peaks to the north (Tioka Puhetini, pers. comm. 1996).[37] Connected to the sleeping platform is Structure 208; it measures 11.3 × 4.0 m in length with some of the stones measuring 2.5 × 1.3 m. A pavement separates the two platforms. The old trail, the construction of which was overseen by the early Catholic missionaries to connect Hatiheu and Taipi, curves along the south side of the *paepae hiamoe*. Many of the stones were removed from these structures to mark the outlines of the trail. A banyan tree is growing between Tohua Tahakia and Structures 207 and 208. Both Tohua Tahakia and Kamuihei are visible from this hill site.

The Hikokua-Kamuihei-Te Iipoka-Tahakia image types are stylistically similar to petroglyphs found on the four tribal ceremonial complexes. But, in this area, the images occur in greater numbers; the panels are more complex; there are a larger variety of anthropomorphic types; and they are most often not associated with any structures (Figure 48). A total of 471 or 58.8 percent depict geometric motifs, while anthropomorphic types represent 257 figures, or 32.1 percent. Zoomorphs is represented by a total of 74 figures or 9.1 percent (Table 17). Some petroglyphs are unique and defy identification (Figure 49). Bird images are not common in the Marquesas, but a boulder with birds and *mata komoe* (331hth 14) is located within the Hikokua-Kamuihei-Te Iipoka-Tahakia area (Figure 50).

Surrounding the tribal ceremonial complexes are extensive and highly elaborate domestic house platforms, terraces, and large communal subterranean breadfruit storage silos. It is possible that some of the domestic platforms may be temple or ritual sites. The same pattern extends to the Ototemoui, a ridge above the tribal ceremonial complexes, and continues southwest to the Kahuvai area.

Table 17. Hikokua-Kamuihei-Te Iipoka-Tahakia area frequencies and percentages of image types

Image Type	n	%
Anthropomorph		
Stick Figure	145	–
Square-bodied	12	–
Double-outlined	10	–
Open-bodied	16	–
Naturalistic	13	–
Profile-squatting Human Figure	1	–
Anthropomorphic-lizard	1	–
Exotic	4	–
Subtotal	**202**	**25.2**
Disconnected Body Part		
Eyes	8	–
Face	47	–
Subtotal	**55**	**6.9**
Fauna		
Dog	27	–
Lizard	3	–
Bird	8	–
Unidentified	4	–
Subtotal	**42**	**5.2**
Sea Form		–
Turtle	21	–
Fish	10	–
Octopus	1	–
Subtotal	**32**	**3.9**
Geometric Motif		–
Curvilinear	314	–
Linear	95	–
Exotic Figure	62	–
Subtotal	**471**	**58.8**
Total	**802**	**100**

Summary

The western section of Hatiheu Valley, probably occupied by the Puhioho tribe and some decent groups, is located between two well-watered rivers. This is a fertile area with more rainfall than the western part of the island. The section more or less forms a traditional pie-shaped Polynesian land unit. The area is archaeological rich, and numerous structures are associated with rock art. Furthermore four of the eight tribal communal complexes or *tohua* are located within the two rivers—Vaiuua and Puhioho. Lastly I examined these *tohua* and their associated rock images. At the time of the field research the area was mostly uninhabited.

Chapter 5

The Archaeological Survey of the Hatiheu Valley

Rock and land were believed to be born and to grow exactly as were man and trees. Stones, islands, trees, streams, the sea, the heavenly bodies, and the heavens, like man had their psychic beings as well as their physical manifestations.

— E. S. C. Handy, *Polynesian Religion*

In this chapter I continue to examine the survey directed at the western section of Hatiheu to see if it is possible to detect an interspatial organization of rock images, rock images within the structures, or rock images within the larger community. While Chapter 4 investigated the archaeology of the four *tohua*, or tribal ceremonial complexes, this chapter deals with the Ototemoui and Kahuvai areas.

Ototemoui Ridge

The land, Ototemoui (or Ototemui) also named Tehoi, is a small, steep, and narrow ridge directly southwest of the Hikokua-Kamuihei-Te Iipoka-Tahakia area. Today the land is the property of the Catholic Church, and a land deed dates to March 27, 1928. *Oto* translates "to interior," while *te moui* means "the area/place behind," or "beyond." *Ototemoui* then, perhaps referred to "the land behind" or "beyond" the four *tohua* in Hikokua-Kamuihei-Te Iipoka-Tahakia area. The occupants of Ototemoui Ridge were probably connected to the ceremonial sites below. A great number of images are found on the Ototemoui Ridge, especially in the lower half. On the slope a total of 268 figures cluster on both large and small individual boulders and on high-status *paepae hiamoe*. Overall, the images depict anthropomorphic stick figures, faces, eyes, dogs, and fish similar to those in the Hikokua- Kamuihei-Te Tipoka-Tahakia area. However, there are variations in the way they were made, their size, and interspatial relationship. For instance, anthropomorphic figures tend to be taller. While the average anthropomorphic figure measures 30–40 cm, on Ototemoui Ridge some of the figures are up to 90 cm tall. Some figures overlap (Figures 51 and 52). The way in which they were made ranges from deeply pecked to lightly bruised, as well as intaglio. In the same area, the architecture consists of complex megalithic *paepae* with multiple components. Several banyan trees grow on Ototemoui Ridge. One *meae* complex, marked by a large banyan tree, is located on the east side of the ridge (Figure 53). A boulder in the walled courtyard has a pecked face motif on one corner stone that wraps around the boulder on both sides (Figure 54). On Ototemoui Ridge, there are more anthropomorphic figures than geometric motifs. A total of 135 figures or 50.4 percent depict anthropomorphs; 41.0 percent or 110 figures are geometric. Zoomorphs are depicted with 23 figures or 8.6 percent (Table 18). The emphasis on anthropomorphs may be related to a belief system that focuses on ancestor veneration and genealogy, the perceived idea that they are closely related to the chief(s), and thus to the ancestral gods.

Two large breadfruit silos (Features 192 and 196), diameters of 5.0 m and 5.5 m, are on the lower part of the ridge between Tohua Te Tipoka and several high-status *paepae hiamoe*. As it is often impossible to distinguish between a high-status *paepae* and a *meae*, it is conceivable that some of the *paepae hiamoe* are ritual places. On the uppermost part of Ototemoui, narrow terraces with platforms cling to the steep hillside. No distinct features characterize them except that they are constructed of small rough boulders and stones collected from the hillside. Level space is restricted to 1–1.5 m, and the terrace could, at the most, support one or two persons simultaneously. Neither images nor grinding stones are associated with these sites. Considering the strategic location with its clear view over the entire Hatiheu Bay, I am inclined to believe that the hill terraces were built as lookout points to detect enemy attacks from the sea and thus served to protect the Hikokua-Kamuihei-Te Iipoka-Tahakia and the Ototemoui area. An alternative explanation is that they represent miniature shrines (Millerstrom 2006a).

Table 18. Ototemoui Ridge frequencies and percentages of image types

Image Type		n	%
Anthropomorph			
Stick Figure		70	–
Square-bodied		2	–
Double-outlined		2	–
Open-bodied		2	–
Naturalistic		16	–
Exotic		3	–
	Subtotal	**95**	–
Disconnected Body Part			
Face		40	–
	Subtotal	**40**	**50.4**
Fauna			
Dog		18	–
Lizard		1	–
	Subtotal	**19**	
Sea Form			
Turtle		2	–
Shark		2	–
	Subtotal	**4**	**8.6**
Geometric Motif			
Curvilinear		79	–
Linear		8	–
Unidentified Figure		**23**	–
	Subtotal	**110**	**41.0**
Total		**268**	**100**

Kahuvai

Located southwest of Ototemoui Ridge in the mountain region, Kahuvai is another discrete settlement occupying both sides of the Puhioho stream. Numerous images and complex architecture were documented and mapped between 1988 and 1989.[38] While a total of 499 images were recorded in this section, only 20 images are located on the east side of the river, within the research area proper. It is uncertain if the stream served as a tribal boundary or if the settlements on both sides of the river were socially related in the past. As the settlement on the west side of the river is located at the base of the mountain near the river, it is likely that the Ati Puhioho occupied that entire area.

The residential unit on the east side of the Puhioho River consists of three megalithic *paepae hiamoe* (Structures 211, 212, 214), two large silos, and a number of stone-faced agricultural terraces. *Paepae* 211 is the largest structure consisting of a number of enclosures and walls (Figure 55). The north-facing wall, with a length of 18.5 m, is 2.5 m high and is built in two courses with small stones in between the boulders in the wall. Four bas-relief anthropomorphic figures (331hth 126, 127), one unfinished, are located on two large boulders in the base of the wall (Figure 56). Several terraces with facing stones and a garden area are situated east of 211. Structure 213 is more elaborate than the rest of the terrace alignments and may or may not be an agricultural terrace (Figure 57). The facing wall varies from 1.4 m to 2 m in height. It is built around several natural outcrops and is not well constructed. Two breadfruit silos (diameters 7.2 m and 10 m) are located north of Structures 211, near Structures 112 and 214. One of the silos (Feature 215) is more than 2 m deep. Structures 112 and 214 are *paepae hiamoe;* abstract geometric petroglyphs (designated 331hth 123 and 125) are found on the west and north walls of 214. The motifs are circular geometries. A shell trumpet (*Cassis* sp.) was found on the surface of 214 (see Chapter 7).

The images on the west side of Puhioho are similar to those of the Hikokua-Kamuihei-Te Tipoka-Tahakia and Ototemoui areas, but the petroglyph panels are not so densely covered with figures. A notable exception is a megalithic boulder where several anthropomorphic figures overlap two dogs (Figure 58). Overlapping figures are relatively uncommon but are found at Tohua Tahakia, Ototemoui, and Kahuvai. As Table 19 shows, a total of 80 percent or 16 petroglyphs depict geometric figures. The four bas-relief figures represent 20 percent of the total.

The megalithic structures constructed with massive boulders, images, stone-lined pits, and a large breadfruit silo, suggesting that a high-status family group occupied the residential unit or hamlet, perhaps in the late prehistoric or early historic period. *Cassis* sp. shell trumpets, bas-relief figures, and megalithic structures are consistent with cultural remains from this time period. Typically, breadfruit silos of this volume are associated not with residential units but with tribal ceremonial places. Today breadfruit trees still grow in the area, and a few kava plants were seen on terraces on both sides of the river. In some legends, the chief's kava plantations, which were all named, were located far back in the valley (Handy 1923). Krusenstem (1968 [1813]) observed kava plantations during his visit in 1804. According to Petard (1986:136 [1958]), the use of kava as a beverage completely disappeared in the Marquesas around 1925 when it was replaced by alcohol.

Table 19. Kahuvai frequencies and percentages of image types

Image Type	n	%
Anthropomorph		
Naturalistic	4 (bas relief)	
Subtotal		**20**
Geometric Motif		
Curvilinear	16	
Subtotal		**80**
Total	**20**	**100**

The Walled Area (the area has no name)

The western section of the valley contains a large number of walls and alignments. A settlement next to the Puhioho River, beginning in the lower part of the valley and continuing into the central part, is hemmed in with stacked walls. The walls are generally attached to house platforms and, presumably, protected household gardens. Some may be boundary walls. Stacked-stone walls, typically 1 m high and more than 0.5 m wide, originated with the notion of private property and the need to keep roaming pigs out of gardens. Many of the walls are depicted on the 1927 Hatiheu land map so they were constructed prior to 1927. Several of the traditional *paepae* within the walled area were rebuilt to form a fully paved platform on which to place a house without a dividing wall. In some cases, stacked walls run over older structures (Figure 59). On a knoll, there are three medium-sized breadfruit pits (Features 37, 38, and 39) in a row (Figure 60). These are the only *ma* silos in the western section built on an elevated area, probably to retrain dryness. Diameters range from 2.5 m to 3.5 m. Stones from a nearby *meae* complex (Structures 52 and 54) have been dismantled to build historic walls. No petroglyphs were found in this section.

Compared to the rest of the research area, a great variety of trees and ornamental plants grow within the walled section, many introduced to eastern Polynesia in the middle of the nineteenth century. In addition to the common coconut trees (*Cocos nucifera*), screw pine (*Pandanus tectorius*) and *hou*/*fau* (*Hibiscus tiliaceus*) are in abundance. The walled settlement contains, for instance, mango (*Mangifera indica*), *vi* apple (*Spondias dulcis*), *mapa* or Tahiti Chestnut (*Inocarpus fagifer*), *kapok* (*Ceiba petandra*), guava (*Psidium guajava*), at least two varieties of vanilla (e.g., *Vanilla planifolia*), croton (*Codiaeum variegatum*), candlenut (*Aleurites moluccana*), tamarind (*Tamarindus indica*), soursop or *coeur de boeuf* (*Annona muricata*), *ylang-ylang* (*Cananga odorata*), and several varieties of flowering red hibiscus (e.g., *Hibiscus rosa-sinensis*). Judging by the sleeping platforms, the breadfruit silos, introduced plants, and the modified house platforms, the walled area was occupied in the late prehistoric through the late historic era up to the 1960s.

The Niuamapu-Tauehua Agricultural Sector

Located on the hillside between Tohua Tahakia and the Puhioho River is a *mape* (*Inocarpus fagifer*) forest with a great number of stone-faced terraces that probably were used for cultivating breadfruit (*Artocarpus altilis*), sweet potato (*Ipomoea batatas*), and other species cultivated by the Enana (Petard 1986 [1959]; Yen 1974:140–144).[39] The lower hillside is generally muddy from natural springs in the area; taro (*Colocasia esculenta*) also was, no doubt, cultivated here in the past. The agricultural systems continue to the north.

The agricultural sections on the west side consist mostly of small residential units with private breadfruit silos (Figure 61). This is in sharp contrast to the size of the communal *ma* pits and house platforms, and the abundance of images found in the area of Hikokua-Kamuihei-Te Iipoka-Tahakia, Ototemoui Ridge, and Kahuvai. Figure 62 shows the diameter of 28 breadfruit pits we recorded and mapped in the research area. The diameters of the pits in the agricultural section measured from 2.5 m to 3.8 m, while the silos recorded in the Hikokua-Kamuihei-Te Iipoka-Tahakia, Ototemoui, and Kahuvai measured from 4 m to 13 m in diameter. There are only 3 petroglyph boulders located in this section.

The 1995 survey brought to light extensive abandoned taro fields in the open valley floor north of Tohua Tahakia. Many could not be surveyed because of the high water level. For instance, in June and July of 1995, the taro fields were under water and mud, and only the outer limits of the fields could be mapped. Archaeological residential units border the swamp area, but several house platforms could not be reached during our survey. Stone alignments from some of the house platforms disappeared into the deeper part of the swamp (Figure 63). Small platforms in the swamps, now mostly under water, indicate that the water flow was previously controlled. One small *meae* (Structure 168) near Puhioho River was surrounded by deep water.

A particularly well-made agricultural system at Tauehua consists of terraces constructed along a tributary stream near the Puhioho River. Located between the agricultural system and the Puhioho River, a megalithic sleeping platforms (Structure 159) is connected to the agricultural system by a historic wall. Part of the stacked wall dissects the taro field system suggesting that the fields were divided at one time. Perhaps one part belonged to the *meae* (Structure 168) to the north. One boulder near the east wall has petroglyphs that depict tattoo motifs (Figure 64). A bas-relief turtle (331hth 47) is depicted on the paved terrace of a megalithic *Paepae* 159 (Figure 65 and Figure 66). Three circular pits are located on the floor of an attached enclosure to the south. No large breadfruit silos were found in this area. Further south, a boulder with petroglyphs is located in the middle of the Puhioho River (Figure 67).

According to local residents, taro grew in large quantities in the "distant past" in both the eastern and western part of the Hatiheu Valley (Katupa, Puhetini, pers. comm., 1993, 1994, 1996); Yvonne Katupa, the mayor of the Hatiheu district remembers that when she was a young girl she saw stone alignments and taro in what today is a swamp. Feral pigs that roam the uninhabited western section of the valley have long since uprooted the taro plants (Katupa, pers. comm., 1994; Millerstrom 1996a).

Addison (2006), focused his research on Marquesan agriculture, and surveyed and mapped the taro pond fields on both sides of the Te Ahu Paaoa River, Atikea, on the eastern section of the Hatiheu Valley. Some 200 pond fields and 400 features, mostly residential and agricultural, were mapped. Irrigated taro pond fields made up 8.66 percent of the surveyed area.

Ethnohistoric and ethnographic accounts suggest that taro cultivation was widespread in prehistoric times (Crook 1790s; Gracia 1843:221; Porter 1970:149, 397 [1822]; Tautain 1897). However, the development of the Marquesan agricultural systems and its sociopolitical dynamics have been archaeologically neglected (Addison 2006; Kirch 1991b, 1994b; Kirch and Lepofsky 1993:190; Yen 1974:140–144). Crook (2007:75, 143 [1797–1799]) wrote that while the sweet potato was common, yams were less so. Taro was cultivated, but it only grew in water. Archaeological surveys have noted the presence of stone-faced terraces for dry-land farming and outlined taro ponds on several islands that demonstrate that many of the Polynesian crops, such as sweet potatoes, yam gourd, and taro also were part of the Marquesan diet (Addison 2006; Allen 2004:147; Bellwood 1972; Kellum-Ottino 1971; Suggs 1961:185).

In general, breadfruit trees were the main force in Marquesan sociopolitical development. The cultivation of breadfruit and the development and control of the tribal communal underground silos for breadfruit paste are seen as key aspects of political control (Hiroa 1938:207; Kirch 1991b, 1994b; 2000; Linton 1925). In contrast, taro cultivation is considered a secondary aspect of Marquesan agricultural production (Kellum-Ottino 1968: Kirch and Lepofsky 1993:190).

Kirch (1984, 1994a) and Spriggs (1986) defined the development of Polynesian agriculture as a three-step process: 1) adaptation; 2) expansion; and 3) intensification. Kirch's (1984, 1994b, 2000) model of prehistoric intensification of agricultural production in Polynesia is closely linked to four basic components: 1) increasing population pressure; 2) human-induced environmental changes and their impact on agricultural productivity; 3) increasing social stratification; and 4) increasing conflict. The Hatiheu case is examined within this theoretical framework.

The early inhabitants who settled in the western section of Hatiheu Valley presumably created swamp areas over time, suitable for taro cultivation. During initial colonization and agricultural expansion, people burned and cleared the upper valley slopes for agriculture (Kirch 1984, 1991b:113–133, 1994b). Anthropogenic activities in the steep volcanic slopes caused soil erosion that deposited alluvium on the valley flats below. Frequent storms aided this process. Crook (2007:89 [1797–1799]) noted in the 1790s that the upper parts of the Marquesan valleys have "a reddish ochrous cast, & [are] clayey." Depleted steep hills with deteriorated soil then caused the need for semipermanent forms of agriculture (Kirch 1991b:120). The alluvium deposit created a rich agricultural environment on the valley bottom where people settled and planted their gardens (Spriggs 1986; Kirch 1994b).

Intensive agriculture in the Marquesas could have included arboriculture, dry-land cultivation, as well as taro irrigation. It involved increased labor with the construction and maintenance of extensive terraces, garden beds, irrigation ditches, mulching, weeding, planting, and harvesting. Intensive agriculture increased the yield per unit area and the amount of land available for cultivation; the production aspects necessary to create a sufficient amount of surplus food needed to sustain the workers during large-scale communal projects. Marquesan customs dictated that, if a chief or a commoner resident needed a labor force to complete a new house, dig a new communal breadfruit silo, or construct a new ceremonial place, feasting occurred both before and after completion (Handy 1923, 1927). In addition, tubers provided food for pigs, a prestigious commodity in most ritual contexts, and one of the main food items consumed during feasting. A strong focus on competitive feasting is indicated by a relative increase in the frequency of pig bones found toward the end of the Hanamiai sequences, Tahuata (Rolett 1998), in the Hane Valley, Ua Huka (Sinoto 1966, 1968; Sinoto and Kellum 1965), and in the Haatuatua Valley, Nuku Hiva (Suggs 1961).

Population pressure was one social dynamic that led to the development of intensive agriculture that in turn gave rise to a complex hierarchical structure (Kirch 1994b:310–312). From approximately A.D. 1600 and onward, population pressure, lack of arable land, a concern with status, and the chiefs' demands for food surplus led to competition for resources and warfare in the Marquesas. It is uncertain if these economic and social dynamics led to intensive agricultural production in the valley, or if intensive agriculture enabled several tribal groups and subgroups to occupy the same area. Thomas (1990) argued that it was the system of competitive feasting, not ecological conditions or demographic pressure that led to agricultural intensification. Close spatial and probably temporal relationships between the irrigated taro field systems and the four ceremonial tribal complexes suggest that taro production, in addition to breadfruit, played an important economic role within Hatiheu's sociopolitical structure.

The swampy area on the western side of Hatiheu provided an ideal condition for taro cultivation and greatly aided, sustained, and provided surplus for the tribes and *ramages* that occupied this part of the part of the valley. For instance, field experiments by Spriggs (1981, 1984) in Vanuatu estimated a potential yield of taro. Under ideal conditions, wet taro cultivation has a mean corm yield of 30 to 60 metric tons/hectare/year for a pond-field system (see also Kirch 1991b, 1994a; Massal and Barrau 1956). In addition to the corm they also ate the leaves. Although taro is subject to diseases, the growing of the tuber has the advantages of a short-time maturity, high yield, and relatively long storage time in the ground. Furthermore, it has the ability to propagate itself. Taro cultivation, however, is dependent on ecological conditions. A high degree of micro-environmental variability in rainfall patterns between the windward and leeward sides of the islands, in addition to prolonged drought, caused agricultural variation (Addison 2007:111–127; Rolett 1998). However, as pluviometric data suggests (Cauchard and Inchauspe 1978), lack of rainfall was less of a threat in Hatiheu and Taipi valleys, Nuku Hiva, than in other valleys in the Marquesas. Yet, even this well-watered section of Nuku Hiva may have been affected. Kellum-Ottino (1971), who discussed the early historic evidence of famine in the Marquesas, wrote that beginning around 1804 or 1806 a serious famine lasted four to six years and ravaged Nuku Hiva. Apparently it was caused by a tidal wave that destroyed agricultural areas in the lower part of the valleys. Two famines, but with less magnitude, took place in Taipivai, Nuku Hiva, in 1820 and 1862. Besides natural destruction caused by tidal waves and drought, intentional destruction of breadfruit and coconut trees, as well as taro, kava, and mulberry gardens by revenging enemy tribes, were an equal threat to the people.

To meet the increased demands by the chiefs for food for the labor force and competitive feasting, the economy shifted from a traditional household-based subsistence to an economy that produced a food surplus.

In Hatiheu Valley, fertile swamps formed in the valley depression and adequate rainfall permitted intensive taro production. In addition to large scale breadfruit cultivation, these agricultural systems allowed the development of the large and closely spaced ceremonial places and the status-oriented occupation of the Ototemoui Ridge and Kahuvai, which are clustered to the north and to the east of the agricultural area.

None of the abandoned taro systems located on the west side have been excavated and dated, thus their age is uncertain. However, the taro ponds with the settlement in the central part date from the late prehistoric–early historic time period. I have argued above that the yield from the taro fields, in part, supported the labor force that constructed and the maintained the tribal ceremonial complexes. Very likely the taro systems formed an important component in the competitive feasting cycle. Due to severe demographic decline soon after European contact, there was no longer a labor force sufficient to tend and maintain the fields. After severe population decline, taro agriculture in Hatiheu Valley ceased to function on a large scale, and the pond fields were largely abandoned. Jardin, a French botanist, noted in the 1860s that although the Marquesans highly esteemed taro, they barely cultivated the plant because of the extensive labor involved (e.g., planting, mulching, and weeding); the less labor-intensive breadfruit trees that grew in sufficient quantities were preferred (Jardin 1862:51). Today, taro cultivation is considered a secondary aspect of Marquesan agricultural production (Kellum-Ottino 1968; Kirch and Lepofsky 1993:190).

The breadfruit crop was favored because of its ability to be stored in silos for a long period, and because it required almost no maintenance. The breadfruit trees could be a liability, however, because they were often favorite targets for destruction during warfare. Capt. David Porter (1970:25–29 [1822]) witnessed the destruction of approximately 200 trees in an afternoon on Nuku Hiva in 1814 during a skirmish between the Hapa and the Taiohae tribes. Thomas (1990) opines that this destruction was part of a strategy to deplete the resources of an enemy group and thus limit their ability to increase their prestige in competitive feasting.

Table 20. Niuamapu-Tauehua frequencies and percentages of image types

Image Type		n	%
Anthropomorph			
Stick Figure		17	–
Double-outlined		2	–
Exotic		1	–
	Subtotal	**20**	**66.7**
Disconnected Body Parts			
Faces and Eyes		2	–
	Subtotal	**2**	**6.7**
Sea Form			
Turtle		1	–
	Subtotal	**1**	**3.3**
Geometric Motif			
Curvilinear		2	–
Exotic Figure		5	–
	Subtotal	**7**	**23.3**
Total		**30**	**100**

The Agricultural Sector and Petroglyph Boulders

Only three petroglyph boulders with a total of 30 individual figures are located in Niuamapu-Tauehua, the agricultural area (Table 20). Of the total of 30 petroglyphs, 22 figures or 73.4 percent, depict anthropomorphs. Seven figures or 23.3 percent are geometrics. The turtle carved in bas-relief (331hth 47) located on the megalithic houseplatform (Structure 159) is said to have belonged to the chief warrior Keikahanui. Across from Paepae Keikahanui, an irrigated taro system, stands an image boulder measuring 1.3 × 2.0 × 1.6 m. The top surface is polished, a unique feature, and the designs are *ipu* figures (331hth 375). The figures are similar to the tattoo motifs at Kamuihei area. While the placement of the boulder suggests a boundary marker, the motifs could also reflect an affiliation with Tohua Kamuihei I. If indeed the warrior Chief Keikahanui and the Chief of Kamuihei were related; as the local history claims, this hypothesis makes sense. Although not common, various forms of the *ipu* motif are found elsewhere in the Marquesas. On a petroglyph boulder known as Hakaiki (chief) near Tohua Koeomai, Hanaiapa, Hiva Oa, several deeply pecked *ipu* motifs are found on the north side (Linton 1925:172–173; Millerstrom 1985c). The *ipu* motif is a common part of the tattoo repertoire and was used in a great number of variations (Handy 1922; P., et M.-N. Ottino-Garanger 1998). Crook (2007:59 [1797–1799]; von den Steinen 1925), who compared the incomplete circle with a "lagoon island," thought it was the most common Marquesan tattoo motif. Both the image boulders, which indicate association with the *tapu* class, are linked to the high-status residential unit.

Further inland, at Niuamapu, a boulder measuring 1.3 × 1.2 × 2.2 m (331hth 46/1 and 46/2) is located in the middle of the Puhiohio River. It appears to have no archaeological association. But located some 26 m to the north is a residential unit, apparently occupied by warrior Chief Keikahanui's mother. The west panel facing the settlement across the stream depicts several anthropomorphic stick figures surrounding a *mata komoe*, a face; the north panel consists of geometric figures, and it faces downstream (see Figure 67). Because of the boulder's location in the center of the stream, in addition to the orientation of the panels, the boulder perhaps served as a tribal boundary marker.

The lack of petroglyphs in the agricultural section demonstrates that the *mataeinaa* that occupied the agricultural section were not directly involved in the making of images on boulders on or around the house sites and the *meae*. It appears they were largely concerned with the food production that sustained the people, in general, and the chiefly class to the east, in particular. The location of the petroglyphs suggests that all activities involved in making the images took place on or around the elite area, the tribal ceremonial complexes and the elite households. Although there is no evidence to show that the petroglyphs were linked to agriculture one can speculate they were part of agricultural fertility and associated with everything related to the fertility of the elite as well as the *mataeinaa*.

Food Storage Facilities

A total of 28 breadfruit pits were documented. There are two types of pits or silos, the communal and the private. The communal breadfruit silos are located in the Hikokua-Kamuihei-Te Ilpoka-Tahakia area, in Ototemoui, and Kahuvai (Table 21). Most of the large silos are connected to ceremonial places, but those in the Ototemoui-Kahuvai area are associated with high-status residential units. This indicates that some families, besides those of chiefs, had considerable influence and were able to organize a workforce for both the construction of *ma* pits and for the filling of massive quantities of breadfruit. For instance, the two silos (Features 192, 196) linked to the high-status residential units on Ototemoui have an estimated volume of 20 m^3 and 88 m^3, respectively. Similarly, in Kahuvai two silos (Features 215 and 216) linked to a high-status residential unit could hold at least 157 m^3 and 47 m^3. While these pits held a large volume of breadfruit paste, they are of medium size compared to the communal *ua ma* east of Tohua Hikokua that held an estimated volume of 226 m^3. Placed to the south of Tohua Tahakia is a 70-×-40-cm, stone-lined rectangular pit (Feature 24) and a circular depression that measures 3.4 m in diameter. It is uncertain, however, if these pits represent *ma* pits.

Conversely, no communal *ua ma* are found in the agricultural sector. The *ua ma* placed near residential units in the agricultural section range in diameter from 2.5 m to 4 m. While the first harvest with

Table 21. Hatiheu Valley, west side locations, measurements, and storage capacities of the breadfruit pits

Location (Agricultural sector)*	Diameter (m)	Depth (m)	Volume (m³)
West (6)	3.2	1.0	8
West (12a)	2.6	0.3	2
West (37)	3.4	1.3	12
West (38)	3.0	1.2	9
West (39)	2.5	1.5	7
West (77)	3.2	1.0	8
West (107)	4.0	1.0	13
West (109)	4.0	0.5	13
West (116)	2.5	0.5	3
West (117)	3.5	1.2	12
West (135)	3.0	0.5	4
West (146)	3.5	0.5	5
Hikokua-Kamuihei-Tahakia, Ototemoui-Kahuvai area	–	–	–
Hikokua area (78)	7.5	0.5	23

continued

Table 21.

Location (Agricultural sector)*	Diameter (m)	Depth (m)	Volume (m³)
Hikokua area (89)	7.1	0.5	20
Hikokua area (91)	12.0	2.0	226
Hikokua area (97)	8.0	0.5	25
Ototemoui (192)	5.0	1.0	20
Ototemoui (196)	5.5	3.7	88
Mutoka (204)	10.0	2.0	157
Kahuvai (215)	10.0	2.0	157
Kahuvai (216)	7.7	1.0	47
Tohua Hikokua (20)	12.0	2.0	226
Tohua Kamuihei I (25)	7.5	2.0	88
Tohua Kamuihei I (26)**	–	–	–
Tohua Kamuihei II (9)	13.0	7.0	1237
Tohua Tahakia (9)	3.5	1.6	15
Tohua Tahakia (12)	3.8	0.7	8
Tohua Tahakia (13)	2.5	0.3	1.5
Tohua Tahakia (24)	3.4	?	?

*Numbers in parenthesis indicate the designated structure number
** Not measured due to collapse.

the largest yield, the *mei nui*, belonged to the chief and went to fill his communal pits, the second harvest was used to fill private family pits (Handy 1923:183).

Several structures have two or more circular or rectangular pits on the *paepae*. At Tohua Tahakia, Structure 10, three circular pits are on the front terrace of the *paepae*; a larger ground-storage pit is located nearby. Some of the pits may have been hearths; others represent *ma* preparation pits.

Ma Preparation Pits

In Willowdean Handy's description of food preparation (Handy 1923:186–202), she explains how to drain the breadfruit: "The flesh is placed in a temporary shallow hole lined with plaited coconut leaves and banana leaves" (see also Linton 1925:102–103). Tautain (1897:549–550) wrote that people in Nuku Hiva used both earth pits and the square stone-lined pits on the *paepae* as *ma* pits. It is uncertain if Tautain referred to storage or preparation pits. Dordillon (1931:304) translates *pakeho* as a stone-lined pit used as a place to put fermented breadfruit. Elsewhere Dordillon gives the term *tiemo*, "to plaster a silo with clay." Linton (in Handy 1923:188–189) wrote that the *ma* pits appeared to have had only earthen walls and were never stone lined. For the *ma* to be preserved well the paste had to be drained, and the pits had to be made in watertight clay soil. Hjalmar Stolpe, the Swedish ethnographer, was photographed in Taiohae standing on a *paepae* next to two circular *ma* preparation pits (Söderström 1937:238, Figure 2). The pits are lined with coconut leaf mats and contain prepared breadfruit paste. The pits are probably similar to those described by Linton (1925) as circular containers where the fresh pulp is stored until it ferments. After approximately 7–10 days, the *ma* was removed and buried in the large clay-lined storage pits (Linton 1925:102). Handy (1923:119) observed round pits lined with boulders on *meae* and suggested that some were used for storing breadfruit paste eaten by the priests and their attendants. Some of these pits were repositories for *tapu* objects (Linton 1925:14). Possibly the *pakeho* was used for the preparation of the chief's *ua ma*. Perhaps the process on the *meae* consisted of placing the *ma* in mat containers that were placed over the pits to allow the juices to drain into the *pakeho*. Kellum-Ottino (1971) wrote that in Hane Valley (Ua Huka) the *ma* storage pits were frequently part of house platforms (Kellum-Ottino 1971). Bellwood (1972) mapped 18 *ma* pits in Hanatekua, (Hiva Oa) most of which were located in the central part of the valley.

Communal Breadfruit Silos

The large communal pits are truly impressive. Linton (1923:103) wrote that the largest pit he saw was located in the large and populous valley of Taipi, Nuku Hiva. It reportedly measured 5.5 m in diameter and was approximately 9 m deep. It appears that Linton inadvertently reversed the measurements. From his description I located the pit, and the diameter was 9 m with a depth of 5 m (Millerstrom 1995). If it is the same *ua ma*, the volume would be approximately 318 m^3. This is consistent with the communal *ma* pits I documented elsewhere. One such pit on the land of Te Tuu, located in the upper part of central Hatiheu Valley, measures 9 m in diameter and has a depth of 6.5 m (Millerstrom 1995). The volume would be 413 m^3. However, unlike the other silos I have documented, this one had smooth sides and appeared as if it was never put to use. A banyan tree grows on the west side of the silo, testifying to the belief that the chief's food was held sacred and that communal production was associated with ritual and strict observances. A *paepae hiamoe* with petroglyphs, perhaps the residence of the caretaker, is located nearby. According to Tautain (1887:550), a guardian always protected the large communal *ua ma*.

Only the men could carry *ma* from the small pit to the communal silos. During this process, the men had to observe sexual abstinence, and no women were allowed to be present when they prepared the paste (von den Steinen 1928:36–37). Söderström (1937:239) reported that raw *ma* was used as an embalming agent. During epidemics, dead bodies were occasionally thrown into abandoned *ma* pits (Linton 1925:54). Infrequently, clay breadfruit pits were incised with figures (Handy 1923; Millerstrom 1990). Many of the silos were faced with a few stones in the upper section to prevent loose soil from falling into the *ma* (Linton 1925:103).

Fermenting and storing breadfruit paste occurred throughout most of Oceania (Barrau 1961; Kirch 1984, 1991a, 1991b; Linton 1925; Ragone 1991:203–220; Yen 1975). But in the Marquesas, the process "reached a technological peak" (Kirch 1991a:128). Large communal subterranean silos with semi-anaerobically preserved breadfruit were developed as a buffer against famine due to prolonged drought and periodic warfare. The production of breadfruit and control over the large storage pits for fermented breadfruit paste (*ua ma*) is seen as key aspect of political control (Hiroa 1938:207; Kirch 1991a, 1991b, 1994b; Linton 1925). Lack of arable land made it useful to be able to store abundant breadfruit crops. The breadfruit trees played a pivotal role in Marquesan sociopolitical development. While the whole clan contributed to the *ma* pit, the tribal silos were under the direct control of the chief. This, however, did not necessarily ensure equal redistribution in times of stress. Robarts, the English beachcomber who lived in the Marquesas from 1797 to 1824, described his experience during a famine on Nuku Hiva in which the chief and his family survived while the poor people starved: "but I was well informed that the enemy was very weak for want of food, and their Chiefs would not give any food out of their stone pits" (in Dening 1974:117 [1797–1824]). During the three-year drought in Nuku Hiva (1800–1803), Robarts pointed out that 200–300 people starved to death in a single valley (in Dening 1974:274 [1797–1824]). However, Robarts' elite establishment suffered the least as they controlled the stored surplus.

To illustrate the importance of breadfruit and the horror of famine, Robarts related two occasions where two women, one from a respectable family, were put to death because they cursed the breadfruit (in Dening 1974:265). One was strangled by her own son, the other by a brother.

Architectural Components

Including the four tribal ceremonial complexes, a total of 67 rectangular and circular pits were recorded on 56 structures (Table 22). Of these sites, 42 were regular *paepae hiamoe*; at a sleeping house (Structure 159), three circular pits were placed in an attached-walled enclosure on the south side of the house. The remaining architectural structures consisted of seven *paepae*, two *paepae* complexes, and two *meae*. The two *paepae* complexes consisted of more than one *paepae* with attached terraces. Without archaeological excavation, it is impossible to determine if the circular house pits were fire pits or *ma* preparation pits. I suspect that, in the cases where there are several circular pits on the same platform, all or some represent breadfruit preparation pits.

Table 22. Hikokua-Kamuihei-Te Iipoka-Tahakia archaeological structures with pits

Site	Structure	Type of Architecture	Form of Pit	Rock Art
Tohua Hikokua	B	*Paepae*	Rectangular	
	D	*Paepae hiamoe*	Rectangular	
	E	*Paepae*	Rectangular	
	L	*Paepae hiamoe*	Rectangular (2)	
Tohua Kamuihei I	3	*Paepae*	Rectangular	
	4	*Paepae hiamoe*	Rectangular	
	5	*Paepae hiamoe*	Rectangular	
	10	*Paepae hiamoe*	Rectangular	Yes
	15	*Paepae hiamoe*	Rectangular	
	17	*Paepae hiamoe*	Rectangular	
	18	*Paepae hiamoe*	Rectangular	
	21	*Paepae hiamoe*	Rectangular	
	22	*Paepae hiamoe*	Rectangular	
Tohua Kamuihei II	2	*Paepae hiamoe*	Rectangular (2)	Yes
	7	*Paepae haome*	Rectangular (2)	
Tohua Tahakia	2	*Paepae hiamoe*	Rectangular	
	6	*Paepaep hiamoe*	Rectangular	
	7	*Paepae*	Rectangular	
	8	*Paepae hiamoe*	Rectangular	
	10	*Paepae hiamoe*	Circular (3)	
	11	*Paepae* Cmplex	Circular (3)	
	14	*Paepae*	Circular	
	17	*Paepae hiamoe*	Rectangular	Yes
	19	*Paepae hiamoe*	Rectangular (2)	
	20	*Paepae* Complex	Circular	Yes
Hikokua-Kamuihei-Tahakia area	224	*Te Iipoka meae*	Rectangular	
	92	*Paepae hiamoe*	Rectangular	
	94	*Paepae hiamoe*	Rectangular	
	100	*Paepae hiamoe*	Rectangular	
	128	*Paepae hiamoe*	Rectangular	
	170	*Paepae hiamoe*	Rectangular	
	171	*Paepae hiamoe*	Rectangular (2)	
	172	*Paepae*	Rectangular	
	220	*Paepae hiamoe*	Circular	
	222	*meae*	Circular	
	223	*Paepae hiamoe*	Circular	
Ototemoui	175	*Paepae hiamoe*	Rectangular	Yes
	176	*Paepae hiamoe*	Rectangular	
	181	*Paepae hiamoe*	Rectangular	Yes
	182	*Paepae hiamoe*	Rectangular	Yes
	187	*meae*	Rectangular	Yes
	189	*Paepae hiamoe*	Rectangular	
	195	*Paepae hiamoe*	Rectangular	
Kahuvai	211	*Paepae hiamoe*	Rectangular	Yes
	212	*Paepae hiamoe*	Rectangular	
	214	*paepae hiamoe*	Rectangular	Yes
Walled section	18	*Paepae hiamoe*	Rectangular	
	19	*Paepae hiamoe*	Rectangular	
Agricultural sector	68	*Paepae hiamoe*	Circular	
	71	*Paepae hiamoe*	Circular	
	129	*Paepae hiamoe*	Circular	
	151	*Paepae hiamoe*	Rectangular*	
	158	*Paved paepae*	Rectangular	

continued

Table 22.

Site	Structure	Type of Architecture	Form of Pit	Rock Art
	159	*Paepae hiamoe w/ enclosure*	Circular (3)	Yes
	163	*Paepae hiamoe*	Circular	
	168	*meae, Tauehua*	Rectangular	
Total	**56**		**67**	**11**

*May be a tomb built around the pakeho.

The *pakeho* are stone-lined pits placed on the front paved section of the sleeping houses. Eleven of the house sites, ten of which are sleeping houses and one *meae*, are associated with petroglyphs. Twenty-eight house sites had one *pakeho*, and five houses had two pakeho on the same structure. These were found on sleeping houses and on two *meae* in the tribal ceremonial complexes Hikokua, Kamuihei I, Kamuihei II, and Tahakia. Two *pakeho* occur on the chief's sleeping platform at Tohua Hikokua and Kamuihei II. At Tohua Tahakia, the tallest house foundations (Structure 19) may have been a chief's sleeping platform in the past.

A total of ten *pakeho* were located on sleeping houses, and one *meae* in the high-status or chiefly areas of Ototemoui, Kahuvai, and the area of Hikokua-Kamuihei-Te Iipoka-Tahkia. None of the houses with two *pakeho* were *meae*. Only five house sites with *pakeho* were found in both the agricultural and the walled sector. These were associated with images. One such rectangular pit was apparently used to expose a corpse (Puhetini, pers. comm., 1997). Except for a broken *koma*-type adze and six flakes, test excavation of two *pakeho* in Structures 176 (TU 3) and 182 (TU 6) did not uncover artifacts; nor were significant fragments of charcoal uncovered. Yet the pits were filled halfway with what looked like construction fill. Whether the fill served to stabilize the *paepae* or the fill was periodically tossed over organic matter is unknown.

Several functional explanations for the *pakeho* have been provided: they were repositories for sacred objects, such as skulls and bones of tribal members; they were larders; they served as sacrificial pits where victims were thrown to please the gods; or they were refuse or *ma* pits. Linton (1923:274) rejected the *ma* pit hypothesis as unlikely because "*ma* can only be preserved properly in pits dug in compact clay soil." Tautain (1897:668) explained that a *meae* had one or two *pakeho* according to its importance. To please the gods, human offerings were thrown into these pits. Linton wrote: "Many large *paepae* in Nuku Hiva have a rectangular pit running down to the original ground level" (Linton 1923:274). In general, pits are found in many Marquesan sacred sites, but they are most common in Nuku Hiva. In Nuku Hiva, the *pakeho* was both a constant feature of the simplest form of *meae*, as well as being found in ordinary house platforms (Linton 1923:275; 1925:14).

Handy (1927:193) stated that consecrated pits and enclosures were set aside "as the reception and protection of such remains as food, hair, [and] clothing that had been in contact with sacred persons." Objects that had been in contact with a sacred person thus "sanctified must be protected against desecration" (Handy 1927:193). Linton (1925:115) observed several pits with human bones, especially skulls, on both *meae* and *tohua*. Several old informants, according to Linton, claimed that they had "seen objects which had become tapu through touching the head of the oldest son thrown into such platform pits" (Linton 1923:274). As already discussed above, Ottino (2005:127–136) excavated human skulls from a total of fifteen individuals at a *pakeho*, near the priest complex at Tohua Kamuihei II. The base of the *pakeho* revealed that the stone-lined pit had originally functioned as a breadfruit storage pit. In a *pakeho* in the vicinity they uncovered another burial with a dozen human skulls, one of which showed trephination. The burials were associated with pig and turtle bones. Linton (1923:274) wrote that in the Hatiheu Valley during an epidemic, the owners used a pit as a burial place; afterward the house was burned and the place declared *tapu*. In the Hatiheu Valley a local informant once told me a story of an event that apparently happened in the 1860s. To break a prolonged drought, a stranded turtle found on the reefs at Anaho was placed as an offering in the *pakeho* at Meae Moeaoko in the upper part of the Hatiheu Valley. The informant remembered seeing turtle bones in the *pakeho* when she, as a young girl, cleaned the funeral *meae* that was located on her family's land. The informant told me that when the four young men carried the turtle over the mountain it started to rain (Millerstrom 1990b).

As we shall see in Chapter 6, not all *pakeho* served as burials. Ottino's excavations illustrate that some *pakeho* had, over time, served different functions. Some originally served as storage for breadfruit, and in the late prehistoric–early historic period, some were used as burial pits.

In Tahiti, to appease the gods, the *marae* weeders would clean the ceremonial site before one of the most important religious ceremonies, assembling and uncovering of the gods. The waste was thrown in pits. Teuira Henry (1928:159) explains:

> Then the marae weeders—the king or high chief, the men of rank, and the gentry—with bodies bare to the waist humbly scraped the accumulated moss from the marae stones and reverently collected and threw it into the sacred pit, blindness or death from the gods being the supposed sure consequence if they scattered it to the winds. Then they swept and weeded the grounds, carefully collecting the rubbish into heaps, which they threw into the pit. They began to work on the west side and faced the east as they went forward.

Old tapa strips, matting, and coconut leaves from the priest's kneeling were thrown into the pit. The old image "coverings" that the images were wrapped in were carefully folded and deposited in a "cell" for that purpose on the *marae* (Henry 1928:169).

Some of the *pakeho* probably served to prevent sorcery.[40] Polynesians were always on their guard to protect themselves against sorcery. They took great care to bury discarded personal objects or anything that had to do with bodily functions in pits at the temple or drop them into the deep part of the ocean (Handy 1923, 1927:237–240; Henry 1928:204). To kill or harm another person, sorcery known as *nati* (*nani*) *kaha* was practiced among the Enana (Chaulet 1853–1888; Delmas 1889–1902; Handy 1923:272; Krusenstem 1968 [1813]) and other Polynesian societies (Handy 1927; Henry 1928:203–214). Both Krusenstem (1968 [1813]) and Chaulet (1853–1888) described the rites on Nuku Hiva. As the name suggests, Te Ahu Kaha, discovered in 2003 (but mentioned by Chaulet), was a ritual *paepae* in Hakahetau Valley, Ua Pou, and was perhaps a place where rituals pertaining to sorcery took place (Millerstrom 2006a). Crook (2007 [1797–179]) translates *nati kaha* as a kind of spell or charm and also the disease that is supposed to be inflicted thereby. Sorcerers, who could be men or women, were called *tuhuna nati kaha. Nati* means to tie up or bind. Sorcery involved working a spell for revenge, protection, transference, and removal of spells (Handy 1923:272–278). The bait *mounu* or *momo*, which was regarded as the victims soul or spirit, involved something that was intimately linked to the victim, such as nail trimming, hair clipping, clothing, saliva, food, excrement, and so on (Chaulet 1853–1888; Delmas 1889–1902; Handy 1923:72). Bait was mixed with a prepared powder and placed in a specially woven bag called *ipu* (Handy 1927:37). Ipu, a tattoo motif, can also mean a bag, a small container and, for example, a bowl or gourd. Apparently the power of this charm lay in the weaving of the sennit bag and the preparation of the powder (Krusenstem 1968 [1813]). Similar figures *ipu* figures are part of the petroglyph repertoire. The term *ipu mata* refers to the eye orbit (Le Cleac'h 1997). As already mentioned, during 1797–1799 when Crook lived in the Marquesas, *ipu* figures, according to Crook, were the most common tattoo motif. The placement of the *ipu* image on stones, and especially the use of the *ipu* motifs in tattooing, may have served as an added protection against sorcery.

Cup-shaped Depressions

Research on cupules or *cups*, referred to by the Marquesans as *mata*, is lacking. They are found in various contexts indicating that they served multiple functions. Linton (1925) believed that they served as dye cups for tattooing. However, they are found so frequently at most sites that they must have had other functions as well. They may be situated on isolated grinding boulders, on portable stones associated with houses, and often in concert with polishing spots and v- and u-shaped polishing grooves. Anthropomorphic images sometimes are found on the same surface. Occasionally the cupules are incorporated into images. Most often they are incorporated into the pavements of *paepae hiamoe* on residential units as well as on structures within the tribal ceremonial complexes. For instance, close to two hundred cupules are found on various boulders associated with Tohua Hikokua. They may be on either vertical or horizontal surfaces. An image boulder in the Eiaone Valley contains several cupules in different elevations that are connected with grooves, some of which lead to two nested basins. Two lines lead from the basins to the image panel below. The images consist of circular pecked meandering lines and, in this case, the placement of the cups, grooves, basins, and the images suggests that some type of liquid was poured into the cups or basins on top of the stones and that the liquid drained into the cups and the image grooves (Millerstrom 1985b).

Generally, the cupule diameters range from 2.5 cm to 13 cm. The depths measure from about 0.3 cm to 8 cm; although some do reach a depth of 10 cm, it is rare. Cupules with various diameters and depths are found on the same boulder suggesting that they had different uses or were made at different times. Most of the *mata* are smoothed with rounded edges. In many cases, the cupules are worn on one side, indicating the direction from which the person who used the cups was situated. Some cupules have sharp edges. In a few cases, a shallow rectangular basin is associated. Frequently, stones covered with cupules are reused in walls and house foundations, presumably after they had exceeded their usefulness. Several of these were fractured during moving or construction. Perhaps the stones were reused because of the *mana* the stone contained, or they were used as construction material for practical reasons.

Occasionally, a boulder associated with a house site has up to 40 uniform cupules randomly placed on the same horizontal surface. These cupules are smaller in diameter, approximately 3 cm. It seems as if these particular stones might have served as a game board, similar to the Hawaiian *papamu* for playing *konane*, a game similar to checkers (Hiroa 1957:369–372; Kirch 1985).

Von den Steinen (1969:83–87 [1925]) mentioned that cupules were used to catch the soot from burning candlenut. The fine black powder was used as pigment for tattooing. Kellum-Ottino (1971:81–82) reported that, in the Hane Valley, Ua Huka, she recorded some 170 to 180 cupules that were distributed on 45 sites; one-third of the sites were residential units. Otherwise, they occurred at all site types. In the Poiotona area, however, they were absent on the house sites, but one cup-stone was found on a *tohua*. This led her to speculate that the Poiotona area may have been the earliest inland settlement. As cowry shell scrapers were often found on the house sites, Kellum-Ottino suggested that the cups were associated with *taro* or breadfruit cultivation and that the cups were used for making the cutting edges on the cowry vegetable scrapers. She tested this hypothesis in an archaeological experiment and found that she could obtain, by grinding a fresh cowry shell of the same species and size, a circular perforation 1.5–2 cm deep. Bellwood (1972) recorded a total of eleven cupules in Hanatekua, Hiva Oa, a small number compared to that of the Hane and Hatiheu valleys.

Cupules are easy to make. E. Edwards, G. Cordonnier, and I carried out an experiment on a basaltic boulder at Hatiheu beach. With a round beach stone, Cordonnier made a cupule that measured 5 cm in diameter and 2–3 cm deep in 15 minutes (Millerstrom 1988). Perhaps some of the cups or basins, in particular those associated with polishing grooves and spots, were used to polish the polls of some adzes, the top of *popoi* pounders, and some wood objects. Others may have served as game boards, to make cowry shell vegetable scrapers, or to grind food, crack nuts, and, as the Marquesans claim, to mix and hold dye for tattooing or body paint, such as turmeric. No cups with residues have been noted however.

Zone III: The Interior Zone

With great difficulty our team followed the Puhioho River to the Kahuvai waterfall. The interior zone was overgrown and the visibility poor. We saw only a few house sites along the river, but around the pool of the waterfall we noted some agricultural terraces. Local residents claim that burial caves are located in the vicinity of the waterfall and that fortifications have been seen on the surrounding mountain ridges. Fortifications placed on ridgetops are common elsewhere in the islands (Bellwood 1979:331; Forster 1777, Part 11:10; Porter 1970 [1822]; Suggs 1961). Except for occasional pig hunters, the interior of the valley is rarely visited today. We found no rock art, and no architectural structures were mapped in the interior zone.

Discussion

A total of 158 structures, including those incorporated in the four *tohua*, were recorded and mapped (Table 23). Of the 78 *paepae hiamoe*, 31 are located on the *tohua*. Among these structures I identified at least four *meae*. The remaining architecture consists of platforms and terraces, some of which may have been sleeping houses.

The sleeping houses on the tribal ceremonial complexes measure from 4 m to 18 m in length (excluding the warriors' *paepae hiamoe*); the average length is 11 m. While the warriors' platform at Tohua Kamuihei I is 18 m long, the length of the warriors' platform at Tohua Tahakia measure 28 m. Sleeping houses outside the tribal ceremonial complexes range from 5.5 m to 10.5 m. The facing walls are from 0.3 m to 2.5 m in height. Two unusually tall *paepae hiamoe* built on the hillside of Ototemoui had a facing wall that reached 5 m; another measured 8 m.

Table 23. Architecture frequencies including those structures part of the tribal ceremonial complexes

Architecture	n
Sleeping Platform (*paepae hiamoe*)	78
Platforms or Terraces (*paepae*)	76
Meae	4+
Total	**158+**

The settlement pattern survey in the western section of the Hatiheu Valley demonstrated that the research area between the main rivers, Puhioho and Vaiuua, formed three different occupational zones: 1) the costal lowland; 2) the central zone; and 3) the interior zone. The majority of the architecture, including four tribal ceremonial complexes, is situated in the central zone. This area was divided into two different settlements: 1) the high-status or chiefly settlement located around the four ceremonial places, on Ototemoui and Kahuvai; and 2) the agricultural area along the Puhioho River. Relatively little is known of the beach and the interior zones. Today no one live on the beach or in the interior of Hatiheu Valley. The costal lowland which is presently occupied by the Hatiheu Valley inhabitants, was probably among the first areas to be settled in the past. Ottino et al. (2003:79) noted that "the earliest important occupation" in the Hatiheu Valley was the seventh century A.D. There is no indication that the interior zone was ever heavily occupied.

Population Estimate

The survey by Kellum-Ottino (1971) and Molle and Conte (2015:253–274) in the Hane Valley (Ua Huka), and Bellwood's (1972:36–39) survey in the Hanatekua Valley (Hiva Oa), found that, in general, the settlements were concentrated in the central portion of the valleys. One tribe inhabited the Hane Valley, yet Kellum-Ottino (1971) located three *tohua*. Kellum-Ottino recorded 176 sites, of which 94 were houses in a 25-ha survey. An additional 16 habitation sites were recorded on the *tohua*. Kellum-Ottino established that, compared to Nuku Hiva and Hiva Oa, the Hane Valley structures appeared to be less varied and complex. The width of the paved section varied from 2–3 m to 6–9.2 m, while the width of the sleeping part in the rear of the house measured within 2 m. The maximum height is reported to be 1.6 m. The house sizes ranged from under 10 m^2 to 140 m^2.

Two tribes were known to have lived in the Hanatekua Valley, but Bellwood (1972) did not find evidence that the settlements were distributed in two zones. The lengths of houses were from 3 m to 33 m, which correspond to historic accounts. Within 30 ha, Bellwood recorded 106 structures of which 87 were sleeping platforms. Both Kellum-Ottino (1971) and Bellwood (1972) estimated the populations of the valleys they surveyed, based on habitation structures. Kellum-Ottino (1971) suggested that perhaps only 75 percent of the houses were occupied simultaneously. She calculated that 5–7 persons would occupy each house regardless of size, thus the total population was between 350 and 500 individuals. Bellwood (1972) went to extraordinary lengths to assess the Hanatekua Valley population but settled on two types of estimates. Bellwood (1972:44–47) suggested, following Naroll (1962) calculations that "prehistoric settlement can roughly estimate one-tenth of the total floor area in square meters." Structural evidence that gave a total 2,400 m^2, indicated a population of not over 240 people." Based on carrying capacity, Bellwood estimated that the food resources would sustain approximately 400 individuals. But, as Bellwood (1972:40) pointed out, to calculate the population carrying capacity of the Hanatekua Valley based on available food resources is challenging because of the many unknown factors. Molle and Conte (2015) who examined the Ua Huka population prior to European contact supported Kellum-Ottino's (1971) population estimate in the Hane Valley.

Founded on Kellum-Ottino's (1971) structural evidence, the 78 *paepae hiamoe* on the west side of the Hatiheu Valley would support a population of about 400–550 individuals, or approximately 300–400 individuals if only 75 percent of the houses were occupied at the same time. However, it appears that the houses in the Hatiheu Valley research area are larger than the houses in the Hane Valley. Thus the population may have been proportionally higher. In addition, it is possible that some of the unpaved platforms and terraces also served as sleeping houses.

Structural evidence of the 78 *paepae hiamoe* gave a total of 7,918 m^2 (including the sleeping platforms on the four *tohua* (3,322 m^2), 1,040 m^2 of which comprise warriors platforms), which indicates a population of just under 800 people. However, it is questionable if the *tohua* were permanently occupied. The warriors, for instance, probably lived elsewhere in the valley but only occupied the men's houses during the day, during festivities and rites, and when they were in a *tapu* state, such as before a war. If the warriors' sleeping platforms (1,040 m^2) are removed from the calculation, 6,878 m^2 would represent approximately 687 people. Nonetheless, if no one lived permanently on the *tohua*, the population would total some 460 people. Anchored in these calculations, the population in the western part of the Hatiheu Valley was somewhere between 300 and 800 inhabitants.

Images in the Study Area

Within the theoretical framework of archaeological rock art research, it appears that two sociopolitically diverse systems were in operation in the western section of the Hatiheu Valley. The area to the east and southwest were occupied by the high-status or chiefly class, while most of the *mataeinaa* settled to the west along the Puhioho River in an agricultural sector. While each Marquesan residential unit was surrounded by a garden, the main agricultural sector consisted of less complex residential units with relatively small private *ua ma*. The *mataeinaa* then were mostly concerned with agricultural cultivation, such as *taro*, breadfruit, and other crops that supported the tribe and created surplus food for the high-status group. The tribal ceremonial complexes were found in the high-status or chiefly area only, but *meae* occur in both sections. However, without markers, such as archaeological art, banyan tree(s) and other special trees (e.g., *temanu* [*Calophyllum inophyllum*]), upright stone(s), red or white cut slabs, fully paved platforms, and other high-status characteristics, it is often challenging to differentiate regular residential houses from high-status or chiefly residential units.

A total of 1,284 individual images were recorded (Table 24). The majority, 691 figures or 53.8 percent, represents abstract geometric figures. The total repertoire of anthropomorphic images are 478 figures and tiki or 37.2 percent. This embodies a great variety of types.

The Hatiheu Valley images are numerous, but the repertoire is limited in motif types. Furthermore, the images are homogeneous and repetitive. Overall, this is comparable to the images in the rest of the Hatiheu Valley, and to the Marquesas in general. The homogenous rock art repertoire is, in part, also due to the fact that the specialists, the *tuhuna*, some of which were highly respected, were able to travel freely despite intertribal and interisland warfare (Crook 2007:95 [1797–1799]). Furthermore, motif repetition in preliterate societies is common. Depending on context, each motif was meaningful and carried information (Conkey 1981; Forge 1991). Conkey (1981:4) writes: "Redundancy and ambiguity are often characteristic of visual forms created in non- or preliterate societies. Ambiguity allows a great deal of potential information to be stored in very few symbols; often, it is the particular context in which the symbols are used that may determine their relevant meanings."

In his analysis of the Hanamiai, Tahuata, fishing implements and adze assemblages, Rolett (1998:238–240) found that there is a "lack of archaeologically visible regional variation within the Marquesas" before approximately A.D. 1300. This trend disagrees with Sinoto's (1979) model that there was a cultural differentiation between the central and southern island groups in the early part of the cultural sequence. According to Rolett (1998), another trend seen in the material cultural complex from the early Marquesan period is the close resemblance to artifacts documented from the island of Huahine in the Society Islands from about the same period (Rolett 1998:236–240). Rolett (1993:45, 1998:240) argued that these interarchipelago similarities in early East Polynesian prehistory strongly suggest that Polynesians engaged in systematic long-distance, two-way voyaging and regional interaction. Imported stones that are frequently found in the Hanamiai, Tahuata, artifact assemblage before A.D. 1300 occurred less frequently after A.D. 1450, suggesting that long-distance voyaging declined over time (Rolett (1998:240; Weisler 1998,

Table 24. Hatiheu Valley, west side frequencies and percentages of image types

Image Type		n	%
Anthropomorph			
Stick Figure		266	–
Square-bodied		14	–
Double-outlined		14	–
Open-bodied		18	–
Naturalistic Human Figure		36	–
Profile-squatting Human Figure		1	–
Anthropomorphic-lizard		2	–
Exotic		8	–
	Subtotal	**359**	
Sculptured Image			
Tiki		4	–
	Subtotal	**4**	
Disconnected Body Part			
Anthropomorphic face/ eyes		115	–
	Subtotal	**478**	**37.2**
Fauna			
Dog		54	–
Lizard		5	–
Bird		8	–
Centipede		1	–
	Subtotal	**68**	
Sea Form			
Turtle		29	–
Fish		13	–
Octopus		1	–
Unidentified		4	–
	Subtotal	**115**	**9.0**
Geometric Motif			
Curvilinear		478	–
Linear		119	–
Cupule*		3	–
Exotic Figure		91	–
	Subtotal	**691**	**53.8**
Total		**1,284**	**100**

* Cupules are directly linked to petroglyphs

2002; Weisler et al. 2004:128–148). If Rolett's model is correct, it may explain in part the similarities between the Marquesan petroglyph repertoire and the petroglyphs located in the Society Islands and other Polynesian islands. It is also possible that the petroglyph assemblage that was part of the overall Marquesan ideology remained relatively unchanged regardless of the decrease in interarchipelago activities. The Hatiheu Valley rock images surveyed here are almost exclusively in the context of the chiefly class. To better understand the contexts in which the images were found, I examine the data on three levels: 1) tribal ideology; 2) personal status; and 3) regional tribal identity.

Tribal Ideology

The settlement of the chiefly class consisted of the four tribal ceremonial complexes in the Hikokua-Kamuihei-Te Iipoka-Tahakia region; it extended south to the Ototemoui Ridge and southwest to Kahuvai. The majority of complex megalithic architecture, all the large-size *ma* pits, and the majority of architecture with *pakeho* were recorded in this area.

Handy (1923:37–39) wrote that while the concept of a distinct chiefly class was definitely developed in Ua Pou Island, less is known about Nuku Hiva. In Thomas' (1990:36, 108) opinion, except for Ua Pou, chiefly authority in the early contact period was lacking but that landholders were more socially stratified in the northern group. In the large valley of Hakamoui, located on the east side of Ua Pou, the chiefly class occupied the west side of a stream—the side most protected from enemy invasion. It was called the valley of the *papa-haka-iki*, a term applied to the chiefs. On the east side of the river, the *mataeinaa* resided. Handy (1923:37–39) speculated, based on early historic accounts, that a chiefly class may also have existed on Nuku Hiva in former times. The spatial distribution of elite architecture and the vast number of examples of archaeological art in the western zones of Hatiheu confirm Handy's suspicion that a distinctly developed chiefly class also existed on Nuku Hiva in the past.

However, the spatial distributions of images on each individual *tohua* are variable (Table 25). On Tohua Hikokua, the focus appears to be on the spectators' *paepae* (50 percent). A single boulder with several images represents this high percentage. It is unknown if these images indicated status, if they were related to the dancing stone placed above, or if they were linked to a small god house that stood, at one time, on the platform. Three tiki (21.4 percent), taken together with other high-status items, such as cut rectangular blocks of red tuff, *papatea*, and the central placement of the *meae*, suggest that the ritual place where the priests officiated was the focal point of Tohua Hikokua. The case is similar at Tohua

Kamuihei I where 12 figures (30.8 percent) of the 39 figures are placed on the *meae*. Only 6 figures (15.4 percent) are associated with the chief's residential unit and another 6 figures (15.4 percent) with the warriors' place.

Table 25. Hatiheu Valley, west side percentages and contexts of the images on the tribal ceremonial complexes

Provenience	Tohua Hikokua	Tohua Kamuihei I	Tohua Kamuihei II	Tohua Tahakia	Total
Warriors' Platform	–	15.4	3.7	63.1	82.2
Ritual Complex	21.4	30.8	3.7	14.3	70.2
Chief's Complex	–	15.4	22.2	–	37.6
War Chief Kekahanui	–	–	–	9.5	9.5
Spectators' Area	50.0	20.5	–	10.7	81.2
Alignment	14.2	–	–	–	14.2
Dance Plaza	7.1	18.0	–	2.4	27.5
Paepae (unidentified)	7.1	–	–	–	7.1
Communal Breadfruit Pit	–	–	70.4	–	70.4

The spatial distribution of 27 petroglyphs at Kamuihei II is different. There the boulder next to the communal breadfruit pit contains 19 figures (70.4 percent) of the images, while the chief's residential unit displays 8 figures (22.2 percent). Considering that the chief generally controlled the communal *ma* pit, it appears that the focus on Tohua Kamuihei II was on the chief. Only the chief's residential complex exhibited red *keetu* and, in this case, one of the sleeping houses contained two *pakeho*. The ritual complex depicts 3.7 percent of the total images on the *tohua*.

Tohua Tahakia is unique due to the presence of warrior Chief Keikahanui. More images, some stylistically different (e.g., large and overlapping images) are found on this tribal ceremonial complex as compared to the other three *tohua*. It is certain that major changes, possibly a shift in the power structure, took place at the site. We also know that it was occupied in the early historic period. Most of the 84 images (63.1 percent) were linked to the warriors' place. If the 8 images (9.5 percent) on warrior chief Keikahanui's *paepae* and the images associated with the warriors platforms are added, the total is 61 figures (72.6 percent). Images, in all 12 figures, at the ritual complex represent 14.3 percent. The different emphases on images within the various *tohua* suggest that power and control were variable, according to the different tribes or ramified descent groups that occupied this part of the valley. The uneven distributions of the images in certain sections can be viewed in light of what Goldman (1970), Thomas (1990), and Kirch (1991a) have written about competition between the different status groups.

Personal Social Status

The majority of petroglyphs are not found on the *tohua* but on and around the high-status residential units. Unlike the *tohua* proper, images significantly increase in motif frequency, complexity, and motif variation with distance from the ceremonial complexes. A total of 1,254 figures (97.6 percent) are found in the high-status section. The remaining 30 figures (2.4 percent) were found in the Niuamapu-Tauehua area, in the agricultural sector. Except for one high-status residential unit, none of the household *paepae* in the agricultural area are linked to rock art (Table 26). It is interesting to note that of the total number of figures in the four tribal ceremonial centers, Hikokua-Kamuihei-Tahakia, Ototemoui and Kahuvai, the high status or chiefly area, abstract geometric figures comprise 684 figures (53.3 percent) and anthropomorphic figures are 456 (35.5 percent). The remaining "Others" category consists of zoomorphs, such as dogs, turtles, and fish. The focus on geometric and anthropomorphic figures reflects personal status and prestige. These figures, associated with late prehistoric–early historic architecture, correspond with historic tattoo designs and carved decorative motifs on elite regalia, such as bone fan handles, war clubs, stilt foot holders, carved turtle shell head-pieces, and so on. The remarkable similarities between archaeological art and motif decoration on historic material objects suggest that, while the emphasis and distribution of motifs changed according to time and place, the fundamental principles regarding the symbolic repertoire remain intact.

Table 26. Hatiheu Valley, west side frequencies and percentages of geometric, anthropomorphic, and other images in the chiefly versus the agricultural area

Sites and Areas	Geometric	Anthropomorphic	Other*	Total
High-status Areas				
Tohua Hikokua	7	6	1	14
Tohua Kamuihei I	16	20	3	39
Tohua Kamuihei II	19	7	1	27
Tohua Tahakia	45	27	12	84
Subtotal (n/%)	**87/6.8**	**60/4.6**	**17/1.3**	**164/12.7**
Hikokua-Kamuihei-Tahakia	471	257	74	802
Ototemoui	110	135	23	268
Kahuvai	16	4		20
Subtotal (n/%)	**597/46.5**	**396/30.8**	**97/7.6%**	**1,090/85.0**
Total (n/%)	**684/53.3**	**456/35.5**	**114/8.8**	**1,254/97.6**
Agricultural Areas				
Niuamapu-Tauehua (n/%)	7/0.6	22/1.7	1/0.1	30/2.4
Total (n/%)	**691/53.8**	**478/37.2**	**115/9%**	**1,284/100**

* This category consists mostly of zoomorphs, such as dogs, turtles, fish, and birds.

If the presence of petroglyphs and stone sculptures is indicators of high status, then the clustering of petroglyphs at Tohua Tahakia suggests that the war chief and the warriors held an elevated position within the tribe. However, at Tohua Hikokua and Kamuihei, where the *meae* are clearly defined, it appears as if priests held prominent positions. At Tohua Kamuihei II, most of the 19 petroglyph images (70.4 percent) occur on a boulder adjacent to the communal *ma* pit. A total of 7 figures (22.2 percent of the images are found in the chief's complex (See Table 25). Chiefs, who by virtue of birth were the head of the tribes, may not always have been effective political leaders of the *mataeinaa* connected to Tohua Hikokua, Kamuihei I, Kamuihei II, and Tahakia. In the Marquesas, in contrast to other Polynesian societies, the chiefs had limited sanctity and did not play an active role in tribal ritual matters (Crook 2007:52 [1797–1799]; also see discussions by Kirch [1984, 199la: 125] and Thomas [1990]). An inspirational priest or a shaman who mediated between the people of the tribe and the spirit world was powerful and could travel in both spheres. The knowledge they brought back from the spiritual world was particularly important and authoritative as a source of morality and social control. Some of the inspirational priests who officiated at the *meae* were greatly feared and had considerable influence over the *mataainaa*. In some cases, the *taua* were so powerful that she or he was elevated to *atua/etua* (a god) while still alive (Thomas 1990). It was thus possible for a shaman, shamanca (female spiritual leader and healer) or successful warrior to usurp a weak *hakaiki*. The presence of a weak or unpopular chief may have provided the shaman/shamanca or a brave warrior with the opportunities to gain control over a tribe.

This data support the argument that, due to several critical environmental and social features unique to the Marquesas Islands, an elaborate form of prestige rivalry and competition between the elite permeated the society, especially during the late prehistoric–early historic era (Kirch 1984, 1991a; Thomas 1990). These archaeological features, already discussed, caused the degradation of arable lands and depletion of certain fauna, such as sea turtles and sea mammals (Steadman 1989; Rolett 1992, 1998; Steadman and Rolett 1996). Demographic increase, particularly in the late prehistoric period, signaled major changes in agricultural production, monumental architecture, and personal prestige (Kirch 1984, 1991a; Thomas 1990). Subsistence in this period was based on arboriculture, animal husbandry, *ma* pit storage, and the gathering of marine mollusks (Dye 1990; Kirch 1991a). Particularly, the large tribal ceremonial complexes played a major part in the intertribal and interisland rivalry. Here, the *mau* (large commemorative funeral feast) was held, some of which were years in preparation. Often, several thousand guests from other tribes participated. Such extensive feasts set up reciprocal obligations between rivaling tribes and influential intertribal members. Intertribal rivalry in the Hatiheu Valley, especially among chiefs, inspirational priests, and warriors was manifested in the proximity of the four *tohua*. Within the tribal ceremonial complexes, the *meae* and the chief's residential unit were often placed at opposing ends of the plaza. Both the *meae* and the chief's residential unit then competed for spatial prominence and attention with architectural embellishment and archaeological art.

The same competition appears to have occurred between elite households, as well as between tribes on the north coast of Nuku Hiva. To increase personal prestige, images on rocks were placed on or around the private residential units. Large private reserves of stored breadfruit paste placed near monumental architecture aided in competitive feasting.

Regional Tribal Identity and Hatiheu Polities

Dog images occur in great numbers only on the north cost of Nuku Hiva. They are particularly numerous in the valleys of Akapa, Haatuatua, and Hatiheu. In the Hatiheu Valley, 162 dog figures were documented. Elsewhere, a few stylistically different dog images were documented in Vaipaee, Ua Huka, and Eiaone, and Puamau, Hiva Oa (Millerstrom 199la, 1991b). It is likely that the dog images served as a visual emblematic symbol expressing solidarity between the Marquesan polities who inhabited the north coast of the island.

In the past, two political divisions were recognized on Nuku Hiva: the Teii occupied the western portion of the island, and the Taipi constituted the eastern division (Crook 2007: 128–129 [1797–1799]; Handy 1923:31; Pechberty 1996:165). Traditionally the Teii and the Taipi were bitter enemies (Handy 1923:31). The matter of political divisions is complicated, however. Marquesan society was fluid and, in the event of war, alliances frequently shifted (Crook 2007 [1797–1799]; Handy 1923:31). Some of the same clan names documented in the protohistoric period by Crook appeared, some eighty years later, to have occupied different valleys. For instance, the Tuuoho, one of the clans inhabiting Hatiheu, also occurred in Hakaui; likewise the Naiki tribe of Pua (north coast), which was once expelled from the Puamau Valley, were also present on Ua Huka (probably Vaipaee), Hiva Oa, and Ua Pou (Kellum-Ottino 1971:43; von den Steinen 1925:17).[41] These connections may explain the presence of a few dog images on Ua Huka and Hiva Oa.

Sometime before 1844, warfare was waged between the Taipi tribes of Hatiheu and Anaho against the Naiki tribe on Ua Huka A truce was made in 1844 when the Naiki people set off for the northeast coast of Nuku Hiva with a "*hameva* (flag of truce), a large *tapu* pig, a large loggerhead turtle, and a gourd shell full of blackfish teeth strung two and two together on stripes of white tapa" (Lawson 1867 cited from Kellum-Ottino 1971:165).[42]

According to Crook (2007:127–136 [1797–1799]) Akapa was inhabited by three groups: the Pua, Naiki, and Ati-toka. Traditionally Akapa was considered part of the Hatiheu district, as the valley is today. The Taipi group comprised the Hatiheu, Hooumi, and Taipi tribes. Anaho and Haatuatua valleys were dependencies of Hatiheu (Handy 1923:32). There is some uncertainty regarding Akapa. Gracia (1843:xii) listed the Taipi people in Anaho, but he registered the Hapa, the traditional enemies of the Taipi, in Akapa (Hakapaa).

Conclusion

Overall, the same image types found in the western part of the Hatiheu Valley mirror the rest of the valley and the Marquesas as a whole. Petroglyphs are found throughout the valley (Table 27). Of the more than 478 boulder with over 3,379 figures the concentrations are found in the vicinity of the tribal ceremonial sites and in high-status residential clusters.[43] It suggests that, as a group, the Marquesans consciously shared a collective ideology that expressed certain social values. As a group, the images appear to have been part of practices designed to promote and reinforce social solidarity in order to maintain a belief system that bolstered the political position of the elite. Maintenance of group solidarity is manifested in occasional collective expression that usually takes on ritual forms (Radcliff-Brown 1952). All rituals require some form of concrete object or objects that then serve as an agency of social control. In the Hatiheu Valley, the images' high visibility and their

Table 27. Hatiheu Valley locations, boulders, and frequencies of petroglyphs as of 2016

Sites	BLDR	Images (n)
Tohua Hikokua	8	14
Tohua Kamuihei I	10	39
Tohua Kamuihei II	5	27
Tohua Tahakia	9	84
Hikokua-Kamuihei-Tahakia area	55	802
Ototemoui	26	268
Kahuvai	70	499
Kahuvai, research area	4	20
Tohua Maikuku	42	139
Maikuku area	36	177
Ivi Maikuku	24+	54+
Tohua Paahaua	2	4
Uhu	20	64
Ihupuaka	20	148
Teuatoki	15	89
Te Haetaei	15	108
Vaipupuhi	14	101
Teumu	9	38
Paitua	6	33
Hae Konini	6	53
Pataha	5	82
Mutoka	5	148
Vaikoa	2	11
Poio	2	32
Niumapu	1	24
Tauehu	2	6
Okiau	2	27
Other	63	288
Total	**478+**	**3,379+**

primary association with high-status architecture in the vicinity of tribal ceremonial complexes may have served as a reminder to the general population of the social order within the community.

From the above discussion it appears as if the lack of images in the agricultural sector suggests that not every segment of the population expressed the chiefly ideology in the same manner or even held the same ideology. As Giddens (1979) argues, only the ruling class is strongly committed to a dominant ideology. While it is likely that the *mataeinaa* participated in the chiefly ideology, consciously or unconsciously, during public rites that took place at the tribal ceremonial complexes, this social doctrine was not manifested in their architecture and associated features.

It is unclear what role, if any, the women played. As reviewed in Chapter 3, it is certain that some women were part of the chiefly social philosophy. High-status women, and particularly the inspirational priestess or shamanca played a part in the social hierarchical system because they performed rites associated with the *meae*. Female figures depicted in the petroglyphs and the sculptures tend to support the hypothesis of the inclusion of some women. Thus, the images conveyed an additional layer of meaning that reached beyond the ideology of social control

Chapter 6

Archaeological Investigation of the Western Hatiheu Valley

The Marquesas Islands offer few opportunities for archaeological research; both the geographic conditions and the character of the native culture are unfavorable.

—— Ralph Linton, *Archaeology of the Marquesas Islands*

This chapter focuses on the results of test excavations, charcoal samples, and datable artifacts that can be used to help place the archaeological art in a temporal, as well as spatial cultural context. Considering the abundance of petroglyph boulders associated with ceremonial complexes and high-status architecture, especially in the vicinity of Tohua Kamuihei I, II, and the area around Meae Te Iipoka, I excavated at Mutoka, situated below Tohua Kamuihei I, and on Ototemoui Ridge, northwest of Tohua Kamuihei II. Each test unit was associated with both petroglyphs and architecture.

In total, four 1-×-1-m units and two *pakeho*, stone-lined pits, with an approximate combined volume of 2.5 m³ of cultural deposit were excavated. In addition, charcoal samples were collected from the base and upper part of an *umu* (earth oven) that was exposed in a road cut during construction. The earth oven is located in the village proper approximately 200–250 m from the ocean. A stone alignment with a petroglyph (331hth 88) is located nearby.

Excavation Strategies

The test units were excavated in a combination of arbitrary 10-cm levels and cultural levels. Depth control was maintained by measuring distance below surface. Only trowel and brush were used in the excavation. Features were excavated as individual units, and their contents were bagged separately. All excavated sediment was visually inspected; except for samples, the sediment was not screened because of the hard clay-like texture. To control for microartifacts, a sample of sediments in a 20-×-18-cm plastic bag was collected from each 10-cm level that was later wet-screened.

Charcoal fragments were collected from all test units. However, due to the poor quality of the charcoal, not all were appropriate for AMS analysis (John R. Southon, pers. comm., 1998). Five samples were processed and dated by the Center for Accelerator Mass Spectrometry, Lawrence Livermore National Laboratory (LLNL). None of the charcoal samples were identified to species. The resulting age determinations were calibrated following OxCal version 2.18 (Stuiver and Becker 1993).

Mutoka, Test Unit 1

Structures 200 and 201 (Figures 68 and 69), two terraced *paepae hiamoe* or sleeping platforms, are located approximately 50 m northwest of Tohua Kamuihei I. Both form part of a house cluster and are separated by an unpaved courtyard. Structure 200 measures 14 × 4 m; the height is two to three courses and it measures up to 1.55 m. In general, the terrace was poorly constructed. The first course in the facing wall consists of both large and small boulders; the second course is built of greater than 1 × 1 m, relatively even-sized, boulders. The boulders that form the western part are roughly of the same size. Few of the stones are tightly fitted; only the eastern portion of the structure is paved. As is often the case with sleeping houses, one of the paving stones in the facing wall served as a polishing stone. The top surface has two cupules (12.5 × 7.0 cm and 15.0 × 8.0 cm), and a 35-×-2.0-cm u-shaped polishing grove. The polishing boulder is situated 1 m directly across from a megalithic petroglyph boulder (331hth 367). The boulder measures 1.8 × 1.4 × 1.3 m. Seven pecked anthropomorphic stick figures are arranged on the south, east, and west sides, as well as facing up/on the top surface (Figure 70). Randomly placed on the horizontal top surface are five cupules; their diameters and depth range from 5 cm to 14 cm, and from 1 cm to 8.5 cm respectively. To the north and across the courtyard, sleeping House 201 forms the seaward end of the house cluster. The north wall of Structure 201, now partly collapsed, is 15 m long and more than 2 m high. As can be seen in Figure 69, an unproportionately large boulder (1.35 × 1.2 m; depth of stone could not be measured) with a petroglyph face (331hth 390) is prominently incorporated into the base of the east wall. Two polishing stones are situated in the west wall. Structure 201 is also poorly constructed. Both sleeping platforms (200 and 20) are soil and rubble filled. Test Unit 1 was placed between the north wall of Structure 200 and petroglyph boulder 331hth 367, a distance of 1.1 m.

Level I: 0 cm to 12 cm

No surface artifacts were found between the house foundation (Structure 200) and the petroglyph boulder. The compact overburden consisted of 12 cm of rich black soil (Munsell color 2.5Y N 2/0). About 7 percent coarse pebbles (> 10 mm) and several fire-altered cobbles were intermixed with the sediment. One basalt flake, wedged between two stones in the first course of the facing wall of Structure 200, was uncovered 4 cm below surface. Four fragments of red volcanic tuff and two flakes, one polished, were uncovered 11 cm below surface. Charcoal flecking was observed in the interface of Levels I and II, toward the center of the south wall.

Level II: 12 cm to 20 cm

In the south upper level, six flakes and two flake fragments were exposed. A large number of fire-altered stones were found in the northwest section. Several flat pieces of stone, 2–3 cm thick, had spalled off of the petroglyph boulder and were uncovered in this level. Increased charcoal flecking occurred at 12 cm below surface. Circular pieces of red pigment less than 1 cm (Munsell color 2.5YR 4/8) were found in the upper stratum in the northwest and southeast corners. A centimeter-long piece of branch coral was exposed 12 cm below surface in the northwest corner. Within the interface of Levels II and III, 20 cm below surface, or 12 cm below the last course of Structure 200, a circular lens of dark soil (Munsell color 5Y 2.5/1) containing charcoal fragments was noted. It was designated Feature 1. Feature 1, an 18-cm deep hearth, measured 35 cm in diameter. Several ash-coated, fire-altered cobbles were uncovered. The hearth extended to 42 cm below the surface. Approximately one-half of the hearth was located below Structure 200.

In the same interface, an additional dark lens (designated Feature 2) was exposed in the southwest corner (Figure 71). More than one-half of Feature 2 was situated outside the test unit and under the petroglyph boulder. It was 20 cm deep. The excavated sediment consisted of loose black soil (Munsell color 5Y 2.5/1), angular pebbles, and numerous fire-altered cobbles. A total of 44 fire-altered cobbles, averaging over 10 cm, were uncovered both in the oven and around the sides. Several of the stones had a pitted smooth outer texture and some were coated with gray and red ash. One basalt flake was exposed 33 cm below surface. Maximum depth of Feature 2 reached 47 cm below surface.

Level III: 20 cm to 32 cm

Both features 1 and 2 were cut into Level III. Level III was approximately 12 cm deep and consisted of dark brown soil (Munsell color 7.5YR 3/4) and numerous fire-altered cobbles. Intermixed yellow paleosol and cultural sediment occurred some 30 cm below surface. One flake was uncovered.

Discussion

Test excavation and dating analysis established that the Mutoka site underwent several occupation phases. Structure 200 overlaps an earlier platform. The original *paepae*, visible in the lower part of the north-facing wall, appears to have been a small platform built of relatively uniform boulders about 50 × 50 cm in size. The *paepae* was then modified and enlarged with larger boulders to its present size. A single course was added to the original *paepae* on the east side, and the *paepae* was then extended with two stacked courses on the west side. Because of the absence of historic material (yet the presence of basalt stone flakes, including one polished adze flake), the last construction phase of Structure 200 probably dates from the late prehistoric to early historic period.

The construction period of Structure 201 is uncertain. None of the boulders were carefully fitted, as were those in the walls of Structure 200. The structure seems to have been deliberately and carelessly constructed around the petroglyph stone to anchor the east wall. I speculate that it is a historic structure.

Features 1 and 2 both predate the earlier *paepae* below Structure 200. A charcoal sample from the base of Feature 1 was analyzed and yielded the following result: Laboratory number LLNL 44511: 100±50 B.P. (Table 28). Calibrated age ranges for this sample are A.D. 1690–1730 (at one standard deviation) and A.D. 1670–1770 (at two standard deviation). With 68.2 percent confidence, the radiocarbon determinations are A.D. 1690–1730 and A.D. 1810–1920. Given that no historic material was found in the area, nor was any excavated, it appears that the hearth was in use in the late seventeenth or early to middle eighteenth century.

Table 28. Hatiheu Valley, west side radiocarbon dates*

Lab No.	Site	Depth (cm)	Context	Uncalibrated	1 sigma	2 sigma
44511	Feature 1	40	Hearth	100±50	1690–1730	1670–1770
44512	Feature 2	47	Earth Oven	340±50	1490–1640	1450–1660
44514	Test Unit 4	30–32	Burnt Surface	240±50	1630–1690	1490–1700
44515	Earth Oven	120	Earth Oven	300±60	1490–1660	1450–1810
44513	Earth Oven	144	Earth Oven	540±60	1310–1440	1300–1460

* Delta 13C values are the assumed values according to Stuiver and Polach (1977). 13C ratio is -25 for all samples. Calibrations for charcoal samples follow Stuiver and Polach (1993).

Feature 2, an *umu*, was approximately 20 cm deep. Although most of the *umu* was located outside the test unit, I estimate that its total length was close to 0.80 m. The length, I believe, falls within the normal range of a family earth oven. Measurements of earth ovens elsewhere in the Marquesan archipelago are, in most cases, lacking or restricted to communal ovens at ceremonial complexes. Suggs (1961:47) described a circular oven about 5 ft. (1.52 m) in diameter, on Tohua Teiviohou, Taipivai, Nuku Hiva. At the north end of Tohua Hikokua, Hatiheu, an earth oven measuring 2.6 × 3.4 m is still visible. Both ovens were probably communal. However, measurements of earth ovens and hearths are available for other Polynesian islands. Kirch (1989b:9–41) excavated 10 hearths and 1 earth oven at Kuolulo rock shelter in the Anahulu Valley, Oahu Island, on Hawaii. According to the measurements provided in his Table 2.1, the hearths measured from 19 cm to 40 cm in maximum diameter with a depth ranging from 3 cm to 17 cm. A large circular earth oven (Feature 10) measured approximately 1 m in diameter with a maximum depth of 35 cm.

One analyzed charcoal sample collected from the base of Feature 2 (LLNL 44512) yielded an age determination of 340±50 B.P. (see Table 28). The calibrated age range for this sample is A.D. 1490–1640 (at one standard deviation and with 68.2 percent confidence) and A.D. 1450–1660 (at two standard deviations and with 95.4 percent confidence). All the radiocarbon determinations are within close range of each other and indicate that Feature 2 dates back to approximately the middle to late fifteenth or middle seventeenth century. The earth oven is the earliest visible evidence of human activity on the site and predates Feature 1.

Interestingly, over half of the oven (Feature 2) was located under petroglyph boulder 331hth 367. Despite the boulder's large size, it was clearly moved to the location of the *umu*, thus collapsing the oven. This event was perhaps contemporaneous with Level II and Feature 1. Feature 1 was situated close to the image boulder, thus the heat that emanated from the oven cracked the lower part of the boulder, spalling off the flat pieces that were uncovered in Level II.

Linking the images with the hearth is problematic. Were the anthropomorphic stick figures already placed on the boulder before it was moved near the house site, or were they made after the boulder was moved? There are two reasons to believe that the images were already on the boulder prior to the move. First, anthropomorphic stick figures are believed to be the earliest petroglyphs types, possibly predating most activities that took place on Mutoka. Secondly, the boulder was perhaps considered special because of the figures. Although moving large boulders took a great deal of effort, it was no obstacle for the Marquesan of the past as is evidenced by the megalithic boulders often used in many house platforms and the odd placement of some image motifs. Some petroglyphs are made on large boulders and then tightly fitted next to another boulder. Thus the image is hidden from view.

Two documented cases support the fact that some boulders were considered so special that they were moved after the petroglyphs were made. Both are located in the vicinity of Tohua Maikuku, in the upper part of the Hatiheu Valley. In 1989 I recorded an anthropomorphic stick figure on a large, irregular megalithic boulder that was deliberately placed on the paved section of a house platform. Another interesting case involved two boulders, one of which had a pecked petroglyph of a dog on a flat side (331hth 307). Measuring 1.83 × 1.67 × 0.97 m, the image panel was placed less than 10 cm apart, facing the flat side of an undecorated boulder. While the image is not visible, it is possible to feel it by placing a flattened hand between the stones. Some of the images were indeed highly charged with *mana*, perhaps so charged with supernatural power that in some cases they could not be displayed. This has been noted for the rock art images of different cultures (e.g., Clottes and Lewis-Williams 1998).

In regards to the cupules and the polishing groove on the top surface of 331hth 367, it is conceivable that by using the *mana*-charged image bolder to sharpen adzes, some of the supernatural power would be transferred to the tool. Polishing stones and petroglyphs are frequently associated. Similarly, randomly placed cupules often appear with a variety of anthropomorphic figures. While the purpose of cupules is not clearly understood, they likely served a variety of functions and some perhaps served to transfer *mana* from the stone to a person or an object.

Mutoka, Test Unit 2

Situated 20 m southwest of Structure 200 is a megalithic platform (202) with four upright stones, a communal breadfruit storage pit (10 ×1.9 m), and a terraced platform (199). *Auti* plants grow on the surface of Structure 202. The *paepae* and a restricted area surrounding the *paepae*, referred to as Hae Konini, was considered *tapu* and shunned by the local people in the late 1940s (Yvonne Katupa, pers. comm., 1992). *Hae Konini*, meaning "house of pleasure," may have been a house built for the *Kaioi*. Thomas (1990) translated the term *Kaioi* to a "sexually oriented adolescent society."

The exterior length of the north wall of Structure 199 is 25 m, and it is intersected by several perpendicular stone alignments (Figure 72). Height ranged from 3 m to 3.5 m, with one to three courses. The surface appears unpaved but numerous water-worn boulders (> 30 cm) are scattered on the northwest portion. Perhaps the northwest portion was previously paved; the south portion is unpaved, and it is level with the ground. Stylistically, the terrace is similar to *Paepae* 200 and 201. Petroglyph boulder 331hth 115 (3.5 × 4.4 × 2.6 m), forming the northwest cornerstone, is prominently situated. Pecked images consist of anthropomorphic faces and stick figures, fish, and a great number of geometric shapes that are arranged on the upper north and south sides (Figure 73). Test Unit 2 was placed directly in front of the south petroglyph panel.

Level I: 0 cm to 10 cm

The first 5 cm covering the lower part of the images consisted of dark gray soil (Munsell color 2.5 N 3/0) mixed with 3 percent angular stone, charcoal fragments, and extensive root systems from nearby coconut and *hau* trees (*Hibiscus tiliaceus*). Next to the petroglyphs at the base of Level I, a circular piece of red pigment (Munsell color 2.5YR 4/8), with a 1-cm diameter, was uncovered.

Level II: 10 cm to 20 cm

The stratigraphic situation in Level II was similar to that of Level I. A flake was uncovered 12 cm below the surface. Minute charcoal fragments mixed with paleosol were observed throughout the level.

Level III: 20 cm to 30 cm

The sediment deposit in Level III was similar to previous levels. Excavation was terminated at 30 cm below surface as it became clear that the deposit consisted of construction fill, that is, black cultural sediment mixed with yellow paleosol, gravel, charcoal fragments, and numerous cobbles.

Discussion

The results of the excavation of TU 2 were informative but somewhat inconclusive. Structure 199 was built around the large petroglyph boulder 331hth 115 and another megalithic boulder underneath. The sediment fill appears to have been removed from the nearby communal breadfruit pit located approximately 5 m to the west. No carbon samples were collected for Structure 199. Similar to Structure 201, the stones were not carefully fitted. But the construction, perhaps late prehistoric–early-historic, was at a time when breadfruit became an economic mainstay, and surplus breadfruit was preserved in large communal subterranean pits. The latter part of this stage is marked by elaboration of status and prestige, which involved cultural practices, such as tattoos, monumental stone sculptures, and the widespread use of cut red tuff for architecture (Suggs 1961:185). The great numbers of geometric figures, such as concentric circles and squares, in addition to faces perhaps reflect the practice of tattooing, as many of the images are identical to tattoo motifs. The piece of red pigment uncovered in the lower part of Level I, similar to those found in TU 1, Level II, may have been used to enhance the figures, perhaps during special rites.

Ototemoui Ridge, Upper Section, Test Unit 3

Paepae 175 and 176 are part of an isolated house complex located in the steep upper section of Ototemoui Ridge. Four petroglyph boulders, several terraces, alignments, walls, enclosures, and six polishing stones in various sizes are part of the site (Figure 74). Tioka Puhetini, my guide and advisor on Marquesan cultural history, believes the house complex was a *meae*, although he was unable to explain why.

Structure 175 is a classic *paepae hiamoe*. Three walls of the terrace measure 9.1 × 8.9 × 8.4 m, and the facing wall measures 1.7 m. While the downhill northeast wall measured 1.7 m in height, the southwest wall was 0.8 m. A stone-lined *pakeho*, mostly collapsed, was located in the front paved section. Except for a 1.10 m wide pavement behind the wall that divides the front open terrace from the sleeping portion of the house, the structure was unpaved in the back. The downhill facing wall is impressive. Seven boulders, with widths from 0.8 m to 1.2 m and a height averaging 1.10 m, form the base; the upper course is laid down with seven long boulders that averaged 1.30 × 0.4 m. Open spaces between the boulders were filled in with round cobbles. The northeast base cornerstone is pecked with a grid-like motif and a *mata komoe* (331hth 392). As is frequently the case with the *mata komoe* motifs, the face wraps around the corner, giving the face a three dimensional form, suggesting that a viewer should walk around the corner stone to see the full figure (Figure 75). A 1-m wide pavement is located in front of the petroglyph boulder and the facing wall. A wall, part of which connects with Structure 176, situated to the west, encloses the area.

Paepae 176 measures 8.3 × 9.7 × 9.3 m. The west wall is collapsed. The maximum height measures 1.4m at the east wall. An unpaved terrace (16 × 10 m) was constructed on the downslope, ending in a northeast wall, 3–4 m high. A modified flake was found in the rubble of the partly collapsed wall. Five polishing stones with both cupules and sharpening groves were found on the terrace and on the *paepae*. Some changes occurred on the site. Originally Structure 176 was a classic sleeping house as can be seen from the two slabs still in place in the east side of the dividing wall. However, at one point, most of the dividing stones were removed, and the sleeping area was paved with flat stones similar to the front section. At least two dressed rectangular boulders that fit the dimensions of the dividing stones are now part of a wall on the south side of the house complex. A relatively intact stone-lined *pakeho*, located in the front paved section of 176, was excavated (TU 3).

Level I: 0 cm to 10 cm

Unevenly shaped, the upper edges of the *pakeho* measured 1.10 × 0.9–1.0 m; it was 0.65 m deep. The first 10-cm layer consisted largely of candlenut or *ama* endocarps and fragments, Tahitian chestnuts, smooth river stones, and boulders (> 30 cm). No candlenut trees are located nearby. Most of the *ama* nuts showed rodent gnaw marks indicating that they were hoarded in the pit by rats. An impressive Tahitian chestnut tree grows on Structure 175. A dense root system from a *hau*, a tree growing on the north side of the same platform, filled a large part of the upper layer. The sediment appeared to be rubble fill mixed with paleosol. The soil color consisted mostly of brown/dark brown (Munsell color HU7.5 YR 4/2) mottled with reddish brown (Munsell color HU5 YR 3/3). Several fire-altered oven stones, some coated with ash, were uncovered. The sediment contained no charcoal.

Level II: 10 cm to 20 cm

Level II was similar to level I.

Level III: 20 cm to 32 cm

Due to the restricted workspace, the time it took to remove boulders, *kiva*, and live tree roots, plus the unproductive yield, testing was limited to a section (50 × 35 cm) in the southwest corner. This level was excavated to below the first course of the *pakeho* wall that extended 97 cm below the platform's surface. The excavated sediment was similar to the above levels.

Level IV: 32 cm to 44 cm

Below the first course of *Paepae* 176, a 10–12 cm thick pavement of fairly uniform, angular stones (> 20 × 10 × 5 cm) was uncovered. The tightly packed pavement rested on paleosol.

Discussion

No separate cultural layers were identified in the *pakeho*. The excavated sediment appeared to consist of oven rake-out mixed with both cultural sediments and paleosol. Charcoal was absent. While the excavation did not shed light on the presence of the images at the site, the angular stone pavement may be an important step in the construction sequence.

The followed scenario was perhaps the order of construction. First, selected boulders were collected in the vicinity and brought to the site. Prior to construction of the platform proper, workers modified rough edges and sides in order for the boulders to fit closely together. The flaking created a large amount of lithic pieces with rough edges. But the use of these uneven pieces greatly facilitated construction. According to a mental template, a level foundation in the form of a house foundation was measured out on the slope. The angular lithic pieces were then laid out as the foundation. Several advantages are apparent. Firstly, the foundation provided a visual architectural plan that was easy to change before the large and heavy boulders were put in place. Secondly, the angular stone foundation helped keep each individual boulder firmly in place. Considering the steep angle of the slope, the rough base served to stabilize the whole terrace. In addition, the layer hedged against erosion due to periodic flooding. Thus excess water had a place to drain rather than undermine the terrace.

Ethnographic information on house construction is fragmented. According to Handy (1923:150–151) the master *tuhuna*, a specialist that directed all aspects of house construction, including the proper rituals, typically laid the two stones at each front corner to indicate the true line of the house foundation. Handy (1923:151) wrote: "The dimensions of the house platforms were set and regulated by spans. A single line of stones laid out on the ground indicated the outline of the platform, the line being called the rat's path (*te a'anui kio'e*). The platform was then built up row by row; the interior being filled with earth as each level of stones was laid."

Linton (1925:7) related that to prevent flooding the first step of house construction "was apparently the digging of a trench across the rear of the terrace to carry off flood water from the hill above. Then the outer side of the trench was faced with stones to prevent the water from undermining the house floor." It is possible that the type of pavement found under the foundation of Structure 176 was used only when constructing house foundations on especially steep hills. For instance, no underlying pavement was visible under *Paepae* 200, Mutoka, a terrace built on a surface with a low gradient.

If the excavated base pavement in TU 3 correctly represents the entire layer of the platform, then this is a significant discovery that has not been mentioned elsewhere in the Marquesan literature. In addition, the excavation confirmed Linton's (1925:13) suspicion that the house pit, the *pakeho*, reached to the base level and was built at the time of construction.

Ototemoui, Upper Section, Test Unit 4

A partly enclosed area is situated directly to the east of Structure 175. To the south, on a level area, approximately 9 m from the northeast corner of Structure 175, a large polishing boulder (Feature 37) is located close to a petroglyph boulder (331hth 394). The polishing boulder is 0.75 m high, and the horizontal surface measures 1.90 × 1.50 m. Twelve cupules, 10 u-shaped polishing grooves (some superimposed), and a rectangular basin (30 × 14 × 6.5 cm) are placed on the surface. Motifs on the petroglyph boulder 331hth 394 (1.1 × 1.1 × 1.75 m) depict two dogs and an anthropomorphic face (Figure 76). Test Unit 4 was placed between Feature 37 and the petroglyph stone. I expected that excavation would reflect activities associated with both the image boulder and the polishing boulder.

Level I: 0 cm to 10 cm

The sediment consisted of dark cultural soil. One flake with some cortex was uncovered in the southwest corner 10 cm below surface.

Level II: 10 cm to 20 cm

At 12 cm below surface charcoal flecking occurred in the southwest corner. At 15 cm below surface, due to time restriction, only the southwest quarter of the unit was excavated. Two flake fragments, one polished, were excavated in the southwest section.

Level III: 20 cm to 32 cm

Charcoal flecking occurred throughout the unit. A 1-cm, thin burnt layer was exposed at 31 cm below surface. Paleosol was reached at 32 cm below surface. Charcoal samples were collected from the burnt surface.

Discussion

Test excavation indicated that cultural residue was distributed toward the southwest area rather than toward the two boulders. The presence of the polishing stone and the basalt flakes and fragments, in addition to one flake with cortex, suggests that stone tools were both honed and made on the site. The cause of the thin burnt layer is uncertain, but it was perhaps related to the clearing of the area before or during occupation of the site.

A charcoal sample from the burnt layer produced the following result: LLNL 44514: 240±50 B.P. Calibrated age ranges for this sample are A.D. 1630–1690 (at one standard deviation), and A.D. 1490–1700 (at two standard deviations). The late date is rejected because no historic artifacts were found at the site. With 68.2 percent confidence, the range of radiocarbon age determination is A.D. 1630–1930. With 95.4 percent confidence the radiocarbon determination extends from A.D. 1490 and beyond A.D. 1910. All the age determinations show a large spread and the date of the burnt layer remain uncertain. However, the radiocarbon determinations in concert with the megalithic architecture type suggest that the site dates to the late-prehistoric-early-historic time period.

Ototemoui, Upper Section, Test Unit 5

An approximately 1-m high wall extends 4 m from the east side of *Paepae* 176, then runs southeast, perpendicular to the steep slope behind the settlement for 14 m before angling off to the east for another 16 m. The last part of the wall is placed along the base of the hillside apparently to prevent erosion. This wall includes a *keetu*-type petroglyph boulder (331hth 393) and a boulder with vertically placed cupules (Feature 43). The extensive wall system is a late addition. When petroglyph and vertical polishing boulders occur in walls elsewhere in the valley, they are often placed in a secondary context. Both 331hth 393 and Feature 43 were at one time removed from their original location. When Feature 43 was placed in the wall, it ceased to function as a polishing stone.

The petroglyph boulder, because of its size and modified shape, had been removed from the dividing wall of *Paepae* 176.

Two isolated petroglyph boulders, one boulder with images depicting an anthropomorphic face with mismatched eyes (331hth 395), and the other a shark (331hth 396), are situated in close proximity to the enclosure (Figures 77a–b). Two additional polishing boulders, each with TU 5 was placed directly in front of petroglyph boulder 331hth 393, located in the east–west wall. The boulder measures 0.65 m in height and has a width of 1.0 m; it is 0.50 m deep. Facing 10° degrees magnetic north, the figures on the panel depict two anthropomorphs (one headless) and one anthropomorphic face with mismatched eyes. As seen in Figure 78 the legs of Petroglyph B were covered with topsoil.

Level I: 0 cm to 10 cm

Yellow paleosol was observed immediately below the surface in the northeast corner. The remaining deposit consisted of a brown clay-like soil mixed with charcoal fragments, angular gravel, and paleosol. In the interface between Levels I and II, a concentration of charcoal fragments was uncovered within a 40 × 35 cm area in the center of the unit.

Level II: 10 cm to 20 cm

Sediment in Level II was similar to that of Level I. A thin wash of paleosol was visible along the east and west edges of the unit. Petroglyph B extended to 15 cm below surface. The dark charcoal lens was not well defined.

Level III: 20 cm to 30 cm

There were no changes in the consistency and color of the sediment. The charcoal lens disappeared at 25 cm below surface. Charcoal specks and angular gravel were observed elsewhere in the unit, but the charcoal specks were too minute to provide a good charcoal sample. The presence of a thin layer of paleosol, especially toward the side of the unit, was perhaps the result of floods. No midden material was uncovered.

Discussion

Excavations demonstrated that the settlement suffered from regular floods during rainstorms. As discussed in Chapter 4, with the exception of Taipivai, rainfall in the Hatiheu Valley is considerably higher than in other areas on Nuku Hiva. Regrettably, TU 5 did not produce a suitable charcoal sample for analysis. The reason for an intentional fire close to the petroglyph boulder is uncertain. Could it be that the boulder was no longer considered special? Its placement in a retaining wall rather than the usual position in a *Paepae* wall suggest that the stone had lost its supernatural power and become *noa* (common). Perhaps the isolated settlement originally was a *meae*, as Tioka Puhetini claimed, but that it was, at one point, transformed to a house site.

Ototemoui, Central Section, Test Unit 6

Sleeping House 182 is truly a megalithic structure (Figure 79). Built on a steep section of Ototemoui Ridge, some 175 masl, the structure is imposing in regard to both construction and proximity of the ceremonial complexes located below. A private breadfruit pit (4.5 m in diameter and 4.0 m deep) is situated approximately 50 m to the northeast of the structure. Additionally, the house site contains an exceptional number of architectural components, as well as associated structures and features. Despite its complexity, the form is a classic sleeping house. While the overall length of the north and east walls are 11.5 m and 10.5 m (120.75 m^2), the west and south walls are less clearly defined.

Facing downhill, the buttressed north wall measures more than 6.5 m in height. The east-facing wall is 4.4 m in height, and the west wall ranges from 1.1 m to 1.5 m high. The south side blends into the slope above. Adjoining Structure 182 at the northeast corner is a 5.5-m high wall extending some 14 m to the east. A megalithic petroglyph boulder, Number 331hth 406 (4.0 × 3.0 × 3.0 m), forms part of the wall. The images, mostly anthropomorphic faces and circles, are seen from both uphill and downhill. Smooth beach stones are intermixed with irregularly shaped, large flat boulders (> 0.6 × 0.6 m) that pave the front terrace. Nine fitted boulders (1.25 m) form the two-course, 1.3-m high dividing wall that separates the front terrace from the sleeping section. A 4.5 m wide area behind the dividing wall is paved. The third boulder from the east part of this wall depicts nine images (331hth 401), five of which are *mata komoe* (Figure 80). The petroglyphs face a stone-lined *pakeho* on the front terrace. Two large breadfruit trees grow near the north and east wall and on the sleeping area of the terrace and these have destroyed part of the *paepae*.

Eleven polishing boulders with a total of 28 horizontally placed cupules and several polishing surfaces are located on top of the on the structure. Five polishing boulders are placed in the dividing wall and in the north and west-facing walls; one boulder is part of the facing boulders in the *pakeho*. Two petroglyph boulders are placed on the *paepae*; three are located nearby. Placed on the front section of the terrace is a polishing boulder with five cupules surrounded by two oblong polishing depressions; one end depicts a small *mata komoe* (331hth 402) figure wrapped around the sides. Oddly, none of the grinding boulders contain polishing grooves. This is in contrast to the many large polishing boulders found just below Structure 182 where cupules, polishing grooves, and circular polishing areas are found in association with a great number of images.

Test Unit 6, the *pakeho*, is situated 0.90 m north of the dividing wall. The sides of the stone-lined pit measure 12.5 × 12.0 cm and are from 0.92 m to 0.97m deep.

Level I: 0 cm to 10 cm

The first 10 cm of overburden consisted of dark brown sediment (Munsell color 7.5YR 3/4) with boulders, cobbles (< 20 × 0.2 cm), circular vesicular basalt oven stones coated with dark reddish brown stain (Munsell color 5YR 3/4), and 13 whole and fragmented *ama* nuts. Three *ama* nuts showed rat gnaw marks. Some charcoal flecking was observed in the southwest corner. One basalt flake was uncovered near the south wall of the pit.

Level II: 10 cm to 20 cm

Some charcoal fragments were collected at 18 cm below the surface along the east wall. Yellowish red soil (Munsell color 5YR 4/6) was mixed with very dark brown sediment (Munsell color 10YR 2/2). One basalt flake fragment was uncovered in Level II.

Level III: 20 cm to 30 cm

Soil condition and color were similar to Level II. Because of the restricted space, due to two large boulders in the northern section of the *pakeho*, excavation was limited to the southwest section. A broken and incomplete adze of the *koma*-type was uncovered 30 cm below the surface, approximately 1.30 m below the surface of the platform. The adze was resting on a circular red ash stain.

Level IV: 30 cm to 40 cm

Red ash stain with charcoal flecking was noted in the southwest corner. Several of the stones in this level were coated with reddish ash and charcoal. A large amount of angularly shaped ground stones are found throughout the *pakeho*.

Level V: 40 cm to 50 cm

Red (Munsell color 5YR 4/6) and brown (Munsell color 10YR 2/2) sediment were intermixed in this stratum. No midden material was uncovered.

Level VI: 50 cm to 60 cm

Red ash-stained sediment mixed with brown sediment similar to Level V continued to be uncovered. For safety reasons, the excavation was terminated approximately 1.60 m below the surface of the platform. Apparently the *pakeho* extended to the base of the platform to perhaps a depth of 4 m to 5 m.

Discussion

Sediment in TU 6 consisted of paleosol, construction fill mixed with oven rake-out, a discarded broken adze, a flake, and a flake fragment. No distinct cultural layers were observed. Although charcoal samples were collected, it turned out that none were suitable for dating.

The megalithic architecture, together with the presence of a *koma*-type adze suggest that the site dates from the late Classic period, perhaps sometimes after A.D. 1600. The *koma*-type adze is associated with increased tuff and wood carving activities (Suggs 1961:111–112).

The residents of *Paepae* 182 occupied an elevated position within the tribe. In fact, settlement patterns research in the area demonstrated that Ototemoui Ridge was an area formerly occupied by a high-status or chiefly class. However, personal status markers associated with domestic architecture, such as worked slabs of red or yellow volcanic tuff, cut slabs of white beach rock and pieces of branch coral, were noticeably absent in the area. Status markers at this residential complex include the location on a small ridge, the size and complexity of the structure, the megalithic boulders used in the construction, the presence of numerous grinding stones with cupules, smooth beach stones, and a great number of petroglyphs. These architectural embellishments, in concert with close proximity to several tribal ceremonial complexes, indicate high status in this part of the valley.

Earth Oven

In the 1987 during the construction of a new road along the Vaiuua River, located in the lower central part of the valley, a bulldozer exposed the profile of an *umu* in the east side of the road cut. In 1996, we cleared the *umu* profile with a trowel and mapped the feature (Figure 81). It was 1.7 m wide at the top and 0.9 m deep with a 0.6 m to 0.7 m thick overburden. Charcoal and ash-stained oven stones filled the pit. A layer of red ash coated both the east and west sides of the *umu*. Charcoal samples were collected from a concentration of fragments found in the upper west section at 1.20 m below surface and at the east side of the base at 1.44 m below surface. Two flake fragments, both of a fine-grained, dense basalt, were retrieved from the central section of the oven. No historic artifacts were found.

A petroglyph boulder perhaps depicting a whale (331hth 88) and three horizontally placed cupules were located some 100 m west of the *umu* (Figure 82). The petroglyph boulder was set in an alignment that may have been part of a *paepae*. Two contemporary houses with gardens presently surround the site. In 1995 a banana plantation was established on the slightly sloping hillside above the *umu*.

Analysis of charcoal fragments found at the base of the earth oven, at 1.44 m below surface, produced the following results: LLNL 44513, 540±60 B.P. The calibrated age range is A.D. 1310–1440 (at one standard deviation), and A.D. 1300–1460 (at two standard deviations). With 68.2 percent, confidence for this sample the radiocarbon determination ranges from A.D. 1310 to A.D. 1440. Overall these dates are in agreement with each other and suggest that the earth oven dates to between the early fourteenth and the middle fifteenth centuries.

Analysis of charcoal collected at the upper part of the earth oven yielded the following results: LLNL 4415, 300±60 B.P. The radiocarbon age determination is A.D. 1490–1660 (one standard deviation) and A.D. 1450–1810 (two standard deviations and with 68.2 percent confidence). With 95.4 percent confidence, the age determination showed a large spread from 1450 to 1940 and on. There is a significant difference between the dated samples from the upper and the lower parts of the earth oven. It is doubtful that the same earth oven was used continuously for several hundred years nor does the stratigraphic section indicate that the oven was used periodically over a long time.

As mentioned above none of the five excavated charcoal fragments were identified before they were processed and dated by LLNL. Later, in collaboration with a colleague, charcoal identification became a separate project (Millerstrom and Coil 2008). Charcoal identification of the wood samples removed from the base of the earth oven shows that *hau* and some kind of bark were burned in the *umu*.

Charcoal samples collected from the upper part of the oven indicated that a variety of species were burnt: Crossostylis, *hau* (*Hibiscus tiliaceus*), *kokuu* (*Sapindus saponaria*) and *mio* (*Thespesia populnea*). *Mio* and *kokuu* were the main species used. Today *kokuu* appears to be among the most common wood used in earth oven. This is particularly the case in Ua Huka where the *kokuu* grow in abundance from the sea to far up in the valleys (Leon Lichtle, pers. comm., 2000). Huebert et al. (2010:61-97) examined over 800 fragments of wood charcoal from seven ovens in Anaho Valley, adjacent to Hatiheu Valley, and provide information on Marquesan earth oven and temporal patterns of fuel use A fair amount of information is now available on the various wood species used as fuel in earth oven (Huebert 2012; Huebert et al. 2016).

It is uncertain how well the wood preserve in the Marquesan climate. It is thus likely that there are some irregularities in our carbon sample and that the lower layer of the earth oven more accurately reflects its age. It is difficult to determine if the whale image, located on the surface in the vicinity of the *umu*, and the earth oven were contemporary. If the date of the earth oven, between approximately the early fourteenth and the middle fifteenth centuries, accurately reflects the oven use, it suggests that the Hatiheu beach area was never completely abandoned. This case then appears to deviate from Suggs's general settlement model, which suggested that, in the Expansion period the Marquesans, due to social instability, such as population increase, interisland and internal warfare, moved to the interior of the valleys. Especially in the Classic period "the beaches were generally shunned and the entire population seemed to gravitate inland, moving up high into the valley heads for protection from increased raiding from the sea" (Suggs 1961:182–187).

Lithic Assemblage

Excavated lithic material was sparse and consisted of 13 whole basalt flakes and eight flake fragments, totaling 247.5 g. Only one incomplete adze was uncovered. Except for two dense, fine-grained, grey basalt flakes collected from the earth oven, the material is course grained and appears to be of local origin. Both metric and nonmetric variables were examined. The metric variables recorded were maximum length, width, thickness, and weight. The nonmetric variables examined were the number of dorsal scars and the presence or absence of cortex and polished surface. The flakes and flake fragments range in weight from 0.3 to 38.8 g with a mean weight of 11.8 g (the incomplete adze was omitted from the calculations). Length varies from 9.4 to 59.6 mm, with an average length of 29.6 mm. Width ranges from 7.7 to 58.6 mm with an average width of 30.1 mm. Thickness varies from 1.3 mm to 14.3 mm with a mean of 6.7 mm. Cortex was only observed on one flake (Test Unit 4). Two flakes from polished adzes were uncovered in TU 1 and TU 6. Ten of the pieces had two flake scars while six had four or more.

The most useful diagnostic characteristics examined on the flakes and flake fragments were the frequency of dorsal scars. In the production sequence the number of dorsal scars increase through the reduction process. Thus, the highest frequency of flake scars occurs in the middle and late reduction stages; the fewer number of dorsal scars, the earlier stage of production sequence (Cleghorn 1986; Kahn 1996; Magne 1985). The Mutoka-Ototemoui lithic assemblage, with the exception of the flake with cortex from Ototemoui (TU 4), reflects the last stages in the manufacturing process. The late production stage is supported by the presence of numerous grinding stones with polishing grooves and spots. With an average length of 38.5 cm, the polishing grooves range in depth from 2 cm to 6 cm indicating both tool finishing and tool maintenance.

The *koma*-type adze uncovered in the *pakeho* in Structure 182, Ototemoui, was of a fine-grained greyish color with highly polished patches on the front, back, and sides. While the poll was broken, the total length was 17.5 cm with a thickness of 6.9 cm and a width of 5.5 cm. Typically, *koma* adzes are hog-backed with a triangular cross-section and range in length from 6.7 cm to 22.2 cm, thickness from 1.8 cm to 5.0 cm, and width from 0.8 to 4.9 cm (Suggs 1961:111, Figure 3b). *Koma* adzes first appear, according to Suggs, in the Expansion period and increase in frequency into the Historic period. Because of the broken poll end, it appears that the adze was intentionally discarded in the *pakeho*.

Conclusion

It is unlikely that any of the five archaeologically tested petroglyph sites represent the earliest petroglyph types in Hatiheu Valley. Ottino et al. (2003:79–100), during his excavation at Kamuihei II, has obtained three dates in areas associated with petroglyphs. Two calibrated dates (2 sigma) fall between A.D. 1420 and 1670; a third date falls between A.D. 660 and 1015. The last date, Ottino believes, may represent the earliest important settlement in Hatiheu, perhaps as early as the seventh century A.D.

The two age determinations from features 1 and 2 in TU 1, situated directly northwest of Tohua Kamuihei I, suggest that the site was settled at least by the middle to late fifteenth century. Test Unit 2 did not contain any datable material, but the petroglyph boulder with the anthropomorphic stick figures and the abundance of geometric figures are probably contemporary with the megalithic structure and the nearby communal breadfruit pit. Red clumps of pigment uncovered in the upper strata of TUs 1 and 2 suggest that the images might have been color enhanced. This may correspond to the early historic practice of applying red pigment to the tiki or anthropomorphic sculptures (Millerstrom and Edwards 1998).

No suitable charcoal samples were collected from the central part of Ototemoui Ridge. But a dated sample from a burnt layer at the upper part of Ototemoui indicated that the residential unit (or perhaps a *meae*) was occupied by the late prehistoric–early historic period.

The *pakeho* TUs 3 and 6 yielded little cultural material, and their functions remain unclear. Linton (1925) called the *pakeho* "skull pits" or "tapu pits." Ottino-Garanger et al. (2003:79–96) and Ottino (2005:127–136) excavated lithics, human cranial fragments, pig and turtle bones in a *pakeho* at Tohua Kamuihei II. In these cases, there were no indications that the two excavated stone-lined *pakeho* served as skull pits. All activities that took place on the platforms, where the pits are located, involved secular rather than sacred activities. It is possible that objects, such as clothing and broken adzes, and personal items like hair and nail clippings, were discarded in the pit so that no one could get hold of the items and use them in magic to harm the occupants. The broken *koma* adze in TU 6 tends to support the idea that objects, which had ceased to be useful, were discarded in the pit.

Both decorated and undecorated boulders were often considered to have supernatural power and were carved, moved, or incorporated into ceremonial and domestic architecture. Although there is yet no firm evidence that petroglyphs were made throughout the Marquesan cultural sequence, anthropomorphic stick figures and *mata komoe* seem to have been made over an extensive period of time. Anthropomorphic stick figures were probably among the earliest figures carved on stones, but they occurred throughout prehistory; they appear to be less common in late prehistoric sites. Excavation indicates that the *mata komoe* were present at least by A.D. 1300. Research in the Hatiheu Valley shows that, while the anthropomorphic stick figures are less common in late prehistoric sites, the frequency of the *mata komoe* figures appears to increase in the same time period.

Test excavations were limited and restricted to a small area. In view of the extensive section surveyed, and the numerous images documented, few sites were dated. Based on the five dates and discussion, it demonstrates that the western section of the valley was heavily occupied by the late prehistoric era and that perhaps the majority of the images were made at this time. While it is difficult to ascertain if the dates from the earth ovens and the hearth reflect associated rock images, together with architectural types and the general data collected in the area, it is possible to place the images within a broad time frame. In the future, it might be useful to try to excavate petroglyph sites that are less complex and in perhaps less densely occupied areas. For instance, sites with only one occupation phase may provide dates that more accurately reflect both the architecture and the associated images.

Figure 83 shows the research area with the 4 tribal ceremonial area, the locations of the six Test Units and the earth oven. The research indicates that the area was generally divided into two general sections. The high-status or chiefly area was centered on the tribal ceremonial units, Ototemoui Ridge and extended west to Kahuvai. This is where most of the petroglyphs, the megalithic house hold unites, and the large private breadfruit silos are found. Four tribal ceremonial complexes are located in this area. The agricultural section is located to the northwest near Puhioho River. Most of the people of the tribe lived here. The house units were smaller and less complex, the breadfruit pits were much smaller and there were few petroglyphs.

CHAPTER 7

Analysis of Surface Artifacts

Archaeology is the only discipline that seeks to study human behavior and thoughts without having any direct contact with either.

—— Bruce G. Trigger

This chapter focuses on the analysis of a small collection of artifacts retrieved from the surface in the western part of the Hatiheu Valley. Artifacts found within an archaeological context can potentially elucidate on the changes in rock art style and function of architecture, and may establish a relative time frame.

A total of one sculpture, one boulder with petroglyphs, two basalt *popoi* pounders, a shell trumpet, three adzes (two unfinished), and a worked flake were discovered during survey and mapping (Table 29). Given the large area examined, relatively few surface artifacts were retrieved. This may, in part, be due to the poor visibility in the forest because of dense undergrowth and the fact that local residents and visitors collect surface artifacts. While whole adzes found in the forest are picked up by the Marquesans; broken artifacts are frequently discarded. In a few households *popoi* pounders found in the valley are still used as kitchen implements.[44] Some Marquesans donate the artifacts to a collection housed at the Hatiheu Museum. What follows is a discussion of each artifact and its archaeological context.

Table 29. Hatiheu Valley, west side sites, frequencies of surface artifacts, and contexts

Location	Artifact	Site Type	Petroglyphs?
West Side (no site name)	Basalt Adze Preform (331hth(A)41)	Platform (174)	No
West Side (no site name)	Basalt Adze Preform (331hth[A]39)	Sleeping Platform (129)	No
Ototemoui	Modified Basalt Flake (331hth[A]43)	Platform (176)	Yes
Tohua Hikokua	*Koma* Adze (331hth[A]42)	Breadfruit Pit (20)	Yes
Tohua Kamuihei I	Pounder (dense light gray basalt). (331hth[A]38)	Pavement (cook house) Part of Chief's Complex (9)	Yes
Tohua Kamuihei II	Pounder, Course Basalt (331hth[A]37)	Dance Plaza	Yes
Kahuvai	Shell Horn (Cassis)	Sleeping Platform (215)	Yes

Images

A portable petroglyph stone (331hth 29) and a sculpture or tiki (331hth T1) were found on the dance floor of Tohua Kamuihei I near the chief's *paepae*.[45] The basalt petroglyph boulder measures 0.58 × 0.52 m and is 0.27 m deep. With large circular eyes, a wide mouth, and a wide nose with nostrils, the anthropomorphic face exhibits "classic" features depicted on most of the Marquesan sculptures (Figure 84a–b). Circular geometric motifs are seen on one of the side panels. An unusual feature of 331hth 29 is that the eyes are formed of two 6-cm deep cupules. Some petroglyph figures do incorporate cupules in the motif. It is uncertain if the face and the cupules were made simultaneously or not. However, several features suggest that the stone first served as a grinding stone and that the face was a later addition; the stone is similar in size and shape to numerous grinding stones recorded in the valley and the cupules had reached nearly a maximum depth; few reach a depth that exceeds 6 cm to 8 cm. The surface is smooth with the edges of the cups smooth and rounded; the petroglyph lines are sharp. The portable boulder, because of its provenience in front of the chief's platform, may have come from the chief's residential complex.

The tiki, formed as a rectangular block with a height of 0.50 m, is carved of red volcanic tuff (Figure 85). Its attributes are "classic," mixed with a few exotic features. Typically, classic features include red tuff as the carving material, circular eyes, arms clasped on the abdomen, and a head size one-third or more the length of the body (Millerstrom and Edwards 1998:61–67). The facial features are atypical; the nose is long and narrow, and the mouth and eyes are small. What appears as arms are placed on the stomach in the classic tiki pose. Most Marquesan tiki are associated with *meae* sites; thus I suspect the sculptures were once part of the *meae* located at the east end of Tohua Kamuihei I. Two petroglyph boulders located at the *meae* depict a total of eleven anthropomorphic faces with circular eyes, stylistically similar to tiki. While the face of this tiki is unique to the Marquesas, the petroglyph faces are the usual type.

Basalt Pounder

Two basalt *popoi* pounders (*kea tuki popoi*) were found at Tohua Kamuihei I and II. Both were types normally used to beat breadfruit, *taro*, and other vegetables into a paste or *popoi* (Söderström (1937:235–242). At Tohua Kamuihei I, Structure 9, a pounder (331hth[A]38) was wedged in a rectangular pavement located in the back of the chief's residence, facing the mountain. The pavement was identified as a cooking area. Several anthropomorphic figures were associated with the chief's *paepae*. The pounder was made of dense light gray basalt, and was 11.0 cm tall with a slightly protruding grip (Figure 86a). The flared base had a diameter of 10.5 cm. On the northeast section of the dance floor of Tohua Kamuihei II, we collected the base of a *popoi* pounder made of coarse basalt (331hth (A)37). It was 0.9 cm high. The round base, with a diameter of 0.9 cm, was only slightly flared (Figure 86b).

With the exception of adzes and flakes, pounders are the most common Marquesan stone artifacts. Yet in-depth research on stone pounders, especially in their archaeological context, is lacking. At this point, only descriptive information and a few catalogue type publications are available (Garanger 1967; Linton 1923; Silverthorne 1936; Von den Steinen 1928). Pounders can be chronologically significant, although with only the base section, dating is problematic. The Marquesans classified pounders into four general types according to use: 1) *popoi* pounders; 2) pounders for infant food; 3) salt pounders; and 4) children's pounders. The most significant implements were the *popoi* pounders (Linton 1923:337–338). These were distinguished on the basis of size and weight into men's and women's tools; the largest tools being used by men (Linton 1923:338). Archaeologists recognized several distinct forms, such as the flattop types called *opu* (stomach), phallic-shaped pounders, conical pounders, and tiki-headed pounders (Linton 1923:338; Suggs 1961:99–103). Garanger (1967:1–82) added supplementary categories. According to Linton (1923:339), the most simplistic form of utensils was the flat-topped *opu* type.

While pounders were made from local materials, the stones with the best qualities came from Ua Pou, Ua Huka (Linton 1923), and Eiao (Von den Steinen 1928:45). Ua Pou and Ua Huka became centers for this type of industry, and finished pounders were traded to the other islands (Garanger 1967; Linton 1923:337). Between 1893 and 1914, a German company established at Ua Huka made a great number of pounders with phallic-shaped heads (Garanger 1967:44).

The Marquesans distinguished between three grades of rock according to the quality of the grain: *ovao*, *puhitea*, and *patako*—the best being the *ovao*, a fine-grained, grey, light, and strong rock (Linton 1923:337). Garanger (1967:44) translated the *ovao*, *puhitea*, and *patako* to mean the quality of the grain, the resistance, and the light weight.[46] All rocks appeared to have been "rather soft, even grained magnetic rocks" (Linton 1923:339).

The characteristics of the Hatiheu Valley specimen from Kamuihei I, Structure 9, indicate that it was either an early *opu* or conical type (see Suggs 1961:99, 102. Figures 30c and d). Both types appeared at approximately to A.D. 1300 and were the earliest known in the Marquesas. They continued to be made until the historic period and are found throughout the island group. The characteristics of the Tohua Kamuihei I pounder were of the qualities described by Linton (1923:337).

The Tohua Kamuihei II pounder eludes classification. However, because of the coarse quality of the stone, it was probably fabricated from local material. Suggs (1961:102), who analyzed 31 pounders, found that over time there was a trend toward a size increase. Garanger's (1967) work on Polynesian pounders from the collection at the Département d'Océanie du Musée de l'Homme, Paris, appears to have followed the same course. I calculate that the 52 Marquesan pounders classified and catalogued by Garanger also increased in height. The earliest pounder, such as the *opu* and conical types, measured from 8.8 cm to 12.2 cm. The Classic tiki-headed pounders ranged in height from 17.5 cm to 23.3 cm. Linton (1923:339), however, stated that the old pounders varied in height from 17.8 cm to 21.6 cm, and the base diameter from 11.4 cm to 14.3 cm.

All except one of the *popoi* pounders excavated by Suggs on Nuku Hiva were found in the context of a *tohua* or *meae* (Suggs 1961:100, Table 10). For instance, four pounders, two of which were the *opu* type, were excavated at Tohua Hikokua (Suggs 1961:68–76). No specific proveniences were given. However, at Moana rockshelter, Haaauai, on the east coast of Nuku Hiva, Suggs (Suggs 1961:57–60) excavated a *popoi* pounder with a carved head. The rockshelter, classified as a habitation site probably served as a fishermen's shelter later in time. Suggs believes that the habitation site is from the Expansion period

Rolett found it strange that no stone pounders were uncovered during excavation at Hanamiai, Tahuata. A few years later, a local Marquesan collected several pounders from the site while it was bulldozed to make room for a soccer field (Rolett pers. comm. 1991). In Hane Valley, *popoi* pounders were only found on ceremonial sites, leading Kellum-Ottino to speculate that their use was "restricted to certain people or to certain sites" (Kellum-Ottino 1971). Von den Steinen (1928:45, quoted in Suggs 1961:102) stated that the tiki-headed pounders were prestige items and the property of chiefs. Similarly, pounders made of red trachyte, because of their color, signified a high-status implement. More research on pounders is certainly needed, especially on pounders that are found in archaeological context.

Lithics

Adze Preform (331hth[A]39)

Regarding lithics, the most interesting discovery was an adze preform found in situ. The adze was lying under a thick layer of vegetation in a circular polishing spot on the outer paving stones of a sleeping platform (Structure 129). The measurements are 11.7 cm in length, 4.5 cm in width, and 4.6 cm thick. It resembles the *teoni*-type illustrated by Suggs (1961:108). Unfortunately the recovered adze then does not add to the chronology of the site since the *teoni*-type adze occurred throughout the Marquesan cultural sequence. Petroglyphs are found at the site complex.

Adze Preform (331hth(A)41

One basalt preform was found wedged between the southwest corner stones of a paved structure (174). It measured 7.2 cm in length, 4.0 cm wide, and 2.6 cm in thickness; the poll was broken. The small *paepae*, probably a *fataa*, measured 5 × 2.9 m and was 0.5 m high. As already discussed, a *fataa* was a *tapu* men's house. Old men who were no longer warriors and young boys spent their time at the *fataa* where the old men drank kava, fashioned tools, and made carvings. It also served as a storage house. Located on the east edge of the Puhioho River, the *fataa* was one of five houses that formed a small hamlet. No petroglyphs occurred in the vicinity.

Worked Flake (331hth[A]43)

At Ototemoui (Structure 176), a modified flake tool was retrieved in the collapsed terrace wall. It measured 7.4 cm in length, 0.3 cm in width, and it was 1.4 cm thick. Although few adze lithics were found on the surface or during test excavation, a total of nine grinding stones in various sizes with 38 cupules, a number of u-shaped polishing grooves, and polishing spots were located on the site. Perhaps adzes were only polished and retouched at the location. While no rock art was associated with this particular terrace, several petroglyph boulders were situated to the east.

Adze (331hth[A]42)
A complete *koma* adze was discovered next to an *ua ma* on the east side of Tohua Hikokua (Information on the adze is missing.).

Lithic Scatter
The ground surrounding a small hamlet, located midway up the central part of the valley and amidst a forest of *kehika*, or "mountain apples" (*Eugenia malaccensis*), had been uprooted by foraging pigs.[47] A great number of basalt lithic flakes under 3 cm to 4 cm in size and preforms were observed scattered around stone Platform 46. The Platform measures 3.9 × 4.7 m; it has a pavement on one side. It is likely that Structure 46 was part of a *fataa*. More stone knapping activities took place at Structure 46 than at any of the other sites examined. The site contained several grinding stones but no rock art was seen in this area.

Cassis Shell Trumpet Horn

In a section of Kahuvai, within a small hamlet, a shell (*Cypraecassis rufa*) trumpet was found on the surface of a megalithic sleeping house (Structure 214). Two large private breadfruit silos are associated with the site (Figure 87). The shell appeared in good condition and may have been of late prehistoric–early historic origin. [48]

The Cassidae trumpet horn was one of two varieties used by the Marquesans. Most shell trumpets were made of the *Triton tritonis*, which was elaborately decorated with sennit cords, pieces of white *tapa*, human hair tassels, and a small, carved-bone tiki (Linton1923:405; Freeman Moulin 1997:250–283). Apparently, the Cassidae trumpet was rare, and Linton could only point to one that was found in a burial cave in Atuona, Hiva Oa (Linton 1923:405).

A Cassidae shell trumpet appears to be both an indicator of time depth and status. What Suggs (1961:48) called a war trumpet of Cassis shell was recovered in a cache on the surface of Meae Peupeu, a megalithic *paepae* in Taipi Valley, Nuku Hiva. Excavation under the stone surface of the temple yielded another Cassidae trumpet along with burials containing whale-tooth earplugs set in shell cups and associated with a European "carving fork" and fragments of a French wine bottle. The wine bottle with one of the burials led Suggs to conclude that Meae Peupeu was a historic site.

Several pieces of corrugated metal found on Structure 214 near the *paepae* were probably used to fence domestic pigs. None of the local residents asked remembered that the area was ever occupied.

Historic Material

Considering the long occupation in the valley, surprisingly few historic artifacts were found. Historic evidence consisted mainly of corrugated metal, barbed wires, traces of cement, introduced fruit trees, and ornamental plants. The *tuu* on Tohua Hikokua was partly surrounded with barbed wire when I first visited the site in 1984. Graves from the Christian era were placed directly outside or on the dance floor of Tohua Hikokua, Kamuihei II, and Tahakia. Structure 22, bordering the Puhioho River, was last occupied in the early 1960s; the name of the owner is still known. Red *keetu*, once part of the dividing wall, was used to pave the entire terrace (Figure 88). In a walled enclosure behind the *paepae*, probably the cooking area, kitchen utensils consisting of enamel pans and aluminum cooking pots were scattered on the surface. Part of a hand-operated Singer sewing machine was also noted. Several ornamental trees, such as croton and red hibiscus, surrounded the site.

Conclusion

In addition to increasing our knowledge of the sites where the artifacts were recovered, some of the surface artifacts yielded information on rock art. The *popoi* pounders from Tohua Kamuihei I and II strengthened the hypothesis that some pounders were both high status and linked to ceremonial sites.[49] While the fine quality of the pounder from Tohua Kamuihei I was consistent with the size and complexity of the ceremonial complex, the locally made pounder from the small Tohua Kamuihei II may represent the ceremonial place of a low-ranking chief. The variations may also reflect different occupation periods. As discussed in Chapter 4, anthropomorphic figures occur on both ceremonial sites, although fewer images are located at Tohua Kamuihei II. Both the tiki and the image-grinding stone found on the dance floor of Tohua Kamuihei I suggest a late prehistoric-early historic period. This time period corresponds with elaborate *tohua* construction, sculptures of red tuff, and emphasis on personal prestige.

Except for the *koma* adze, the lithic artifacts were less useful as chronological markers. One would expect to find *koma* adzes associated with the expansion of large ceremonial centers, as the adze increased in frequency during the Classic period from about A.D. 1600 and onwards.

The Cassidae shell trumpet, in concert with the geometric figures situated on a megalithic *paepae*, indicate that also these types of images were made in the late prehistoric-early historic period. In sum, the trumpet and the images, in addition to the presence of two communal size breadfruit silos, imply a high-status household. However, there is no indication that the *paepae* was the residence of a chief. In fact, it is more likely that an achieved-status elite, such as a warrior or a chiefly family member, occupied the residential unit.

Relatively few surface artifacts were retrieved in the research area. The limited collection is, however, consistent with the number of objects found in the remaining part of the valley during survey. All excavated artifacts and artifacts collected during survey were labeled and placed in the Hatiheu Mairie (town hall), presently part of the Hatiheu Museum.

Chapter 8

Marquesan Rock Art in the Polynesian Context

The anthropological study of art seeks to understand more than just the products of art, but also who makes it and why, the role of art in society, and its wider social meanings.

—— Barbara D. Miller, *Cultural Anthropology*

In this chapter the Hatiheu Valley images are compared with those addressed in the studies by Kellum-Ottino (1971) for the Hane Valley, Ua Huka, and by Ottino (1985) for the Hakaohoka Valley, Ua Pou. The information is then evaluated along with images on rocks from other island and island groups.

Hane Valley, Ua Huka, and Hakaohoka, Ua Pou

Kellum-Ottino's (1971:83–84) settlement pattern survey in the Hane Valley brought to light five new structures with petroglyphs and tiki: Vaiapa 1, Havaiki 21; Puikau 2; Keetupu 1; and Taaoa (Table 30). Meaiaute (Meaeautea?) already a well-known location (Linton 1925) was also described. A ritual site, Meaiaute was built on top of a ridge and it contains four sculptures and a bas-relief figure on a rectangular cut slab ([Figure 89] see Linton 1925; Kellum-Ottino 1971; Edwards 1985b; Millerstrom 1985c). One tribe occupied the Hane Valley, and yet three tribal ceremonial complexes were mapped (Kellum- Ottino 1971:129). At Vaiapa 1, a *paepae* with a circular and a square pit, one corner has petroglyphs on both sides indicating that at least one panel was made prior to construction of the *paepae*. Circular motifs are on the outer panel, and two tiki figures are placed on the inner panels.

Table 30. Hane Valley, Ua Huka, descriptions of rock art sites

Site	Feature
Vaiapa (VA 1 A)	Raised *paepae* with circular and square pits. The image boulder with polishing areas depicted small circles and three face motifs. Two *koma*-type adzes were found on the site.
Havaiki 21 (Hav 21)	Unclassified site. A pair of concentric circles is horizontally placed on a stone slab.
Puikau 2	*Paepae* with circular pits. The motif consists of a small circle. An adze, a sacrificial slab, cowry, and turbo shells were found on the site.
Meae Keetupu 1 (Ke 1)	Anthropomorphic stick figure horizontally placed on a stone slab.
Meae Taaoa	Stone tiki figures.
Site Meaiaute	Four tiki figures in red tuff and one worked slab with relief figures anthropomorph.*

*In the 1990s a fourth tiki was found downhill from Meaiaute. While smaller in size it is similar in form to a tiki in a horizontal lying position, face down at Meae Iipona, Puamau, Hiva Oa.

The Havaiki 21 site was not classified, but Kellum-Ottino (1971) speculated that it might have been a *meae* or a *tohua*. An adze, *ii* (cowry scraper), and a sacrificial slab were uncovered at the site. Motifs on a horizontal slab consisted of a pair of concentric circles connected with a line to a single circle. Six cupules were placed on the same slab, and many turbo shells were found scattered in the area. Kellum- Ottino suggested that the motifs represented tattoo designs, such as a face and that the cupules served as dye cups. A corner stone with petroglyphs and polishing areas was in a secondary context on a *paepae* at Vaiapa 1 A, Puikau. A square pit was on the front terrace; circular motifs were on a facing stone. When the stone was removed, depictions of human faces were found on the inner sides. At Meae Keetupu 1 (Ke 1), Kellum-Ottino discovered a petroglyph stone, 1 × 0.75 m that was reused in an outer wall. An anthropomorphic stick figure had been pecked on it and placed so that the figure was horizontal. When the Ke 1 anthropomorphic stick figure was compared to figures photographed by Linton in Haanaipa, Hiva Oa (1925:Plate XIIIA), and at Haatuatua, Nuku Hiva (Suggs 1961: Plate 11), they were nearly identical.

Kellum-Ottino (1971:85) wondered why a petroglyph stone would be reused in the foundation wall of a *meae* in a place believed to be one of the last areas to be settled. But, when a new *meae* was constructed, it was the custom to take stones from an older structure, thus providing continuity between the old and the new, providing the *meae* with *mana*, and visually tying the temple to the tribe and their land. It could also have been a practical matter.

While the Hane Valley motifs are similar to petroglyphs found elsewhere in the Marquesas, the *mata komoe* are stylistically unique to the Hane Valley (Kellum-Ottino 1971: Plate 14). No other recorded *mata komoe* heads have long necks. Normally, anthropomorphic figures and tiki have heads placed directly on square shoulders (Millerstrom and Edwards 1998).

In Ottino's (1985) survey of the Hakaohoka Valley Ua Pou, only a few petroglyph boulders were found. We recorded one, an isolated boulder with turtles, humans, and *mata komoe* in the same valley during the 1987 field season. At Vaioheiko, high-status megalithic *paepae hiamoe* stones with elaborate linear geometric designs were documented (Ottino 1985; Millerstrom 1987). Few petroglyph boulders were found during Ottino's survey in Hakaohoka Valley, Ua Pou, and Kellum-Ottino's survey in the Hane Valley, Ua Huka. None were reported by Bellwood (1972) from Hanatekua Valley, Hiva Oa. Information from settlement pattern surveys by Ottino (1985) and Kellum-Ottino (1971) provides the only comparable data to that which we have collected for the Hatiheu Valley.

East Polynesia

Large-scale and long-term rock art projects have been conducted in the Hawaiian archipelago (e.g., Cox and Stasack 1970; Lee and Stasack 1999); New Zealand (e.g., Trotter and McCulloch 1971; O'Regan 1998, 2003, 2007, 2008); and on Rapa Nui (e.g., Lee 1992). For an overall Polynesian rock art review see Lee (1996); Millerstrom (2003a, 2006a, 2012); and Hoerman (2016). Despite these efforts, there are interterpitational gaps, particularly in the Society and the Austral islands such as Tahiti. The following sections describe the known rock art from each island or island group.

The Society Islands and the Austral Islands

Since the early 1980s, rock images have been scientifically documented on several French Polynesian islands by archaeologists with SCP, Tahiti. Combined with previously noted sites, a large corpus of rock images now exists. Tahiti includes: (Emory 1933; Edwards 1988; Millerstrom and Baumgartner forthcoming); Moorea (Millerstrom 1997b; Millerstrom and Baumgartner 1998); Bora Bora, Huahine, Raiatea (Inventaire Archeologique de Polynésie Française 1989); and Maupiti (Emory 1933). The Australs include: Tubuai (Candelot 1980, 1987; Edwards 1991; Millerstrom 1991b); Raivavae: (Edwards 1998, [2003] 2005; Marshall 1955, 1961; Millerstrom 1991b;); and Rapa (Smith 1965).

Despite the large motif inventory, cultural, social, and temporal context is generally lacking. For instance, spatial distribution, temporality, architectural and environmental association, among other factors, need to be more closely considered.

Considering the large size of the island surprisingly few petroglyph sites have been located on Tahiti, the largest island in the Society group (Table 31). However, anthropomorphic faces and figures, turtles, and concentric circles occur at Faaa, Papenoo, Tipaerui, and Arue, Tahiti. Some interesting images, perhaps depicting "masks" are found at Vaiote, Tautira (Emory 1933:171–179; Garanger 1980;

Lidin 1932). Emory believed that the "masks," unique to Tahiti, Raiatea, and Bora Bora, represent headdresses related to burial ceremonies of important people (Emory 1927:236–239, 1979:200–221) (Figures 90). A priest, in full mourner's costume, inhabited by the deceased's spirit, would perform rites at the appropriate *marae*. At Tevaitoa, Raiatea, a funeral headdress is depicted on a corner stone, identified by Emory as a chief's house platform (Emory 1922:Plate19b; Inventaire Archeologique de Polynésie Française 1989). In the Vaiati Valley, Bora Bora, petroglyphs on a large image boulder of tuff, commonly referred to as "Ofai honu," (turtle rock) depict a headdress, four turtles, and at the base a post-European inscription (Figure 91). All were made with the same technique (Inventaire Archeologique de Polynésie Française 1989), and apparently are of relatively recent origin. Rev. J. M. Orsmond, of the London Missionary Society, who resided in the Society Islands from 1817 to 1856, described the priests' funeral costume in detail (Henry 1928:293–294).

Table 31. French Polynesia, relationship between the number of sites, boulders, images types, and land area as of 2016

Island	Site	Boulder	Number of Images	Area (km²)*
Tahiti	12	23	133	1,040
Moorea	12	12	128	132.5
Bora Bora	3	10	41	38
Huahine	3	5	30	74
Raiatea	7	29	217	238
Maupiti**	1	6	27	12
Total	**38**	**85**	**576**	-

* From Oliver (1974:8, Table 1)
** The petroglyphs at Haranae, Maupiti have not been properly recorded therefore the number of boulders and individual images are estimated. The numbers derive from pictures kindly provided to me by Heidy Baumgartner-Lesage, August 11, 2016.

A chief mourner in a full funeral dress, drawn by William Hodges, the artist on Captain Cook's second Pacific voyage (Barrow 1979:48, Plate 50), was portrayed in a headdress similar to the Vaiote "masks." Interestingly, when the British Museum, in 1966, dismantled a Tahitian mourning dress (collected by Captain Cook, perhaps the same costume drawn by W. Hodges) for cleaning, a Tahitian wooden image was found inside the head section (Barrow 1979:46–49, Plates 48 and 51). While the Vaiote, Tevaitoa, and Vaiati figures were perhaps a realistic rendering of funeral headdresses, symbolically the images were interwoven with the concept of *tapu* related to the sacredness of the heads of the chiefly class. This included the notion of ancestral spirits, genealogy, lineage, and fertility. Although petroglyphs, especially anthropomorphs, were not a common occurrence in Tahiti, human images were depicted on other material. For instance, the sacred red painted *unu* (wooden boards), that were placed in the center of the *marae* had carved human figures representing god and goddesses (Henry 1928:134). Anthropomorphs were also made of stones; other Polynesian motif types were depicted on less durable material, such as *tapa* and human skin (Green 1979).

Anthropomorphs with forked tails frequently referred to as "lizard men," occur in Raiatea, Tahiti, as well as in the Marquesas (Figures 92 and 93). An unusually fine specimen with two faces is depicted on a large smooth boulder from Tipaerui Valley, Tahiti.[50]

The Papenoo Valley, the largest valley on Tahiti, is located in an ancient caldera in the interior of the island. Considering the extensive settlement, albeit late in time, it is surprising that only two petroglyph boulders were discovered during a large-scale survey in the 1980s (Eddowes 1998; Edwards 1988). A large boulder, associated with a *marae* and an archery platform, has petroglyphs depicting anthropomorphs, concentric circles and geometric figures, all resembling general Polynesian motifs (Figure 94). Despite the link to both a temple and chiefly architecture, no turtles are depicted.

With the exception of Tubuai and Raivavae, turtles are part of the image repertoire from the islands where rock art occurs, although they are sparsely distributed. Nowhere are turtle images depicted as frequently as on Raiatea where they are often associated with anthropomorphs and geometric figures and placed on boulders unassociated with structures.

Rolett (1986:86), relying largely on publications from the 1930s and 1940s, argued that, in the Society Islands and Rapa Nui, turtle motifs were most commonly placed on stones that formed part of the exterior wall of *marae* or on boulders associated with structures of religious importance. Field research conducted by archaeologists from the SCP, Tahiti, has provided new data about the sites. Quantitative data from Raiatea, Bora Bora, and Huahine shows that 37 of the 41 boulders recorded contained turtles (Inventaire Archeologique de Polynésie Française 1989). Ten were linked with *marae*, and the remaining 31 boulders were not associated with any visible structures. However, the sites are often situated next to water, especially rivers.

Moorea Island

Roger C. Green's pioneering settlement pattern archaeology carried out in Opunohu Valley in the early 1960s (Green et al. 1967) was followed by Dana Lepofsky's (1994) work on prehistoric agricultural intensification in the same valley. Jennifer Kahn (2005, 2006, 2010, see also Kahn and Kirch 2014) has conducted an archaeological investigation of households also within R. Green's study area. Except for Opunohu, limited systematic archaeological survey has been undertaken on Moorea. Generally, archaeological inquiry has been limited to documentation and analysis of visible elite architecture (e.g., Decantes 1993; Emory 1933; Wallin 1993). While restoring Marae Nuurua, a costal ceremonial temple located on the west coast of the island, three turtle motifs and a pair of eyes were discovered on an upright stone near the *ahu* (Eddowes 1991). Eddowes speculated that each of the three turtles represented one of the lineages that are known from ethnohistoric sources to have resided on *marae* in the past (Pers. comm., 1992). Sharp et al. (2010) dated three of the largest costal *marae* on Moorea, including Marae Nuurua and 19 *marae* located inland in the Opunohu Valley. They used 230Th/U dating of coral used as architectural features, such as for facing of structures, dressed stones, and offerings. The dates suggested that the *marae* were constructed rapidly over a period of some 140 years immediately before initial Western contact in A.D. 1767. Marae Nuurua was dated to 1761. While the age determination does not directly date the images, it suggests that the turtles were associated with monumental temple architecture and the highest-ranked chiefly lineage.

In collaboration with SPC, the first research with a focus on petroglyphs began on the island in 1989 (Millerstrom 1996b, 1997b; Millerstrom and Baumgartner 1998). In addition to Marae Nuurua, sites with geometric figures, anthropomorphs, and turtles were found in the Tefaarahi (Figure 95) and Paopao valleys (Figure 96) and were associated with agricultural terraces and rivers. A total of 128 individual images were recorded on 12 boulders. The majority of the figures (n = 110) were found at Tefaarahi, an area located inland on the north side of Moorea. Geometric motifs dominate in the Moorean inventory, comprising 63 percent of the images (Millerstrom 1997b; Millerstrom and Baumgartner 1998). This probably relates to the practice of tattooing, which was extensively practiced in the Societies (Ellis 1969:262–267 [1831]; Henry 1928:287–289). One of the words for the art of tattooing, *naonao*, translates to "checkered," suggesting that many geometric tattoo motifs were used in the Societies Islands.

Raiatea, Huahine, Bora Bora, Maupiti, Tubuai, Raivavae, and Rapa

As Table 31 shows, 576 petroglyphs have been recorded in the Society and the Austral Islands. No doubt there are others that have not yet been documented. Abstract geometric motifs are prevalent on Raiatea. Out a total of two hundred and seventeen individual petroglyphs 64 images or 30 percent were circular figures, a few are angular geometries. Some circles occur on Bora Bora and Huahine. Most of these probably represent incomplete or eroded turtle figures. In contrast to the prevalent curvilinear motifs discussed above, the Tubuai repertoire consists mainly of repetitive linear motifs and only a few anthropomorphs occur (Figure 97). Most of the figures are placed on upright stones that form part of the *marae* court (Millerstrom 1991b). Few images have been documented on Raivavae but three human figures, eight anthropomorphic faces, and a few lizards in bas-relief (site RRA 160) were discovered on a group of boulders in 1991 (Edwards 1998:31–58). The boulders surrounded a rectangular cist (Figure 98). Facial characteristics such as narrow eyes and a small mouth are identical to the faces features of sculptured tiki on Raivavae (Millerstrom 1991b). Sadly the bas-relief figures are now destroyed due to a fire in the area (E. Edwards, pers. comm., 2006). At Marae Maunaoto, two motifs are placed on different *marae* stones that enclose the court (Edwards 1991; Millerstrom 1991b). One is a sun or sea urchin motifs that appears to be the most common figures on upright stones at *marae* complexes (Marshall 1961; Millerstrom 1991b). These motifs are frequently depicted on *tapa* cloth, ceremonial paddles, and spears (Barrow 1979; Dodge 1940; Kooijman 1972; Meyer 1995). But distribution of this motif was not restricted to Raivavae. The same design motifs that are depicted on cultural material from Raivavae were also placed on a fisherman's canoe-god from Rarotonga (Peabody Museum; see also Barrow 1979:70, Figure 76). Sun motifs are similarly seen in tattooing (Allen 1992). The other figure, also part of an upright stone on Marae Maunaoto, is exotic and defies identification (Figure 99).

A bas-relief figure, linked to "the depot of the kings," was located on Rapa or Rapa Iti (Smith 1965:90) where a tomb had been carved into the mountain. Smith speculated that, because of the location and the position of the figure, the figure depicted "a supernatural guardian" (Smith 1965:92). Both Stokes and Smith concluded that the tomb was either the drying vault (Stokes 1921:830–832) or the resting place (Smith 1965:92) of each "successive king."

Petroglyphs, apparently mostly turtles, occur on Maupiti, but they have not been scientifically documented since briefly described by Emory (1933:174–175) (Figure 100). No images have been reported from the Tuamotu Islands or from the Cook archipelago. However, this does not rule out the presence of similar motifs on other media. Emory (1947:29–37) observed that Tuamotuan ancestral gods were made of wood, coral, and sennit with attached human hair and red and yellow feathers. Furthermore, turtle effigies were present in the Tuamotus on *marae*.

Overall, the images of the Society and the Austral Islands are closely linked to those of the Marquesas, although local variations and emphasis exist, such as the astonishing number of turtles on Raiatea and Maupiti, and the exotic funeral masks of Tahiti, Bora Bora, and Raiatea (Millerstrom and Baumgartner, forthcoming).

Although images appear in other media, the reason why image making was so prevalent in the Marquesas and not in Tahiti is unknown. Factors such as environmental conditions and geological characteristics, like access to raw material, were similar, and both regions offer similar human habitats. The specific features regarding Tahiti's initial settlement remain uncertain, but the two localities were probably both settled within a short time span of each other. While the Tahitian chiefdoms were more highly stratified than the chiefdom systems in the Marquesas, that factor alone does not explain the difference. The main reasons may lie in different cultural trajectories. Images in tattoo, wood, and *tapa* played an important part of the culture, especially in mortuary contexts. Perhaps dancing, storytelling, and acting served as similar forms of expression as the images of the Marquesas?

Generally, the motif types of the Marquesas Islands fit nicely within the Polynesian imagery already noted on Tahiti, Moorea, Bora Bora, Huahine, Maupiti, and Raiatea. Variously shaped anthropomorphs, turtles, fish, double-outlined anthropomorphs, and linear and curvilinear geometric figures are part of this repertoire. Spatial distributions differ from those from the Marquesas and temporal frameworks remain uncertain. What is common are motif types and, in many cases, associations with chiefly architecture and mortuary practices.

The Hawaiian Islands

Our knowledge of Hawaiian rock art images is derived from published works, including that of Emory (1922, 1924), Cox and Stasack (1970), Kirch (1973a; Lee 1988, 1989, 1990/91, 1995, 2002), Kwiatkowski (1991), Stasack et al. (1996), E. A. Stasack and D. S. Stasack (1997, 1999), Lee and Stasack (1999), and Millerstrom and Kirch (2002, 2004); McBride 2004; Hoerman (2016). Although all the islands have rock art, numerically speaking, Hawaii Island has the greatest number of petroglyphs. A count of individual figures recorded at five locations (Puuloa, Anaehoomalu, Puako, Kalaoa Cave, and Kaupulehu) is well over 30,000 units and is the result of recording projects headed by Georgia Lee (Lee and Stasack 1999). At Puuloa (hill of long life) alone, a total of 23,566 figures were recorded on the rock surfaces of a vast lava field (Lee 1993a).

Quantitative analysis suggests that Hawaiian petroglyphs are site specific. For instance, a *puu* (high lava dome) was the focus of a concentration of cupules and circular geometric motifs at Puuloa. Only 1.9 percent of the recorded images are anthropomorphic (Lee 1989, 1995; Lee and Stasack 1999). Historic references relate that Hawaiians visited Puuloa to place the *piko* (umbilical stump) of a newborn into *puku* ([man-made depressions] Lee 2001). According to Lee (1993a:7), they likely were protective symbols relating to birth, children, and to family. As geologists have dated Puuloa's lava flows to ca. A.D. 1200–1450 (Holcomb 1987:269; Lee and Stasack 1999), the petroglyphs postdate that time.

Anaehoomalu is on the west coast of the island of Hawaii. Here, geometric images constitute 81 percent of a total of 2,126 petroglyphs. Human stick figures and triangular-bodied figures make up 5.8 percent (Lee and Stasack 1999). Unfortunately, most of Anaehoomalu was destroyed by the construction of large tourist hotels and golf courses; only the central portion of the site remains. The recorded figures are but a small sample of what was there originally. At both Puuloa and Anaehoomalu, the petroglyphs appear to be related to a trail (Lee 1995:47–52).

At Puako, geometric motifs are rare and anthropomorphic figures dominate (Lee and Stasack 1999:13). This site falls into four major components, each different and apparently functionally specific (Lee 1995:50, 2002:79–92). Lee and Stasack (1999:13) argue that the focus of petroglyphs at these sites is on family concerns, with a strong sense of continuity. Lee (1995:50) wrote: "At Kaeo 1, there appears to be a strong thread of continuity, lineage, family identification, and succession, as human figures emerge from or branch off from a central image." In this location, some 2,275 petroglyphs were documented. Some depict human feet, and particularly, baby feet, perhaps referring to family or clan concerns (Lee 1995:50). Lee (2002:79–92) also discusses several sites on Hawaii Island that suggest that the ancient Hawaiians selected places in the landscape that already had a particular meaning, and perhaps sociopolitical concerns.

Edward and Diane Stasack launched a long-term field project in 1995 in Hawaii Volcanoes National Park (HAVO). In 1996 they received a three-year grant from the Hawaii Community Foundation to continue their work in HAVO and eventually expanded their research project to West Hawaii with additional funding from other national parks, private developments, foundations, and firms. A huge number of final reports and unpublished manuscripts, more than can be included here, have been produced. These are all archived at University of Hawaii, Manoa, Hawaiian Studies and University of Hawaii's Hilo Hawaiian Studies libraries. Approximately 80 sites and over 30,000 petroglyph units have been recorded, which are in addition to those recorded before 1995, the largest single concentration of which is at Waikoloa Petroglyph Preserve. Some of these sites will not be taken to final reports that will be publicly available but the information will be used in their final analyses. Their research goal is to examine the stylistic rock art elements in an island-wide context.

In their analyses they will be able to assess rock art differences across their survey districts (Puna, Kau, South Kona, North Kona, South Kohala, North Kohala, and some smaller land divisions) and to examine how rock art on surface lava differs from that in caves. In addition they will examine all the images across regions and how they may reflect functional site differences, as well as recorded history. They will use these analyses to produce a series of topical books or monographs on Hawaiian rock art. The final island-wide analyses, according to the Stasacks, will support the idea that the broad function of Hawaiian petroglyphs was essentially that of commemoration and that the petroglyph makers depicted their belief system and cultural values for the benefit for future generations (Edward A. Stasack and Diane S. Stasack pers. comm., October 17, 2010 and January 2016).

Lanai

The small island of Lanai ranks second in the number of documented petroglyphs; Maui places third (Emory 1924; Cox and Stasack 1970). Major sites on Lanai, first examined by Emory (1924), were reexamined in 1987–1988 (Lee 1988). Her study resulted in the discovery of numerous additional images; 1,151 figures were recorded at Luahiwa, Kukui Point, Keomoku, and Kaunolu (Lee 1988; Lee and Stasack 1999). Most of the Luahiwa figures, 571 images on 32 boulders, were located on a steep slope near a rain *heiau*. Anthropomorphic stick figures dominated, with 36 percent, in contrast to triangular-bodied figures, at 25 percent. Twelve percent were dog images (Lee and Stasack 1999).

The Kukui Point site, on the north side of the island, is an outcrop of reddish-brown basalt that may have been an ancient quarry; chips were removed from some of the image stones. Seventy-five percent of the total figures (n = 246) are anthropomorphic; 19 dog motifs were recorded (Lee and Stasack 1999).

At Keomoku, petroglyphs are on boulders that form part of Kahea's *heiau*, a site known to have served as a place where human sacrifices took place (Emory 1924; Lee and Stasack 1999). It is rare to find petroglyphs placed directly on architecture in Hawaii.

On the south side of Lanai is Kaunolu, an impressive site. The area contains a canoe shed, fishing shrines, a *heiau*, and house foundations. Kaunolu's *heiau* is one of the most imposing ceremonial sites on Lanai, and was a retreat for Kamehameha I (1758?–1819), the powerful *alii*, the chief who controlled most of the Hawaiian Islands by 1795. Emory noted 60 petroglyphs here; Lee recorded 276 (Lee and Stasack 1999:114). Triangular-torso forms and anthropomorphic figures particular to the Hawaiian Islands dominate, and some have bird-like characteristics, such as spread-out, wing-like arms. According to Lee and Stasack (1999:114), the Kaunolu muscled human figures appear to be a later development than the same type of figure recorded elsewhere on Lanai.

Kahoolawe

The first Hawaiian petroglyphs to be radiocarbon dated are from Kahoolawe, the smallest island in the Hawaiian chain. Organic material was collected from the rock coating of several stylistically different figures. Ronald I. Dorn, Geography Department, Arizona State University, analyzed 13 Accelerator Mass Spectrometry (AMS) dates. An anthropomorphic stick figure (K23, Site 485, Panel 2-2) yielded the oldest age determination of A.D. 983–1168. A triangular figure (K19, Site 121G, Panel 8), the specific motif of which developed over time in Hawaii, yielded an AMS age determination of A.D. 1650–1950 (Lee and Stasack 1999:202; Stasack et al. 1996). The calendar ages are minimum dates, for it takes several hundred years for a silica coating to form over the petroglyphs.

The dates demonstrate that both anthropomorphic stick and triangular figures were in use at the same time. It should be noted that dating organic material trapped in the silica coating on rock art from desert areas is still in the experimental stage and more research is needed. Nevertheless, all the dates are consistent with archaeological findings and well within the timeframe for occupation on Kahoolawe (Lee and Stasack 1999:201–203).

Discussion

The petroglyphs on the islands of Hawaii, Lanai, and Kahoolawe depict both geometric and anthropomorphic images, but there are motifs that represent fauna, cultural artifacts, and historic elements. While the majority of faunal petroglyphs, such as dog, are found on Lanai Island, many cultural motifs, like sails, *papamu*, or game boards (for playing *konane*) dominate at Kaupulehu, a high-status site on the Kona coast, Hawaii Island. The highest concentration of historic petroglyphs (lettering and script) occurs at Puuloa. Kahoolawe also has a large number of historic petroglyphs, (goats and lettering), which demonstrates that the tradition of image-making continued well into the historic period (Lee and Stasack 1999).

Maui Island

Archaeological sites on the island of Maui are less well studied. However, Patrick V. Kirch directed a comprehensive settlement pattern survey of Kahikinui, a leeward district on the southeastern side of Maui (Kirch 2014). Several new sites were discovered during fieldwork in 1997–1998 (Millerstrom and Kirch 2002, 2004). A rockhelter within the Kipapa and Nakaohu *ahupuaa* was analyzed. Six petroglyph panels were associated with a shrine, plus a significant number of lithics, grinding surfaces, and anvil stones. A dog image was pecked on the wall directly above a hearth (Figure 101). Charcoal samples collected from a hearth place the site in precontact and early postcontact eras (Millerstrom and Kirch 2002).

In rockshelters and cliff faces there is evidence of freshwater springs or seeps (Millerstrom and Kirch 2002, 2004). In the arid environment of Kahikinui, freshwater was a precious resource and thus the petroglyphs may have served as territorial markers, signs of individual ownership, or rights to access.

A small archaeological team surveyed and excavated at Nuu Mauka Ranch, Maui, and investigated seven rock art sites (Millerstrom 2006c). Figures include pecked anthropomorphic stick figures, triangular anthropomorphs with and without muscles, realistic anthropomorphs, dogs, one turtle, historic lettering, and a large number of parallel or random incised lines, most superimposed over older types of petroglyphs (Figures 102 and 103). Although none of the rock art sites at Nuu Mauka Ranch have been excavated, based on the Kahikinui dated site, it appears that the figures span the late prehistoric and early historic period. Superimposition illustrates that the older petroglyphs at Nuu are the traditional Hawaiian motifs. Native Hawaiians in the post-European period used the same cliff faces to incise triangular and stick figures on top of the pecked figures, perhaps because of the *mana* they contained, adding crisscrossed parallel and randomly placed lines. The early post-European contact period is characterized by Roman lettering that reflects early Missionary efforts at literacy. Documented sources speak of a "frenzy" of interest on the part of Native Hawaiian to learn to read during the 1820s and 30s (Millerstrom and Kirch 2004:125; Sahlins 1992; Schütz 1994).

As various Hawaiian research projects have demonstrated, petroglyphs were not randomly placed but were associated with ceremonial architecture and caves, and were linked to sacred land; unusual rock formations, hills, water sources, and rockshelters—all places that were considered sacred and had significance for the Polynesians.

The Marquesan image repertoire is more similar to those recorded on the Hawaiian Islands than on any other Polynesian islands or island group. For instance, nowhere else are the anthropomorphic figures as diverse and numerous as in the Marquesas and in the Hawaiian Islands. Interestingly, the anthropomorphic faces so prevalent in the Marquesas are practically nonexistent in Hawaii. Only at Anaehoomalu, on the island of Hawaii, two faces are located on the same stone. However, they were depicted in the Marquesan fashion with the faces wrapped around the edge of the stone.

Pitcairn

The only known petroglyphs on tiny Pitcairn are at Down-Rope, on the eastern side of the island. A panel depicts an anthropomorph with the arms in an upright position; a stick figure; a square anthropomorph; and geometric shapes ([wheel spoke, star] Edwards, pers. comm., 1988; Heyerdahl and Skjølsvold 1965:3–7; Lavachery 1936). While these images have been described and photographed, they have yet to be scientifically documented. All the figures resemble several of the anthropomorphic types recorded in the Marquesas. Anthropomorphic stone sculptures were also found on Pitcairn in the past. Three or four standing images were said to be on a platform when the Bounty arrived in 1790. Unfortunately, the statues were broken and thrown into the sea. One headless statue from Pitcairn has hands across the stomach (Heyerdahl and Ferdon 1965: Plate lc). It is now in the Otago Museum.

Rapa Nui

Rock images have been documented on Rapa Nui since 1920 (e.g., Lavachery 1939; Lee 1986, 1992, 1993b; Metraux 1937, 1940; Routledge 1919). In 1981 Georgia Lee began a multiyear rock art survey project. Lee documented some 3,995 petroglyphs (excluding more than 4,600 cupules) and rock paintings over a seven-year period. Sea forms with human faces and images of bird-men were highly developed (Lee 1992). The birdman is characterized by a profile of a crouching human figure with the head and beak of a frigate bird. These figures cover the rocks at the ceremonial center of Orongo. Lee examined stylistic development by studying superimposed motifs and found that the early birdman motifs, associated with the birdman cult, developed at the Orongo ceremonial center ca. A.D. 1500. The deterioration of the birdman cult coincided with social unrest and ecological changes. Incised *komari* (vulva figures), supposedly symbolizing fertility, are found carved over the birdman motifs. Thus, changes in the island's cultural and sociopolitical ideology are marked by changes in the rock art (Lee 1992).

The birdman motif can be traced across the Pacific to Southeast Asia (Bellwood 1978; Lee 1992:5). Although a significant part of the Rapa Nui repertoire, only a few birdmen figures are found in the Hawaiian archipelago and are practically absent from the art system of the Marquesas. However, one bird with human legs (331hth 128/1) was documented on a boulder in Kahuvai, Hatiheu Valley, Hiva (Figure 104). Ethnohistoric accounts indicate that carved wooden bird figures may have been used in funeral contexts (Delmas 1927). In 1996, a three-dimensional bird figure in red volcanic tuff, 30 cm long and approximately 16 cm high, was uncovered in the Atuona Valley, Hiva Oa (Antoni 1997:88–90). It had large a mouth and eyes, characteristics of Marquesan anthropomorphic petroglyphs. Thirty-three turtle images are distributed around the coastline of Rapa Nui although most are found within the Tupahotu or Miru clan areas. Several images were associated with local legends (Lee 1992). However, the architectural association was more difficult to ascertain. Although turtle motifs are on an outcrop along with *makemake* faces, fishes, and cupules at Tongariki (Lee 1992:83, Figure 4.65), a large ceremonial site situated on the south coast, few petroglyphs are associated with architecture. This may be due to geology; the fine-grained lava flow that makes such excellent surfaces for petroglyphs may or may not occur near *ahu* (shrines).

Sonia Haoa and her team have documented several new rock art sites during their archaeological survey project in the north and central parts of the island. According to Sonia Haoa, many of the images are geometric and different than those recorded in other parts of the island (Sonia Haoa, pers. comm., February 6, 2011).

No human stick figures have been found on Rapa Nui. The reason for a lack of human stick figures, so abundant in the Marquesas and the Hawaiian Islands, is unclear. We may look to the recent research into how Easter Island was settled. Previously it was believed that initial settlement came from the Marquesas, but today it appears that Rapa Nui may have been settled from the Mangareva-Raivavae area (Green 1998, 2000a; Kirch and Conte 2004). Based on stylistic evidence and the presence and absence of similar figures and motifs, migration(s) from Raivavae to Rapa Nui appear to be a likely possibility. Although there are few petroglyphs on Raivavae, they do bear some affinities to those of Rapa Nui. In addition, large stone tiki were found on Raivavae; these are quite close in concept to the famous statues of Rapa Nui.

There were more interisland connections in the past than previously believed (Green 1998, 2000a; Green and Weisler 2000; Weisler 1998). Weisler (1998:521–532), who applied wavelength dispersive x-ray fluorescence technique on rocks found at a basalt source on Eiao Island, Marquesas Islands, demonstrated that the same characteristics occurred in adzes from Moorea, the Society Islands, and from Mangareva during the twelfth through fifteenth centuries. Green (1998:87, 110, 2000a:71–76) argued that, based on language, culture, and biological information and interpretation, that initial settlers to Rapa Nui arrived from southeast Polynesia, meaning the Southern Cook Islands, Australs, Mangareva, and the eastern Tuamotus. Elements from contact with the west included among others, the anthropomorphic stone sculptures, the *ahu* complex, and one-piece fishhooks. Occasional postsettlement interaction with South America included the sweet potato and the bottle gourd (Green 2000b). Archaeological research has shown that the Polynesians reached the territory of the pre-Hispanic Mapuche people in central-south Chile (Jones et al. 2011). Rapa Nui chickens share the same DNA as chicken bones excavated in well-controlled archaeological context at El Arenal site on the south-central Chile (Storey et al. 2011:139–170). Cultural parallels are linguistic cognates (e.g., the word *toki* for a polished stone axe), *maata* or volcanic stone tools, and hand clubs (Ramirez-Aliaga 2011:95–109).

Other than in the Marquesas, anthropomorphic faces are only found in significant numbers on Rapa Nui. On Rapa Nui, the full face, eyes-nose, and "eye-mask" motifs totaled 517 examples (Lee 1992). While comparing similarities in motifs is often subjective, parallels between the Marquesan and the Rapa Nui face/eye figures are their style, placement on the boulders, and manufacturing technique. Some figures both from Rapa Nui and the Marquesas are placed on edges of a stone with an eye on each side giving a three dimensional appearance. Other commonalties are faces with mismatched eyes (Lee 1992:57; Millerstrom 1997a). While no anthropomorphic faces were documented on Moorea, Raiatea, Bora Bora, or Huahine, they occur in New Zealand, Tahiti, Raivavae, and a few in Fiji (Millerstrom 1997a).

The Marquesan double-outlined anthropomorphs also appear on Rapa Nui. A petroglyph boulder in a small valley adjacent to Tahauku Valley, Hiva Oa, has double outlined anthropomorphs (Linton 1925:149–151. Plate XV, A, B, D; Millerstrom 1985c), leading Handy (1943:22–32) to speculate on a cultural connection between the Marquesas Islands and Rapa Nui. Handy specifically compared the motifs to similar images on the Rongorongo boards. Although not found in great frequency, double outlined anthropomorphs form a part of the anthropomorphic repertoire. In the Hatiheu Valley 25 such figures out of 3,379 (0.74 percent) have been documented. While *komari* are found in great numbers (n = 564) especially at Orongo (n= 334), Rapa Nui (Lee 1992:31), only one that may or may not depict a *komari* was recorded on a large and well-known petroglyph boulder (331hth 1/1) at Meae Te Ipoka in the Hatiheu Valley. Also at the Orongo ceremonial site the man-bird combination is prevalent. A total of 459 figures are recorded on Rapa Nui, while 370 of them are found at Orongo. The man-bird combination is rare in the Marquesas, but a few man-dog, man-fish, and man-lizard combinations occur. Turtles, octopus, and fishes are found on both Rapa Nui and the Marquesas. However, the marine creatures documented by Lee (1992:38) are exceptionally complex and find few parallels in the Marquesas.

Aotearoa (New Zealand)

Compared to rock art research in the Marquesas Islands, rock art research and recording in Aotearoa has an extensive history dating to back to 1852. This is when the surveyor Walter Mantell, one of the first to record New Zealand rock art, sketched Maori figures painted on a cliff wall in North Otago (www.teara.govt.nz). Since then a great amount of literature related to rock art has emerged and various innovated approaches used (e.g., O'Regan 2003, 2006, 2007; Allingham 1990; Ambrose 1970; Thompson 1989; Anderson 1988; Bain 1982, 1985; Allingham 1990; McCulloch 1971). For reviews on Polynesian rock art see (Lee 1996:163–172; Millerstrom 2003:147–164, 2006b:213–225; 2012:235- 243; Hoerman 2016).

Although Maori rock art is recognized as an important component of the archaeological record in New Zealand (e.g., Davidson 1984), continued uncertainty surrounding its archaeological context, how it may relate to other Maori art, and the historically sporadic approach to its research has seen it marginalized in synthesis of New Zealand archeology (O'Regan pers. comm., 2010). A key reference for the stylistic chronology of southern Maori rock art is the publication of the late Toni Fomison interpretation (Fomison and Fyfe 2014).

The work of Michael Trotter and Beverly McCulloch (1971) is a comprehensive description of Maori rock art. It specifically attempts to place the rock art in an archaeological context. However, their conclusions, that Maori rock art is local innovation largely unrelated to other Polynesian rock and that it was only an incidental pleasurable pastime for the Maori engaging in these activities is not in line with what we know about Polynesian rock art today.

Several hundred sites with carved, painted, or drawn charcoal figures are known throughout the country. The great majority is concentrated in the South Island's North Otago and South Canterbury (Davidson 1984:214–218; Trotter and McCulloch 1971). A marked difference in spatial and temporal distribution between the islands has been widely assumed. It is based on the idea that South Island Maori rock drawings were not continuous throughout the pre-European period and that the North Island rock art was similar to late period wood carving found in proximity to late period settlement. This emerged from the summary observations by Trotter and McCulloch (1971).

Many of the images from South Island follow a relatively clear pattern in style and execution, some sites on the South Island depicted *moa* and extinct giant eagles, representing the early settlement period (Trotter and McCulloch 1971, Figures 4c and 35). Trotter and McCulloch argue that the images reflect the work of a culturally homogeneous group, a small population living a nomadic existence between the coast and the interior. Dates from several rock shelters throughout the island ranged from A.D. 1100 to 1500. Sometime after ca. A.D. 1500, perhaps due to environmental changes and because people moved, rock art activities ceased (Trotter and McCulloch 1971:80).[51]

Images from North Island differ from the majority of those in South Island both in motif types and technique. While black charcoal figures are common in the South Island they are rare in the North Island. It is speculated that this may be due to a later settlement in North Island where the population lived in permanent villages with less communication between communities. This, however, is still not clear (O'Regan pers. comm. 2003). In general, the image types fit within what is termed the Classic Maori. Late forms consisted mainly of spirals, curvilinear designs, and anthropomorphs with facial features (Trotter and McCulloch 1971:82). One associated carbon date yielded an age of ca. 200 years B.P. However, in wood carving human figures and spiral motifs may have been used for some 400 years (Davidson 1984). The earliest dated woodcarving in New Zealand was preserved in the Waitore swamp. A carved board with "punched incised" spirals and a human head were believed to be from a period of between A.D. 1380 and 1500 (Davidson 1984:212).

Due largely to the deterioration of the supporting rock, much of it limestone, charcoal, and ochre Maori images in the South Island are continuing to be lost. In 1990 Atoll Anderson and Brian Allingham initiated the South Island Maori Rock Art Project (SIMRAP) with the goal to photographically record all rock art in Te Wai Pounamu (South Island). Initially the project was set up with the assistance of the New Zealand 1990s commission, New Zealand Historical Places Trust, and the University of Otago. Allingham conducted a pilot study in the Oamaru area, North Otago, that resulted in a 300 percent increase in the discovery of new rock art sites and a large increase in the number of unrecorded figures in previously known sites (O'Regan 1997).

At the conclusion of the pilot study, South Island Maori Rock Art Project was adopted as a tribal project by the Ngai Tahu tribe. Over many years Allingham continued to work for the tribe in an effort to survey all the rock art in their South Island territory (www.ngaitahu.iwi.nz). This work has provided support and opportunity for some student research to precede, notably Blanshard (2005), Butikofer (2005), and O'Regan (2007). More significantly it provided a foundation from which Ngai Tahu tribe has set up a tribal rock art charitable trust to promote research, preservation, and education of rock art. Amanda Symon, a local archaeologist, was appointed by the Trust as the inaugural rock art curator in 2003. The trust established Ngai Tahu Maori Rock Art Visitor Center in Timaru, which opened in December, 2010, (www.ngaitahu.iwi.nz/RockArt/). The remaining recording part of the project continues. Brian Allingham and Dr. Yann-Pierre Montelle undertake this work, and a database for the field results is in development (www.odt.co.nz).

The need to complete the survey work and develop the database is recognized as necessary to aid conservation work, but also to enable research that continues to be limited by accessibility to the data. For example, a recent attempt to formulate a stylistic chronology for southern Maori rock art (Blanshard 2005) has met with limited results. A tribal review of the research by Allingham, Symon, and O'Regan noted the chief concern is an inadequate sample size of both motifs and the theoretical basis for determining age (O'Regan, pers. comm. 2006). The presumption is that the Maori rock art most similar to other East Polynesian images is the earliest, while those figures not occurring on other island, for example human figures with acutely angled legs, will be later. The problem is that it is unclear that the East Polynesian samples used are of the same age as the Maori motifs, and the lack of some style in other Polynesian Islands does not itself date that style in New Zealand (O'Regan pers. comm. 2006).

Drawing on Allingham's survey and supported in field work by Montelle, O'Regan (2007) used an intrasite landscape approach to see how the Maori viewed space and if the concept of *tapu* is detectable and influenced placement of rock art figures at two Otago rock shelters. A digital model was constructed of the site using the first 3D laser scan of a rock art shelter in New Zealand. The results indicated *tapu* was a factor in the placement of rock, typically with the later artist looking to make a "respectful" connection with earlier art. Another project undertaken in partnership with the Ngai Tahu Rock Art Trust was a nine-month conservation project that took place in a Takiroa rock shelter by Butikofer (2005). She investigated the dust in the limestone shelter. Butikofer (2005) discovered that the salts on the panel surface includes nitrate that is believed to come from cattle and sheep dung at the site.

While less rock art research has been conducted in the North Island than in the South Island several recording projects continue to take place. Tauranga Museum working with the local tribes Ngati Pikiao and Ngati Makino has recorded a series of red canoe figures located along the foot of an embankment in the Pongakawa Valley Island from the Bay of Plenty (O'Regan, pers. comm. 2006). Apparently there are no traditional knowledge of the art itself but traditional knowledge associated with the site demonstrates how rock art features in sacred or special places in the region (O'Regan, pers. comm. 2006). This apparently is also the case in the central North Island around Lake Taupo where archaeologist Perry Fletcher, in cooperation with local Maori tribes, has recorded over 150 rock art sites during 30 years of site recording. Most of the sites contain "dots" and "dashes" of red ochre, but some canoes and anthropomorphic images occur. In contrast to the South Island, black pictographs are rare while deep relief carvings are the most common.

Stylized human footprints, similar to those found in Hawaii, have been reported from three sites in North Taranaki, North Island. For instance, a panel in a cave from Tongaporutu depicts 83 footprints; several were paired, and one was part of a human face (Delph 1939; Trotter and McCulloch 1971:47, Figure 50). While the majority of the feet showed five toes, three, four, six, and eight toes were also shown.

How do the Maori images compare with other regions in Polynesia? While Maori rock art include anthropomorphs, dogs, birds (especially in the South Island), canoes, and geometries, such as spirals and concentric circles, few of the images stylistically resemble figures from other Polynesian islands or island groups. Monnin and Sand (2004:218) in their extensive publication on New Caledonian rock art did a study were they compared various published image types from several Oceanic islands to try to see how the figures were similar or how they differed. They found that only spiral and concentric circles occurred in most of the other islands were rock art occurred.

Western Polynesia

Except for American Samoa no petroglyphs have been found in Samoa, Wallis, or the Futuna islands, but petroglyphs occur in Tonga and Fiji. Although some petroglyph sites have been discovered since the early 1930s, the reason for a lack of petroglyphs in these islands has not been examined closely.

It is likely that shared Polynesian rock art motifs occurred on other media, such as on *tapa*, wood, bone carvings, or in tattooing. Carved human, fish, and bird figures commonly occurred on Tongan clubs, and to a lesser extent on ones from Fiji; but it is doubtful that the Samoans ever used them (Hiroa 1930:611). Extensive geometric motifs, especially rectilinear, were carved on clubs in Samoa, Tonga, and Fiji. According to Gell (1993) Samoan male tattoo designs also include geometric motifs.

American Samoa

In American Samoa, Kikuchi (1964, 1967) reported on three petroglyph sites. Figures, supposedly of pre-European origin, were found on the coastline behind Fitiuta village, and on the island Tau. One site was found in the lagoon of Leone village (Site ST 128) and one post-European site at Leata, on Tutuila (Kikuchi 1964:163–166, 1967:372–373). In the lagoon of Leone village "an intrusive shelf of ash called Papaloa" depicts images of two octopi, a turtle, an anthropomorph, "outer rings of encircling holes," a cluster of carved lines, a paddle, and a human foot (Kikuchi 1964:164–165). Neither architecture nor any local traditions are associated with the sites. At high tide the outcrop is partly under water.

The shelf is slippery, and it is difficult and dangerous to walk on (Personal experience January 31, 2004). However, Kikuchi has, I believe, underestimated the importance of the megalithic outcrop. The Papaloa site holds a unique position in the lagoon and although it is difficult to see and record the petroglyphs, the outcrop is covered with circular figures. The site is certainly associated with the old settlement of Leone. At low tide the lithic scatter can be seen in the lagoon between Papaloa and the village. The lithic may have been washed out from the inland sites; it is also possible that the present sea level has increased and the sites are now under water.

Te Rangi Hiroa ([Peter H. Buck] 1930:612) wrote that stones representing gods were situated near villages. Offerings were made to these gods at appropriate times or when people were passing. Hiroa claimed that due to the early acceptance of Christianity and to the extensive religious teaching by the Samoans themselves, it was difficult to collect information on ancient religious practices.

The Kingdom of Tonga

In 1920 in Tonga, McKern (1929:79) reported linear motifs, such as vertical and horizontal bars of lines and dots on the facing stones at four different *langi* (royal tombs) at Lanai Katoa, Tauhale, Nukulau, Malae Lahi. He wrote that the "divinely descended kings were entombed" in the monumental architecture, and that the places were so sacred that the Christian Tongans avoided the sites (McKern 1929:80). Similar linear motifs were applied to a frontal stone at Malae Lahi, Uiha Island, located in the northern part of the Haapai Island group. On the same royal tomb Burley (1994) discovered one carved human foot motif (right plantar view) on a facing stone forming part of the northwest corner. Burley hypothesized that the image was linked to the act of *moemoe* a gesture of respect reserved for the highest chiefs (Burley 1994:511). At the time of Capt. J. Cook's visit in the 1770s, the greatest mark of respect and reverence in Tonga was to touch the soles of the chief's feet (Beaglehole 1974:175–176, 953–954; Gifford 1929:118). Burley (1994:512) noted: "The symbolic connotations of a foot on a tomb would demand the ultimate in respect for those interred." Footprints, similar to those at the royal tomb of Malae Lahi, have been found in New Zealand, Hawaii, Fiji, and Samoa but not in the Marquesas. The Tongan images occur on monumental architecture directly linked with royalty and the elite establishment. Their placement and cultural context suggests a conspicuous display and a call for respect and reverence.

It appears as if Tongan petroglyphs are associated both with royal mortuary traditions, and chiefly activities. An unusual site was uncovered at Houmaleeia, Foa Island, in the Haapai Group. The Houmaleeia site is a unique discovery in Tonga since it is the only one of its kind in central Polynesia (Egan and Burley 2009:209–232). Turtles, dogs, triangular anthropomorphs, and footprints, both single and in pairs, are clustered on two main panels on a beach rock (Figures 105, 106, and 107). During high tide the petroglyph site is completely submerged in water indicating a sea-level change. The panel was covered over with coral sand (Egan and Burley 2009:4). Interestingly, the triangular anthropomorphs, thought to be unique to the Hawaiian petroglyph assemblage, also occur at Houmaleeia. Anthropomorphous figures are rarely seen in the Tongan, Samoan, and Fijian limited rock art repertoire. This led Egan and Burley to postulate that the figures were carved either by Hawaiians visiting or living in Tonga, or Tongans that had visited Hawaii and become familiar with the rock art of the Hawaiian Islands. The authors seem to lean toward an indirect "Hawaiian/Tongan interaction prior to European contact" sometime around the fifteenth or sixteenth century A.D. (Egan and Burley 2009:22–23). But, as the authors point out, triangular anthropomorphs also occur on some historic Tongan wooden clubs. It is likely that these figures were part of the Tongan image repertoire used in other mediums, and perhaps, we will see the same images on rock art sites that are yet to be uncovered.

Figures were noted by McKern, but not documented, on a small island in the far south Haapai Group. In 1957 Wordsworth sketched two anthropomorphs, a sun motif, and some circular figures (Egan and Burley 2009:218, Figure 9). Two anthropomorphs, one of which was a triangular figure, were identical to those at the Houmaleeia site. According to Egan there are reports of other sites on isolated islands to the west (Shane Egan, pers. comm. 2009). No doubt there are other Tongan petroglyph sites covered by sand that will be discovered in the future. Interestingly, the Houmaleeia site was again buried in sand after tropical cyclone Rene ravaged the island on February 15, 2010 (Burley pers. comm., April 2, 2010).

Fiji Islands

Limited rock art survey and research has taken place in the Fiji Islands. Furthermore, Fijian rock art is marginalized in the literature dealing with archaeology and Pacific art. The few scattered publications are, for the most part, descriptive, sketchy, lacking in interpretation, and outdated (e.g., Hill 1956; O'Reilly 1954; Paine 1929; Palmer and Clunie 1970; Parry and Phillipps 1951; Snow 1950; Vogan 1937; Watling 1988). Except for Ewins (1995), who recorded and analyzed a rock art site on Naura Beach, Vatulele Island, no one has focused on an in-depth analysis of any of the other sites. Many published comments are contained in a single sentence, while others have sufficient descriptions that allowed us to relocate the sites. While archaeologists had visited some sites, others had only been reported in personal communications to the Fiji Museum. None of the sites have been excavated. There is some confusion as to where some of the sites are located or even on what island they are supposed to be found. Some of the reasons for lack of research are perhaps, because the islands are widespread and the terrain is often rugged with dense vegetation.

Thomas Williams, a Methodist missionary in Fiji in the 1840s, illustrated in his notes two "round, black milestone[s], slightly inclined," one with several carved concentric circles, the other stone dressed with a *liku*, a women's dress (Stringer Rowe 1982:220). According to Williams, particular stones were sacred and, like certain clubs and animals, they served as the residence of deities. God stones could be both gods and goddesses, to which the natives would *tama*, a shout of reverence, in the same manner as to a chief (Stringer Rowe 1982:37, 219, 221). Consecrated stones are found on many religious sites in Fiji; some mark the place of a particular god or goddess. One such stone, named Lovekaveka, was regarded as the home of a goddess, and food was prepared for this deity (Stringer Rowe 1982). Palmer and Clunie (1970:3) were skeptical about the concentric circles engraved on one of the god stones noted by Thomas Williams. They are of the opinion that the circles were drawn on the surface before the stone became a god stone. As evidence they pointed out that numerous god stones have been documented by the Fiji Museum, none of which were marked with concentric circles.

Millerstrom and Cruz conducted three field seasons of rock art survey that took place in 2007, 2008, and 2009 (Millerstrom and Cruz 2009:40–47, 2010:27–35). The focus was to: 1) relocate and record the sites mentioned in the literature; 2) note the archaeological, cultural, and environmental context for each site; 3) evaluate the potential of developing a project on rock art in Fiji, taking into account that the knowledge of rock art in the archipelago is considered a "blank spot"; 4) enter the data in a computer-generated classification system based on a typology already created for the Marquesas Islands, the Hawaiian Islands, and Easter Island; and 5) interpret the information. Our 2007 survey focused on the island of Moturiki, in the Lomo Viti Group. A large boulder with seven concentric circles and several polishing groups was discovered in 2005 in a taro swamp (Figure 108). We estimated that we surveyed approximately 70 percent of the island. Though the island contains numerous fort villages, ring ditch villages, and other archaeological remains, such as an abundance of pottery fragments, no additional rock art site was found.

In total, twenty sites were mentioned in the scattered literature. Nineteen of these sites were visited during the three field seasons. While we located two new sites, some of the sites mentioned in the literature could not be relocated (e.g., the Naboro site and the panels in Bukusia cave, both in Sigatoka Valley, Viti Levu), or they were destroyed (e.g., Bouma, Tavoro Waterfall, Tavioni). Because of the great distances and transportation logistics, we were unable to reach one site in the mountains on Ovalau, the Koroiemalu Cave, Sigatoka Valley, Viti Levu; and the dome-like cave on Sawa-i-lau, Yasawa Group. Furthermore, there are sites on both Totoya and Moala in the Lau Group, but the islands are difficult to reach as transportation is irregular. On Totoya Island there is one boulder with some pecked figures (P. Nunn, pers. comm., 2009), and on Moala pecked circles and geometric designs are depicted on two large boulders (S. Matararababa, pers. comm., 2009).

Some sites may never have existed. For example, a cave site on Yanuca Island visited by the late G. T. Baker (or Barker) but reported by Palmer and Clunie (1970:3–10) was supposed to contain approximately 50 circles, some interlocked. We visited Yanuca Island, shown on Palmer and Clunie's map to be located next to Beqa Island (Palmer and Clunie 1970: Figure 1), and spoke to the elders in the village. They had no knowledge of any such site. They guided us to both the caves on the island, one of which was a chiefly burial place, but we were unable to find any evidence of any circles. But there are several Yanuca Islands in Fiji. We had the opportunity to visit the small island of Yanuca located next to Moturiki. Only the owner and his family live on the island, and they had no knowledge of the site. However, we examined a large rockshelter and observed numerous wasp-nest-stained circles, many linked together covering the wall. Could these wasp-nest stains be the markings reported by Baker/Barker? Another Yanuca island is located on the northeast side of Viti Levu. The island is private, and we had no opportunity to visit. However, we talked with a few people that knew the island, and they had never heard of a rock art site there. Likewise a turtle petroglyph at Nanau-i-Ra mentioned by Hiener (undated, Fiji Museum Archaeological Survey Records) may or may not exist. We visited the island and a diver told us there is a stone of a turtle image out in the sea (Papoo, pers. comm., 2009).

A total of 12 sites were recorded. The Fijian images consist of painted, incised, pecked, and charcoal-drawn figures. Charcoal drawn turtles and pecked painted figures would include ochre handprints (Figures 109 and 110), pecked concentric circles, cupules, and unidentified geometric figures (Figure 111 and 112). Only one pecked anthropomorphic face and two anthropomorphic stick figures have been documented (Figures 113 and 114).

From the sites recorded it is clear that the images are painted, pecked, and incised in four distinct situations: 1) outcrops; 2) boulders; 3) cliff faces; and 4) caves and/or rockshelters. Most of the sites are associated with water, such as the sea, springs, rivers, and a taro swamp. Ridge-crest locations are also important features.

It is noteworthy that at Qaravonu rockshelter a pecked turtle is wrapped around the edge of a boulder giving it a three-dimensional look. In the Marquesas numerous face figures and some turtle figures are depicted on both sides of a boulder. This unique placement of figures also takes place in great numbers on Easter Island.

Petroglyphs and pictographs occur in Fiji but not in great abundance. Although to date limited survey has taken place in the islands, it is doubtful that a large number of images on rock, such as for example, the great number found on the Hawaiian Islands, Rapa Nui, Aotearoa, the Marquesas, and New Caledonia, will turn up in the future. Our 2008 survey on Moturiki Island was an attempt to archaeologically survey a manageable island to see if indeed an entire island survey would reveal additional rock art sites (Millerstrom and Cruz Berrocal 2009, 2010). This case study demonstrates that in Fiji, even if one site occurs in a particular region or island, it is not predictable that additional sites will occur on the same island. It is likely, however, that in the future, isolated sites will be found accidentally, as in the case on Moturiki.

The study of Fijian rock art is interesting in two senses. First, there seems to be a great deal of island and interisland variations. Second, its location in the Pacific as an "in-between" archipelago between Near Oceania (Melanesia) and Remote Oceania ([Western Polynesia] Kirch 2000:155) places it in a special position in cultural terms. The study of traits that may or may not be shared between both regions is especially significant in this context. One of these relevant traits is rock art. In this sense, the lack of information regarding rock art in Fiji is a problem; as it has been shown by our work, Fiji is an exception in the context of the Pacific that needs clarification. While some of the images are closely related to those found in Polynesia both in terms of site situation and image types, the reddish handprints show an affinity with those of, for instance, New Caledonia. Where the close relationship to Near Oceania is concerned, it has been suggested, based on stylistic evidence, "that the earliest paintings may have been made during the early Lapita period of settlement in Fiji" (Ewins 1999:67). Iconographic connections therefore exist, as have been highlighted above, between both Fiji and Near and Remote Oceania. The problem we face now is to try and determine the origin and development of these cultural traditions and their genetic (or lack of) relationship.

Discussion

Nowhere in Polynesia are petroglyphs as prevalent as in the Hawaiian Islands, Rapa Nui, the Marquesas, and Aotearoa. However, more commonalties exist between the Marquesan and the Hawaiian image assemblage than between any other islands or island groups in Polynesia. Lee (1992) has already noted this link. The practice of making images on stones was a strong cultural component in both island societies, lending support to the general theory that the Hawaiian Islands were originally settled from the Marquesas (Bellwood 1978; Emory and Sinoto 1964; Irwin 1992:164–167; Kirch 1984, 1985:68–87, 184–187; Matisoo-Smith et al. 1998; Sinoto 1967). Despite numerous variations, there is an emphasis on circles, cupules, and shared features, such as dogs and an abundance of anthropomorphic stick figures. Over time, some culturally unique figures were developed, such as the Hawaiian triangular-muscled anthropomorph and the Marquesan anthropomorphic faces (and the birdman figures on Rapa Nui). Although the cultural context appears to be site specific in both island groups, as a whole, the images suggest a common underlying principle. While we will never be able to "read" each single image type, central to their complex belief system, as known from many different sources, were their strong concerns for ancestors, lineage, clan relation, mortuary practices, recognition, and succession.

Some island societies never engaged in petroglyph activities due to perhaps a lack of suitable surfaces; instead images were depicted on other media. Different artistic expressions, such as oratory, poetry, music, and dances, expressed similar or certainly related social matters as the petroglyphs were concerned with.

Problematic Issues and Common Assumptions

While the Hatiheu Valley study indicates that the images were of a late prehistoric–early historic period, this may not be the case elsewhere in the archipelago. Given that the Marquesan petroglyph repertoire is extensive and redundant, and that the same types are found in several islands and island groups, the image repertoire probably represents the total sum of the decorative system in the archipelago.

Several common assumptions regarding the type of designs, such as spirals, rectangular geometric motifs, and anthropomorphs, need to be addressed. Barrow (1956:313 in Davidson 1984:209) pointed to the presence of curvilinear elements, including spirals in the Marquesan art system. In addition, many anthropologists and art historians believe that the early Polynesian art was rectangular, and that rectilinear styles were introduced by the first settlers (Gathercole 1979:98; Hiroa 1949:326–327; McEwen 1966:409; Mead et al. 1975:206; Simmons 1977). This does not appear to be the case in the Marquesas, the Society Islands, or in the Austral Islands. While it is true that a large percentage of images represent abstract curvilinear figures, except for New Zealand, spirals are rare. If rectangular geometric motifs were introduced to East Polynesia with the first settlers, there is little evidence seen in the Marquesan petroglyph record. In addition, it is highly unlikely that the pervasive Marquesan art system would have changed so drastically (see Chapter 2).

It is assumed that human figures are the most numerous Marquesan design element. This may be solely due to the fact that human figures are more recognizable. Quantitative analysis of the 3,379 petroglyphs recorded in Hatiheu demonstrates that abstract geometric figures prevail (see Table 4). As a whole, 998 anthropomorphic figures account for 29.5 percent whereas 2,121 petroglyphs or 62.8 percent depict abstract geometric motifs. In the painted rockshelters of Eiaone Valley, Hiva Oa, anthropomorphs are represented by 20 figures (18.2 percent), while 50 (45.5 percent) are geometric figures (see Table 5). The situation in the west section, the research area of the Hatiheu Valley, Nuku Hiva, show a similar pattern. Of a total of 1,284 images, 691 or 53.8 percent are geometrics. A total of 478 figures or 37.2 percent are anthropomorphic figures (see Table 26). On Moorea the most common motifs are also geometrics.

According to Mead et al. (1975), a common Polynesian idea was to work in small design fields. Again this is not the case with the Marquesas petroglyphs. As more research on the Polynesian art systems is conducted, we recognize that a great deal of local variation existed. Several questions remain for future archaeologists and art historians to ponder. Were images temporally restricted in different areas, or were they made continuously throughout the pre-European period? Do the image's placement in the landscapes suggests a significance held in common across tribes and eras?

Almost four decades have passed since the late Roger C. Green wrote in his article, "Early Lapita Art from Polynesia and Island Melanesia: Continuities in Ceramic, Barkcloth, and Tattoo Decoration." Although Green (1979:3) was not specifically referring to rock art, his statement raises important issues: "In recent literature on Oceanic prehistory, there are few archaeologists who refer to investigation of its art or write much about the implications that such analysis may have for the culture history of the area" (Green 1979:3).

In my view, since Green penned the above words, the situation has not significantly changed in regard to Oceanic archaeological art. Hopefully in the future archaeologists will develop models and theories in Oceanic archaeological art and that these theories will form part of discussions on cultural process and change, because to use Green's words "a potential useful line of evidence is ignored" (Green 1979:3).

Let us not forget that the creation of the images on rocks was only a small, but durable, part of a total art system. Other forms of what we may refer to as art including, for example, chanting, dancing, drumming, storytelling, and especially tattooing, were all interconnected with the creation of mental and physical images. Thus, what we call rock art cannot be studied in isolation but must be placed in a wider cultural context that embraces all aspects of the society.

Research in the Marquesas Islands, specifically in the Hatiheu Valley, illustrates that the settlement pattern survey, as a method of archaeological research, is particularly well suited to the study of rock art. Settlement pattern research can incorporate several aspects of archaeology. This venture is significant in that it is the first extensive research project in French Polynesia that focuses explicitly on rock art and helps promote public awareness of the pre-European period. Based on the spatial distribution of petroglyphs, pictographs, tiki, megalithic domestic units, public and sacred architecture it was established that the research area was divided into the chiefly or elite settlement with rock art, complex megalithic domestic units, public complexes, and sacred structures and those of the commoners with small-house domestic units, agricultural terraces and only three boulders with petroglyphs. The research will establish a solid foundation from which to begin productive discussion on Marquesan archaeological art and architecture. The Marquesan data can ultimately be used in a pan-Pacific comparative study on archaeological art and the symbolic value of space in relation to architecture. It further provides an inventory so that we can monitor changes in the preservation practices and recommend cultural conservation programs. This research project represents a significant contribution and useful tool for future lines of inquiry in Pacific archaeology. Finally, it adds to the cultural patrimony of the Marquesas Islands and provides knowledge for future generations.

However, the Hatiheu model needs to be tested in other valleys with similar environments, including Haatuatua or Aakapa (Nuku Hiva), or perhaps Puamau (Hiva Oa), in the southern group. In addition, more rock art sites should be excavated to get a better understanding of the spatial and temporal context and changes over time.

Future rock art research could include archaeoacoustic techniques. Waller (2011) has since 1987 measured acoustics and reverberation at over 300 rock art sites in France, Australia, and the United States and found echoes and/or reverberations at almost all of them (sites.google.com/site/rockartacoustics/). For instance, Meae Iipona, Puamau Valley, Hiva Oa, a large and complex ritual site that has several boulders and stones with rock images in addition to 18 stone sculptures (5 stone statues, 10 heads and 3 sculpture fragments) has exceptional acoustic (Figure 115). This was once pointed out to me by a French musician/conductor.[52] It is quite likely that Meae Iipona, as well as other complex archaeological sites in the Marquesas Islands, may have been chosen just because of this phenomenon.

APPENDIX A

Figures

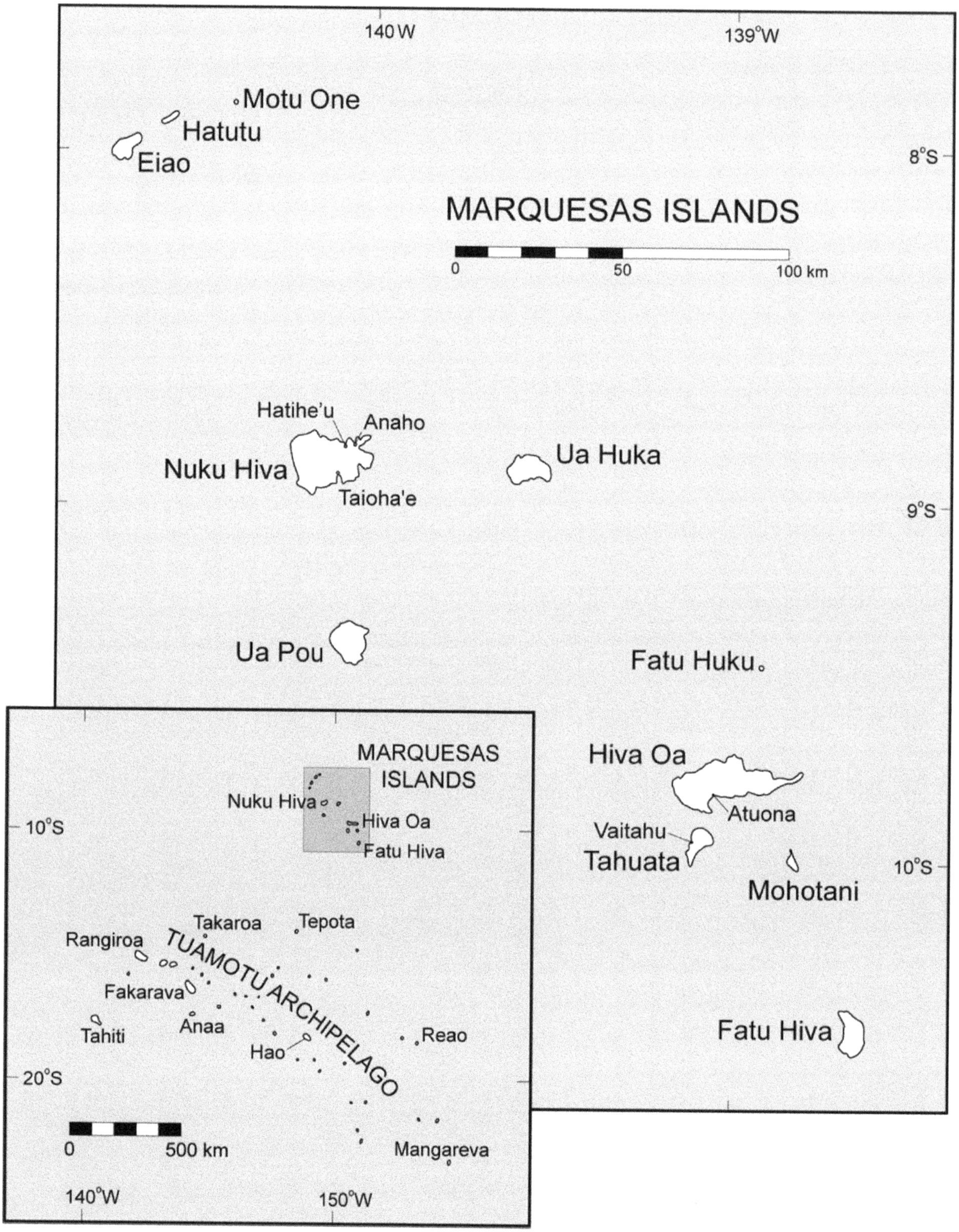

Figure 1. The Marquesas Archipelago, French Polynesia (map courtesy Melinda Allen, 2016).

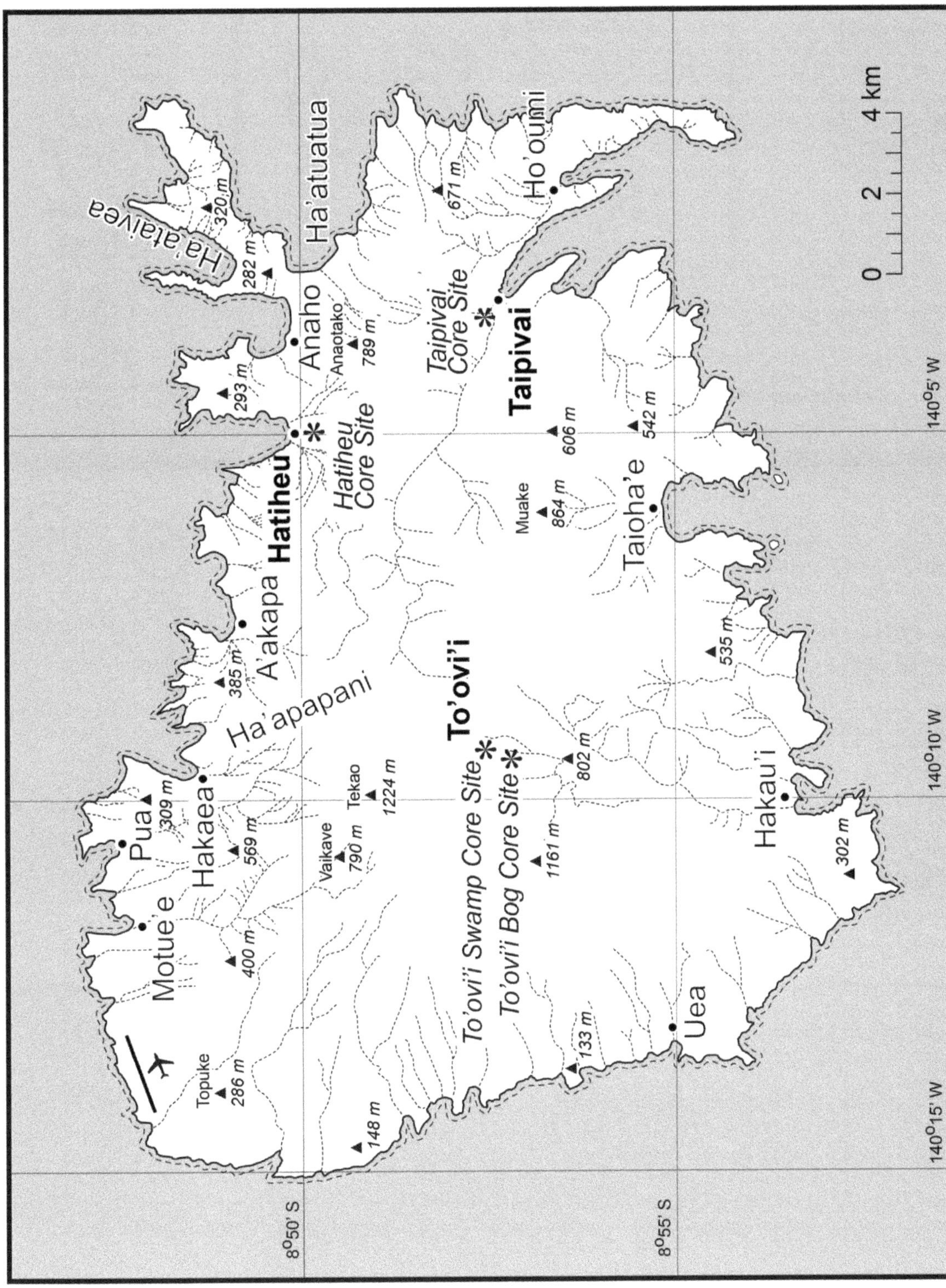

Figure 2. Nuku Hiva (map courtesy of Melinda Allen, 2016).

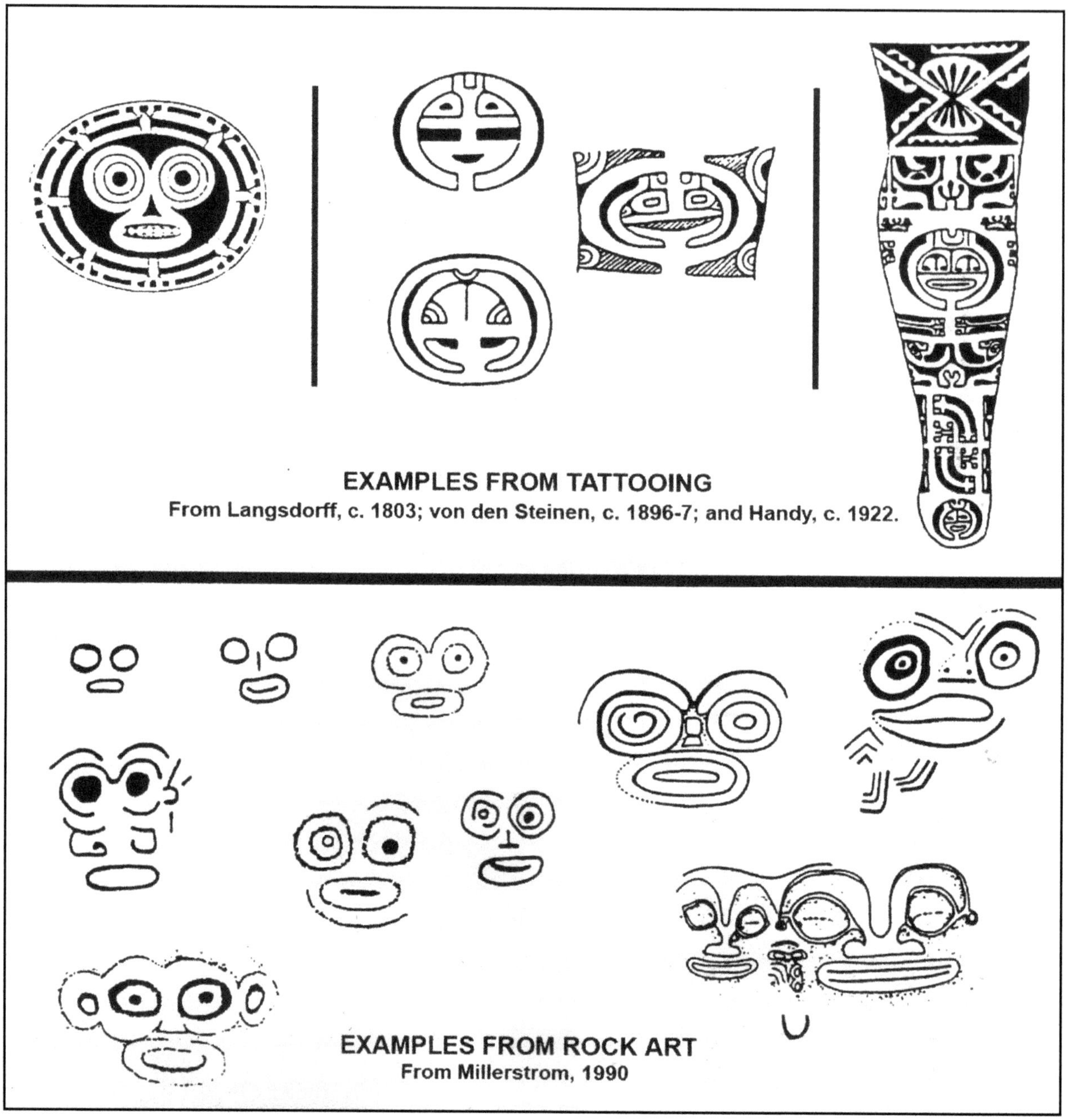

Figure 3. Examples of face motifs seen in tattoos (above) and petroglyphs (below) (Millerstrom and Allen 2006).

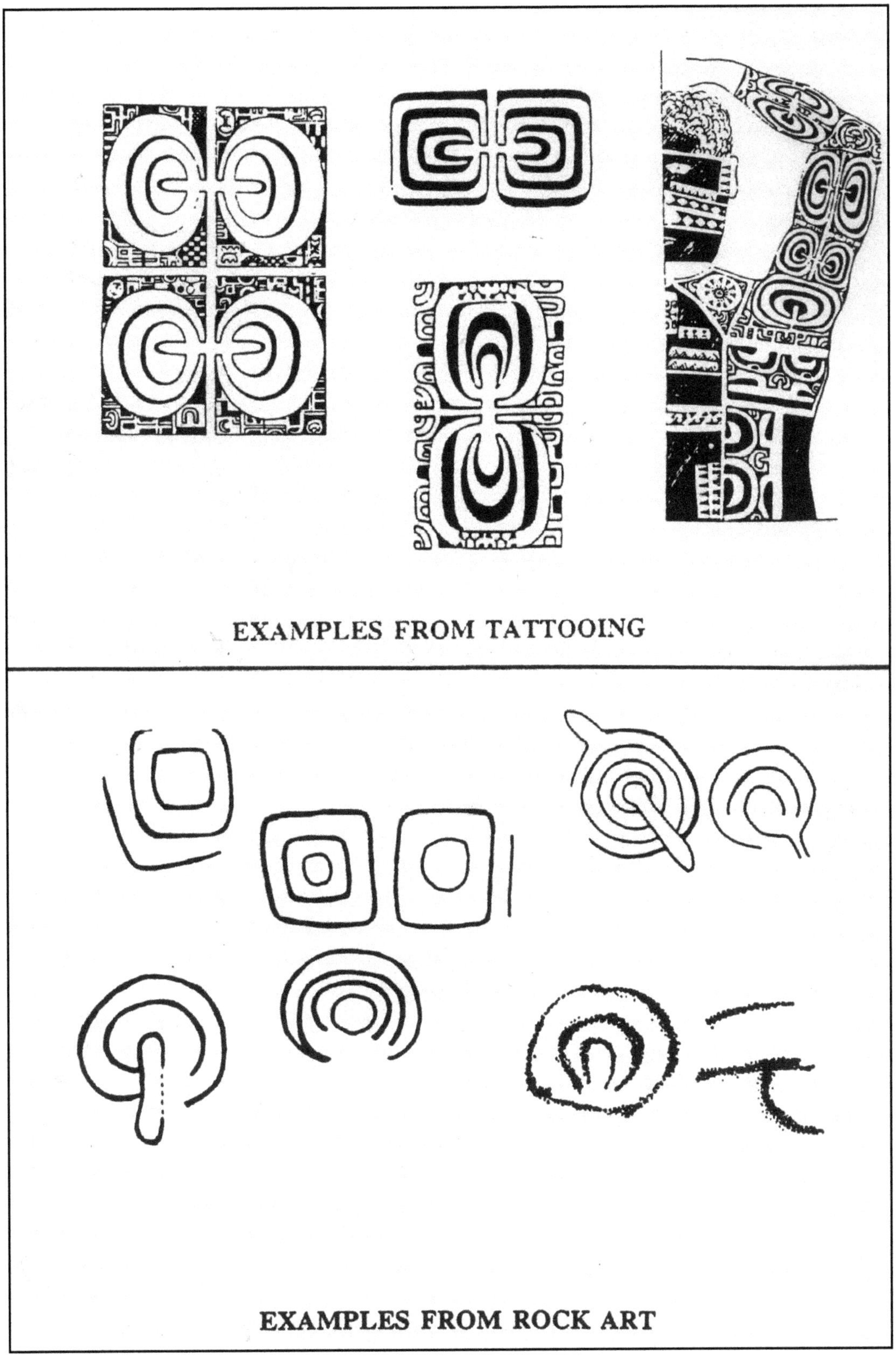

Figure 4. Examples of geometric figures seen in tattoo motifs (above) and petroglyphs (below) (Millerstrom and Allen 2006).

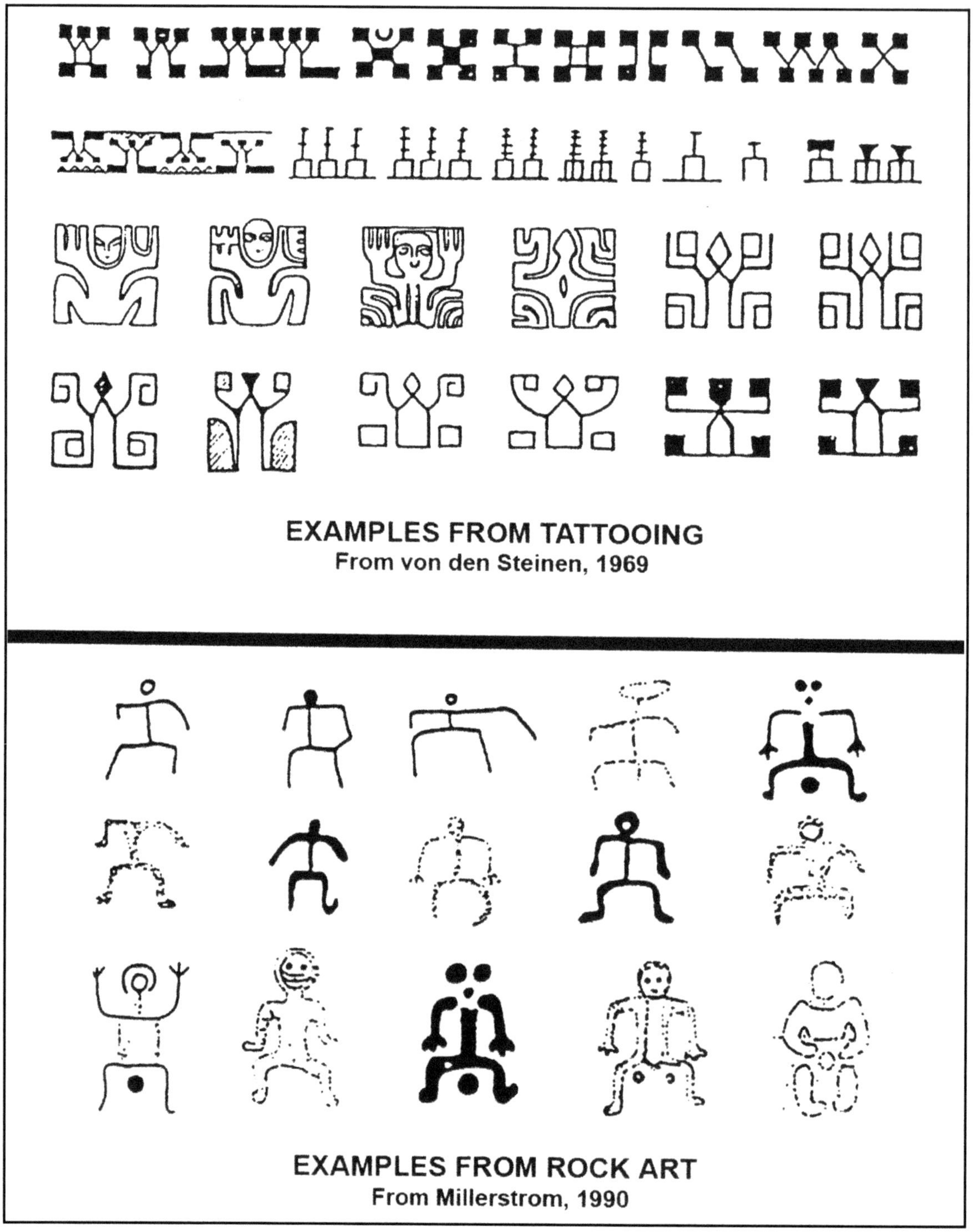

Figure 5. Examples of anthropomorphic tattoo motifs (above) and petroglyphs (below) (Millerstrom and Allen 2006).

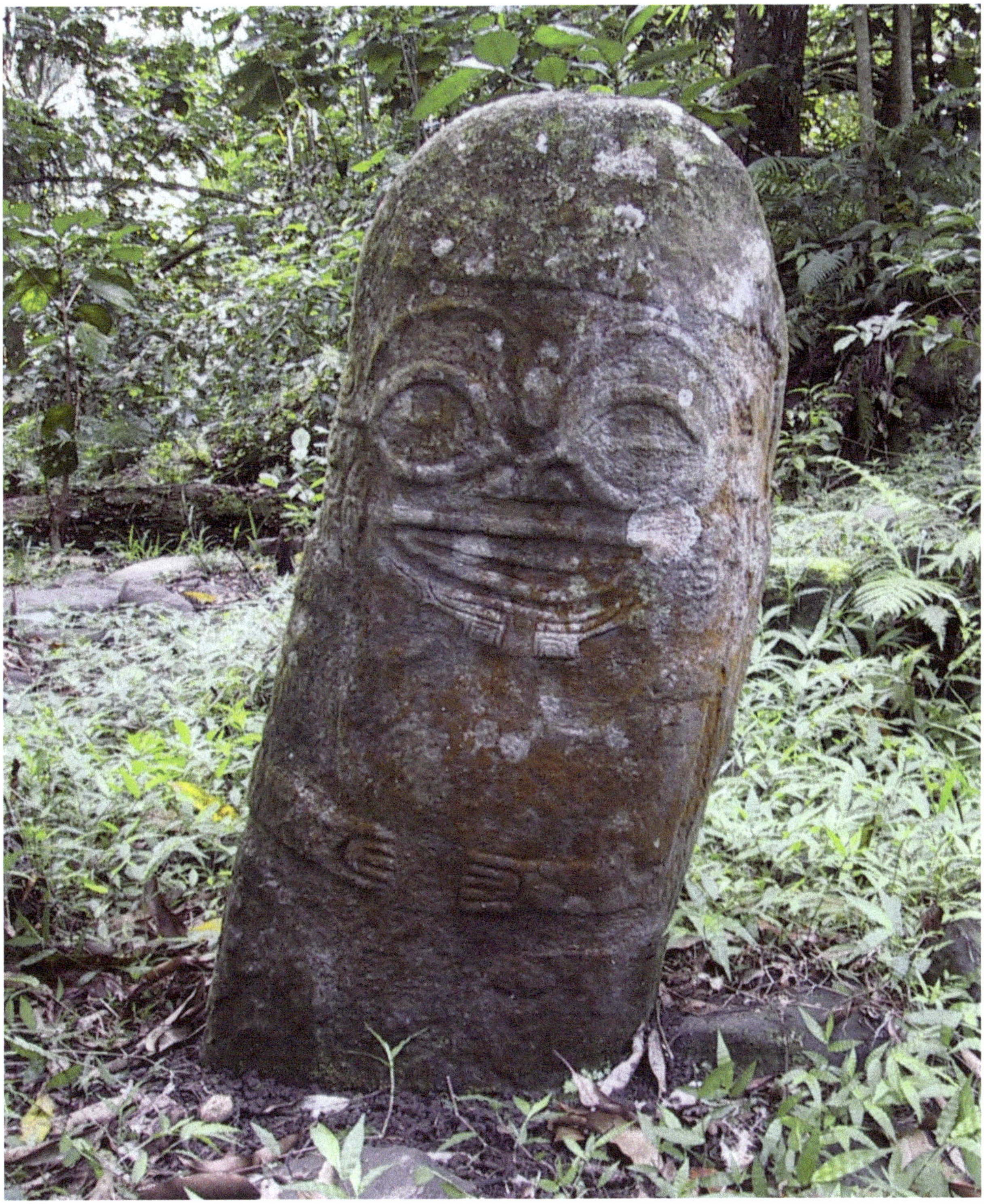

Figure 6. Meae Utukua, Punaei Valley, Hiva Oa (photo courtesy of C. Chavaillon and E. Oliver).

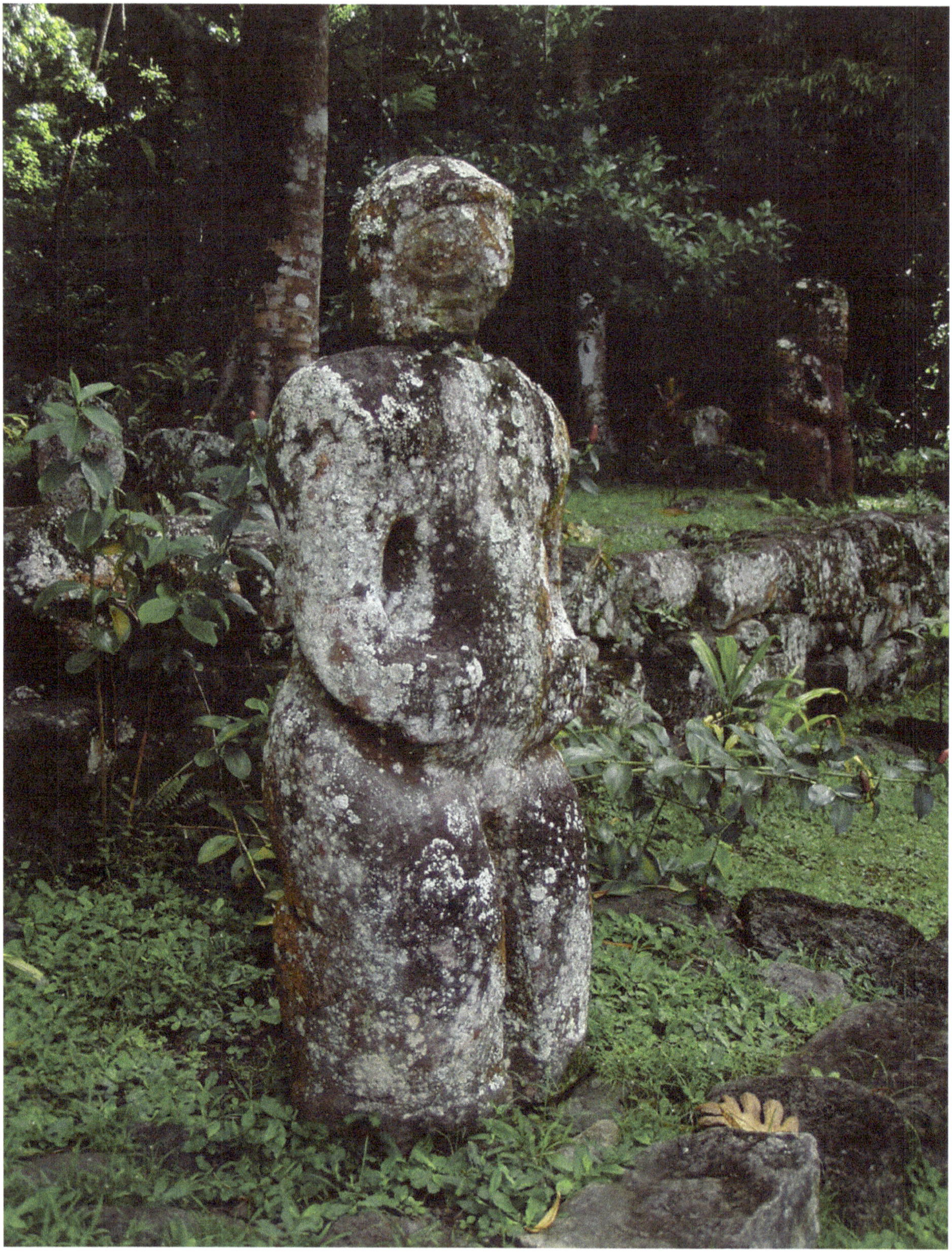

Figure 7. Meae Iipona, Puamau Valley, Hiva Oa.

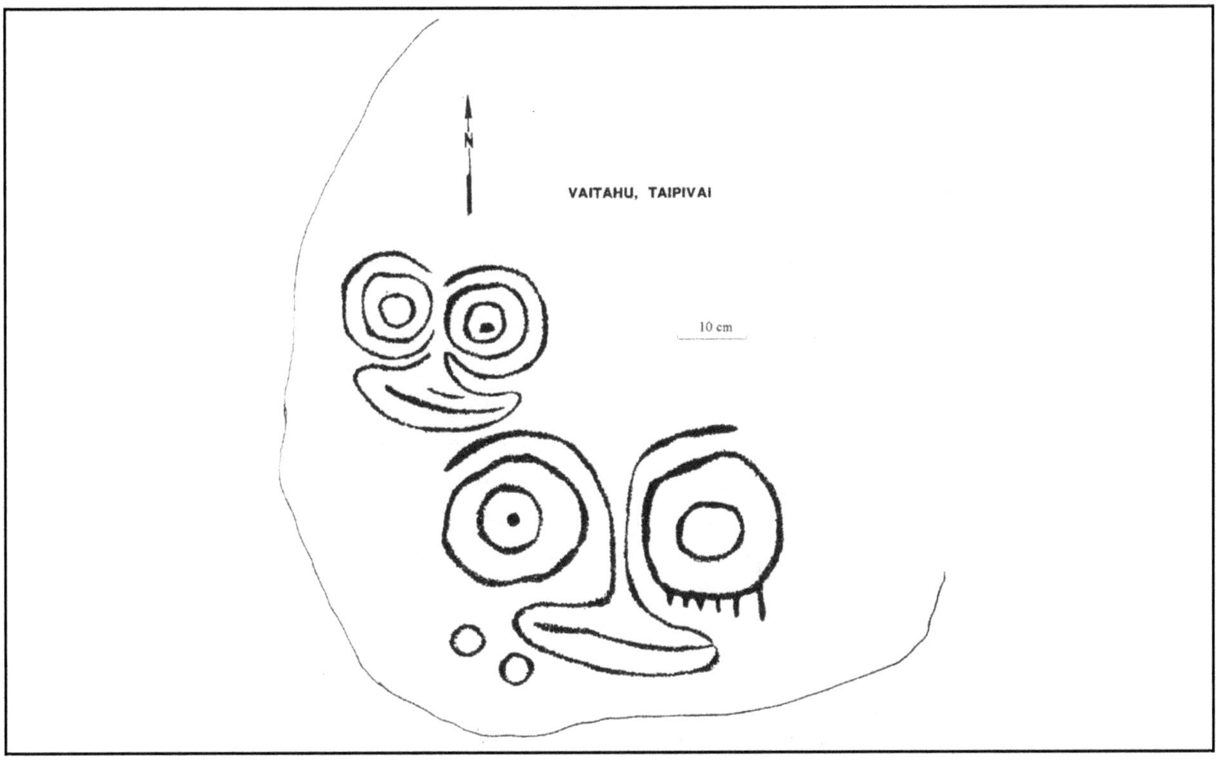

Figure 8. Vaitahu, Taipivai, Nuku Hiva. Two face petroglyphs placed on a boulder (333vai 1).

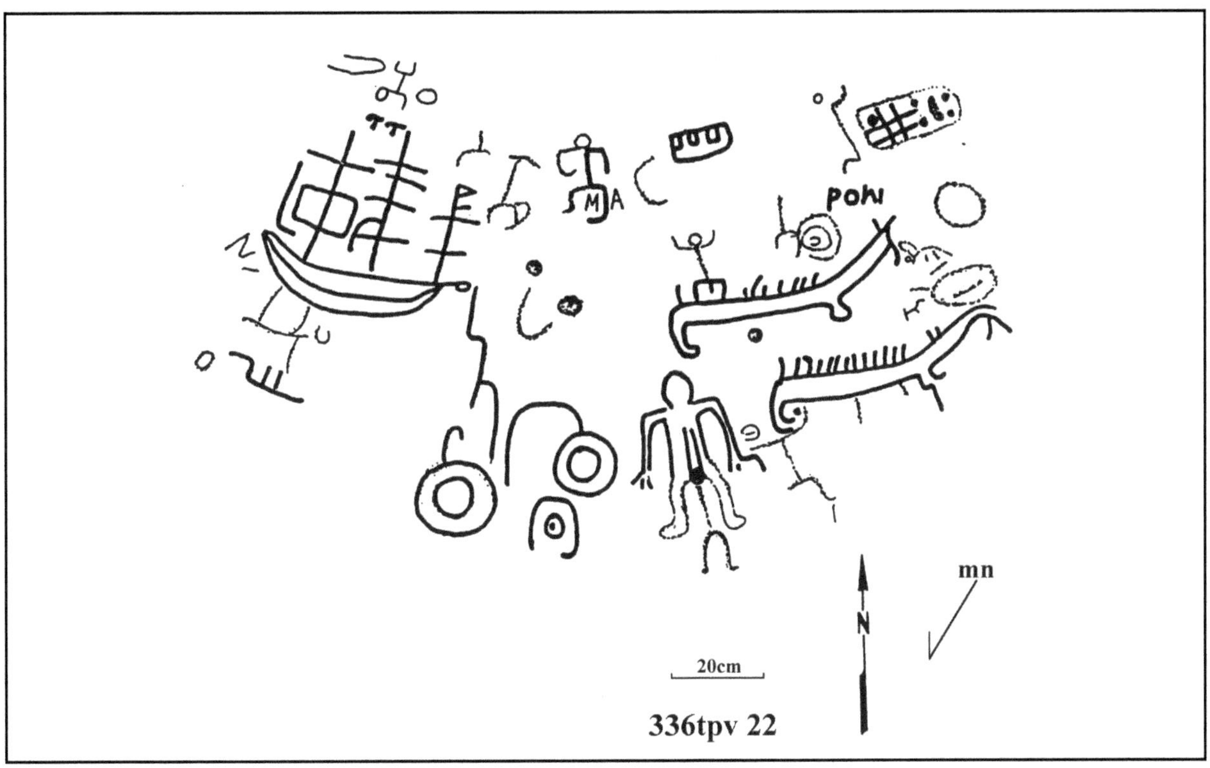

Figure 9. Vaikihi, Nuku Hiva. Incised historic sailing ship, two horses (poni), lettering, and traditional pecked figures on a megalithic boulder (331hth 22).

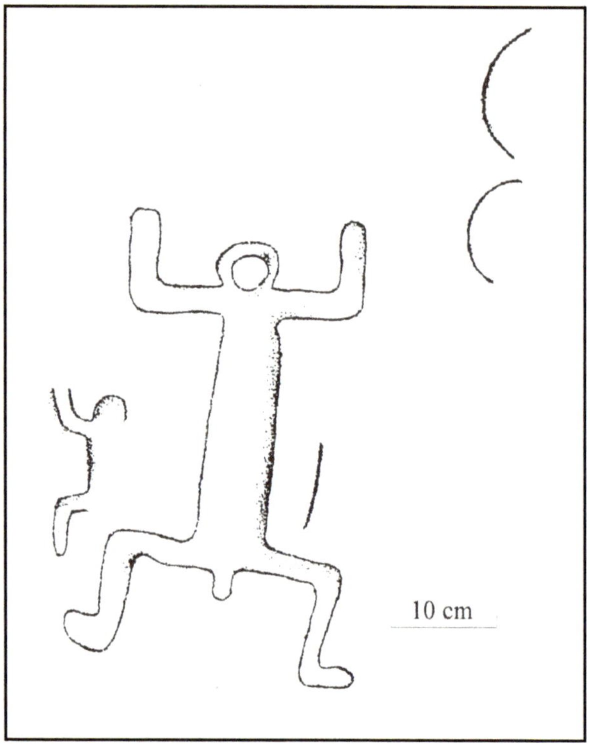

Figure 10. Anaho Bay, Nuku Hiva. Petroglyphs pecked on a boulder next to the sea (333vai 1).

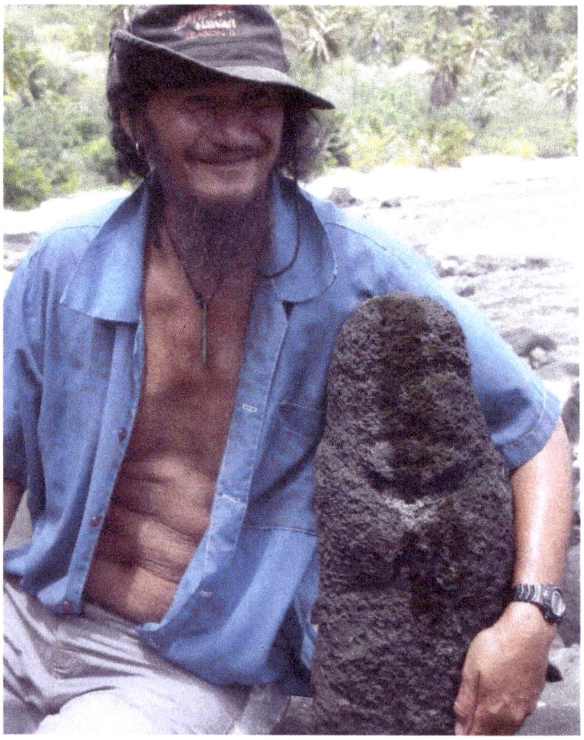

Figure 11. Anaho, Nuku Hiva. Tioka Puhetini with a tiki discovered on Meae Atatai, a fishermen's shrine.

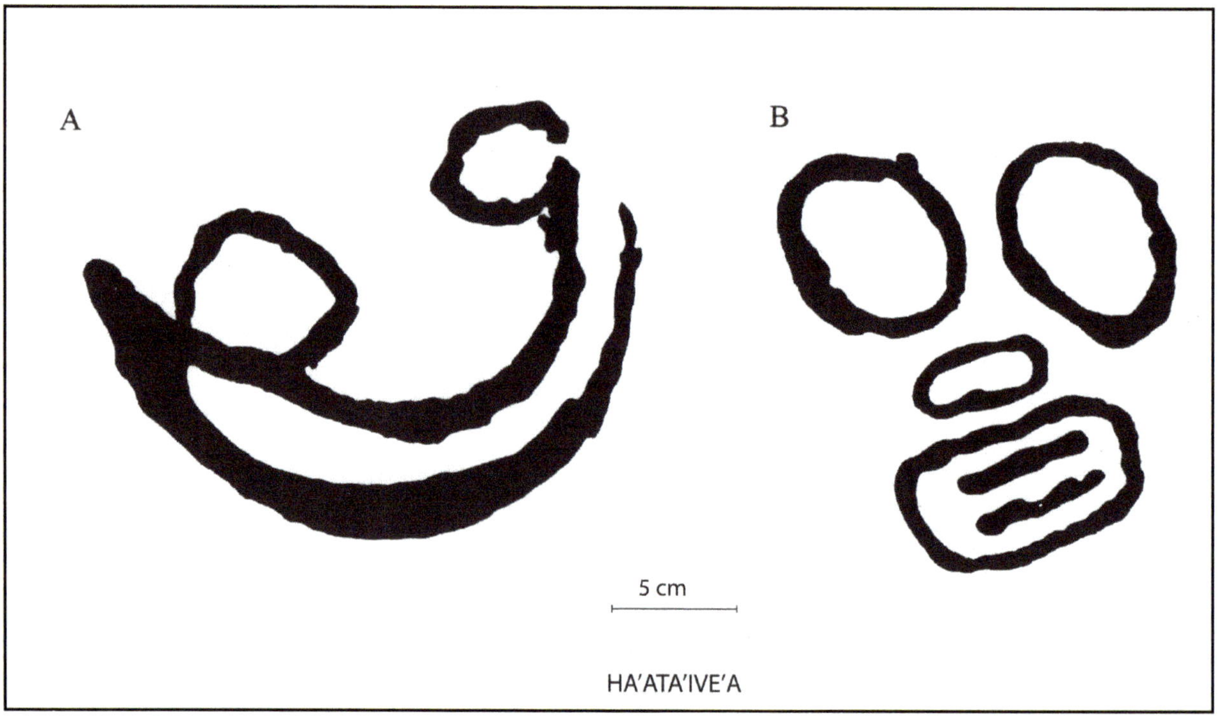

Figure 12. Haataivea, Nuku Hiva. Two petroglyph faces placed on the floor of a rock shelter (331haa 1).

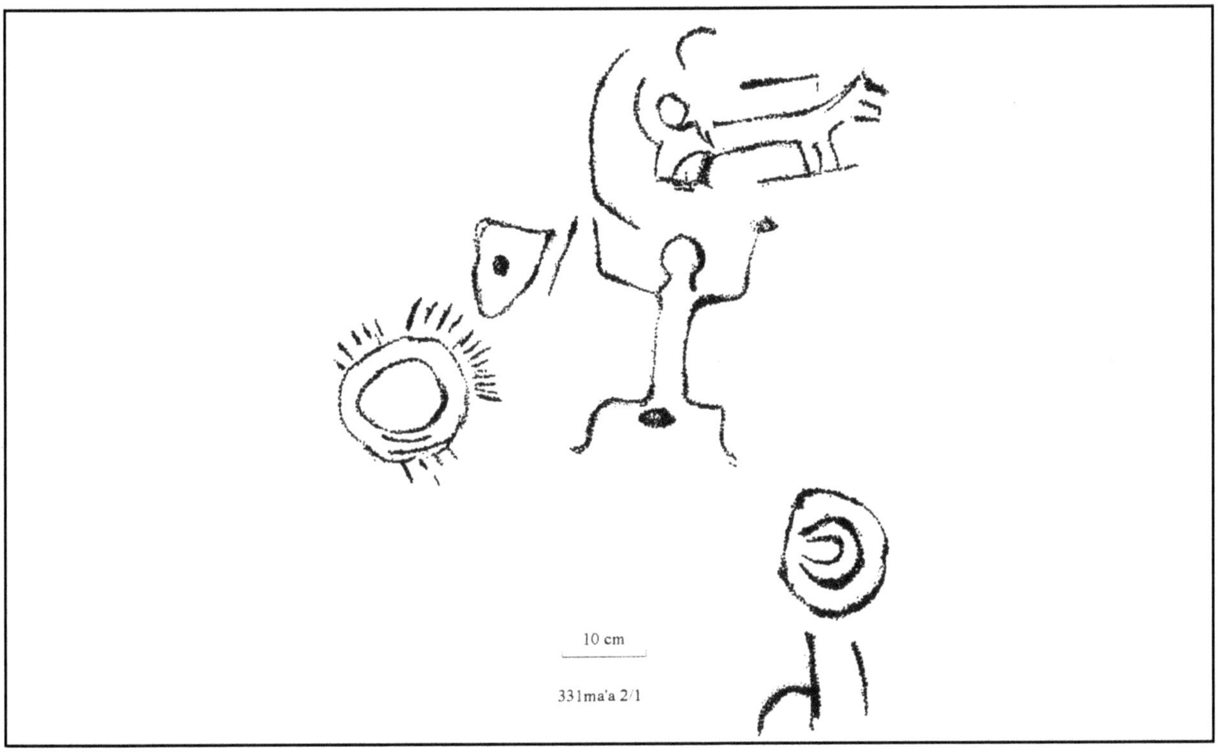

Figure 13. Maatea, Nuku Hiva. Images pecked on a sandstone slab (331maa 2/1).

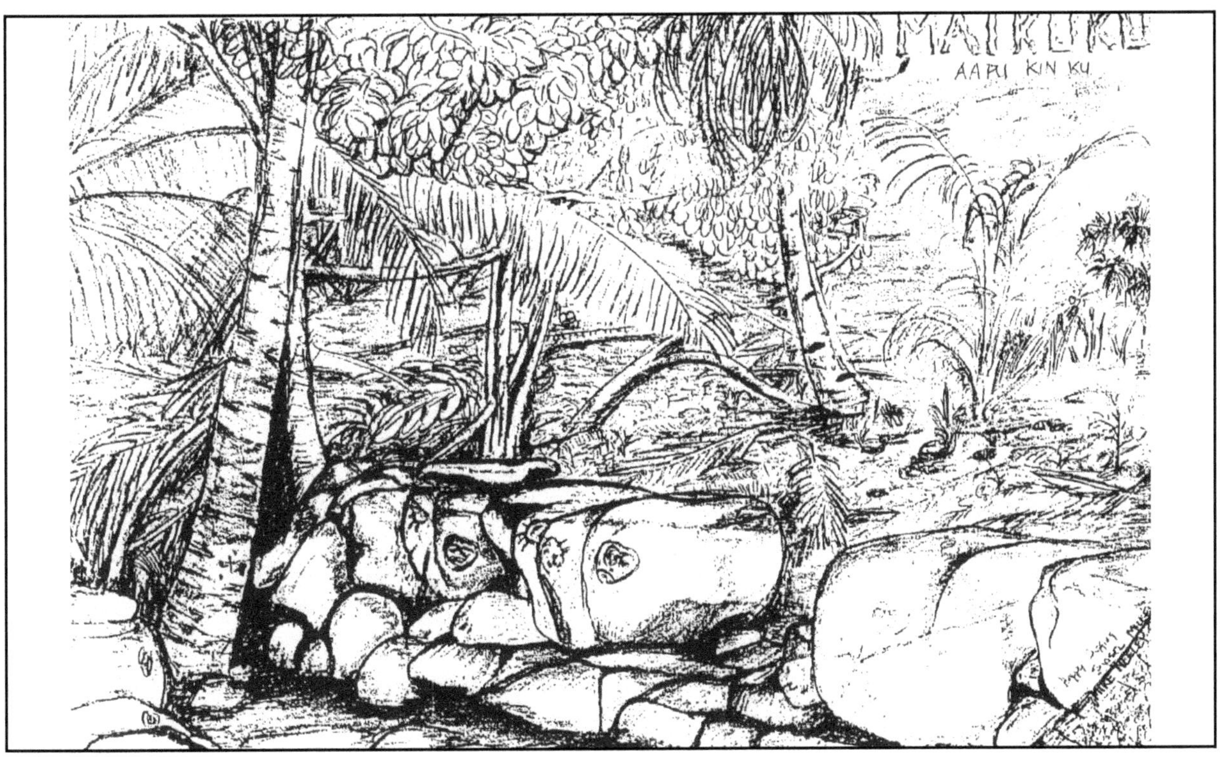

Figure 14. Tohua Maikuku. A narrow alley with images placed on both sides of facing walls. Drawing by Mike Neeley (July, 1993).

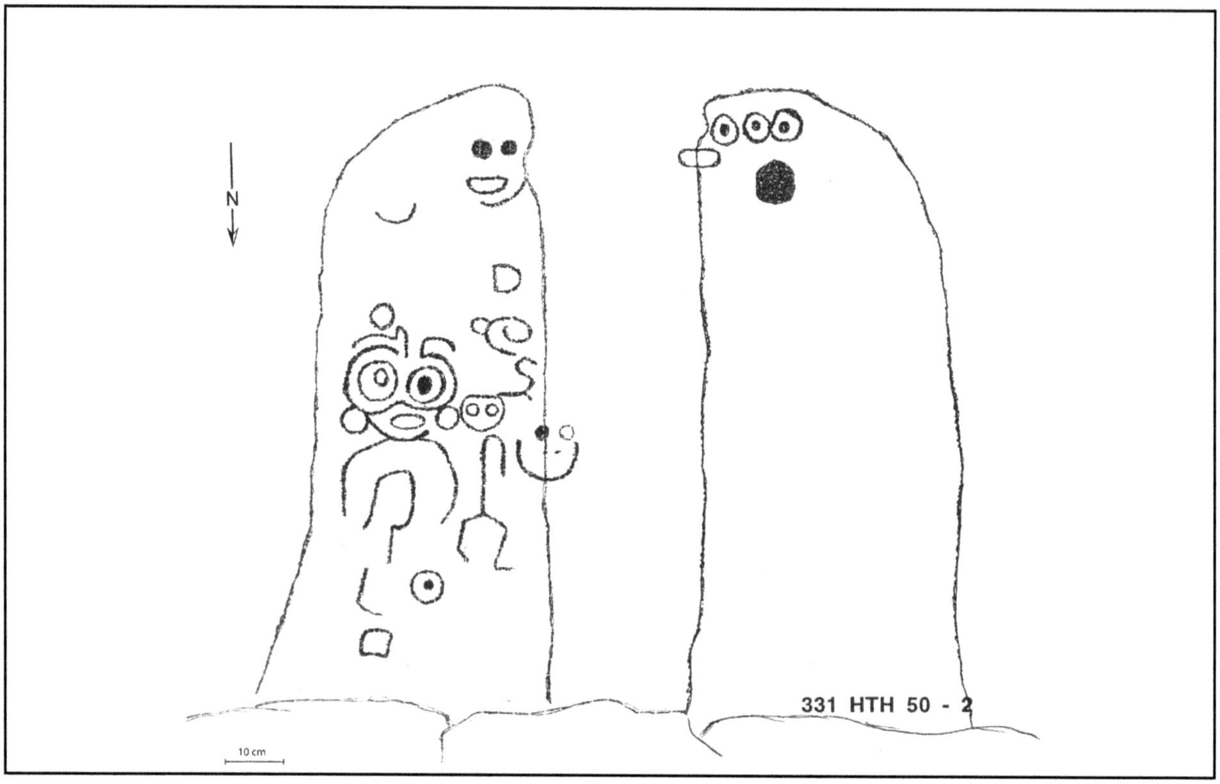

Figure 15. Tohua Maikuku, Nuku Hiva. An upright boulder placed on a small pavement (331hth 50).

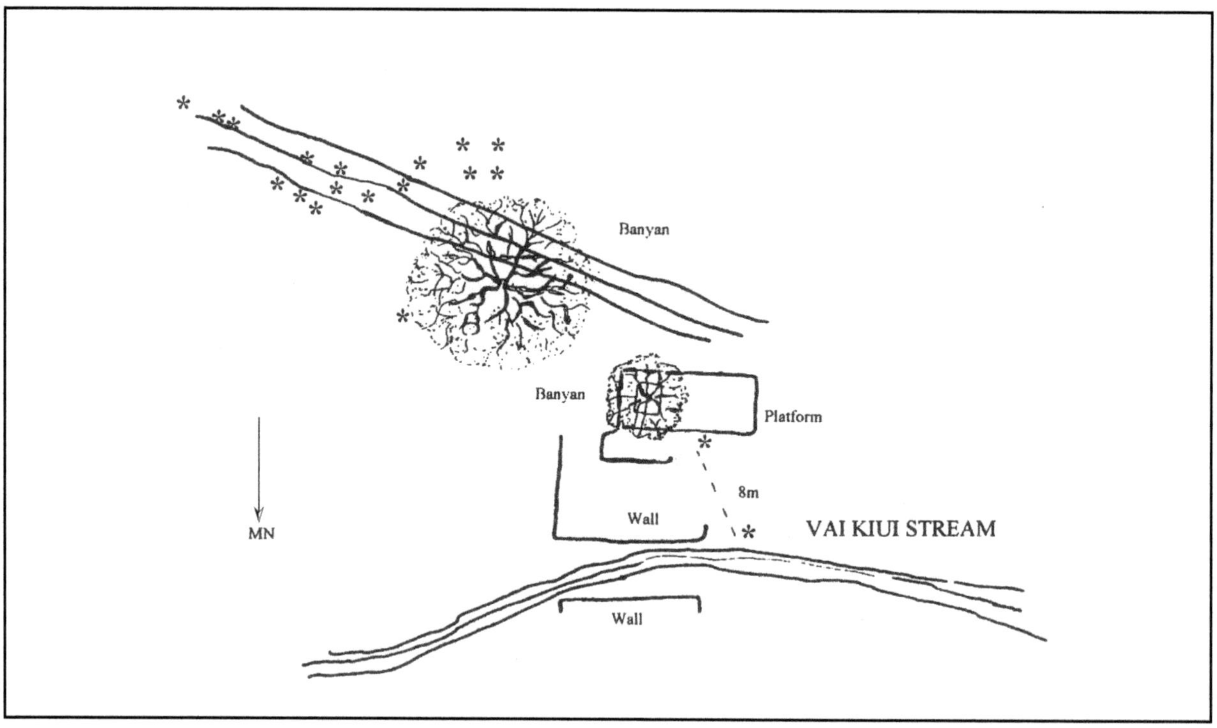

Figure 16. Meae Vaikivi, Ua Huka. The asterisks indicate panels with petroglyphs (map by E. Edwards, 1985).

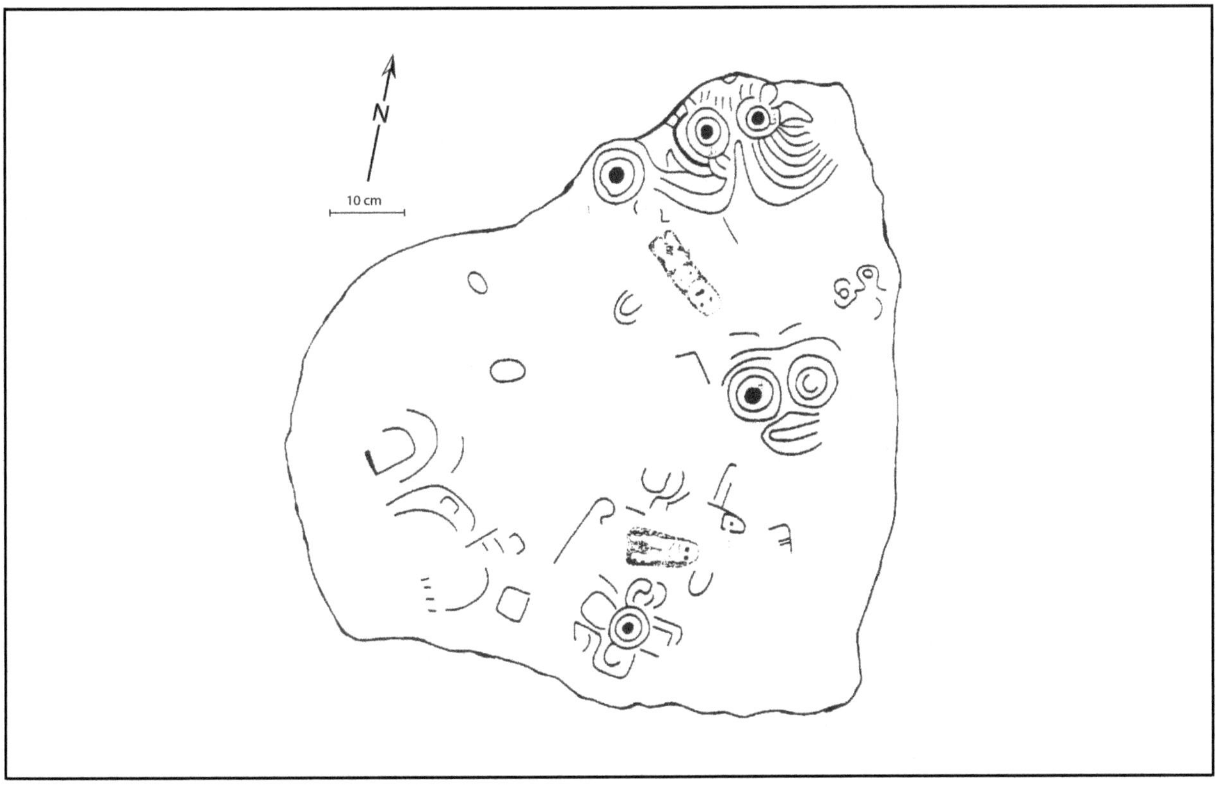

Figure 17. Meae Vaikivi, Ua Huka. Petroglyph boulder located next to the river (35ovai 1).

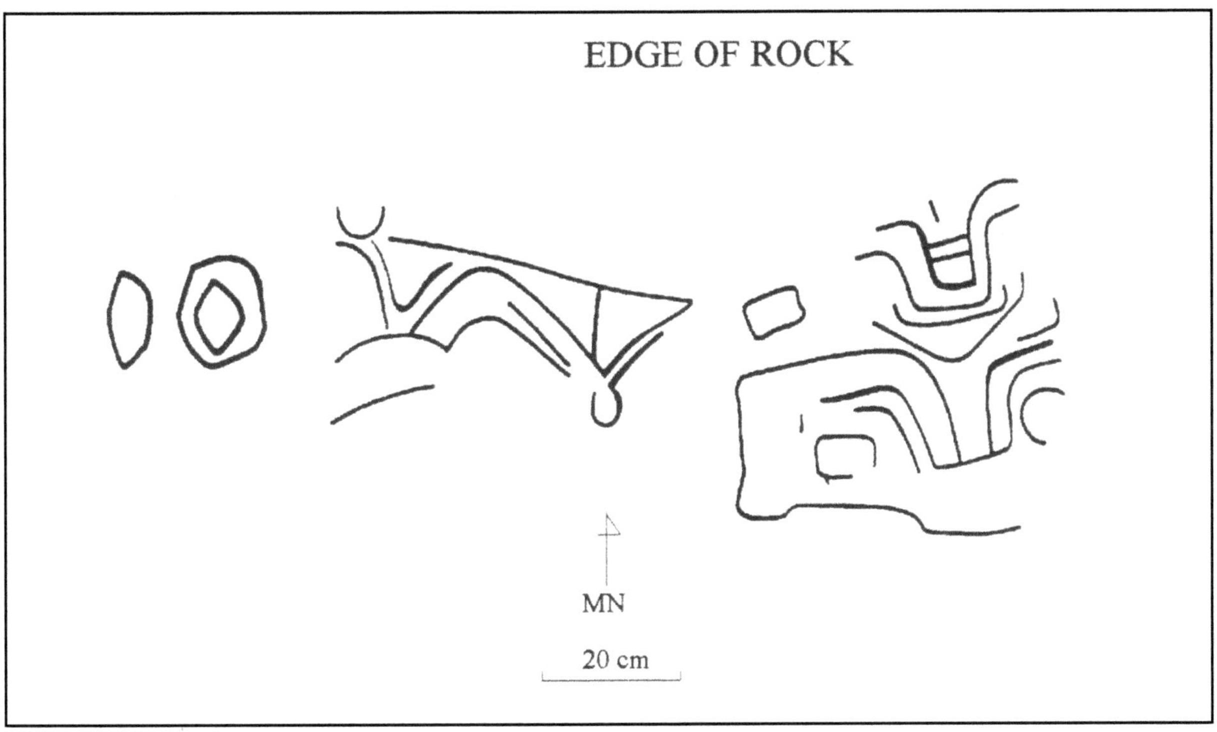

Figure 18. Vaikivi, Ua Huka. Incised figures on clay walls of a rectangular pit (35ovia 4).

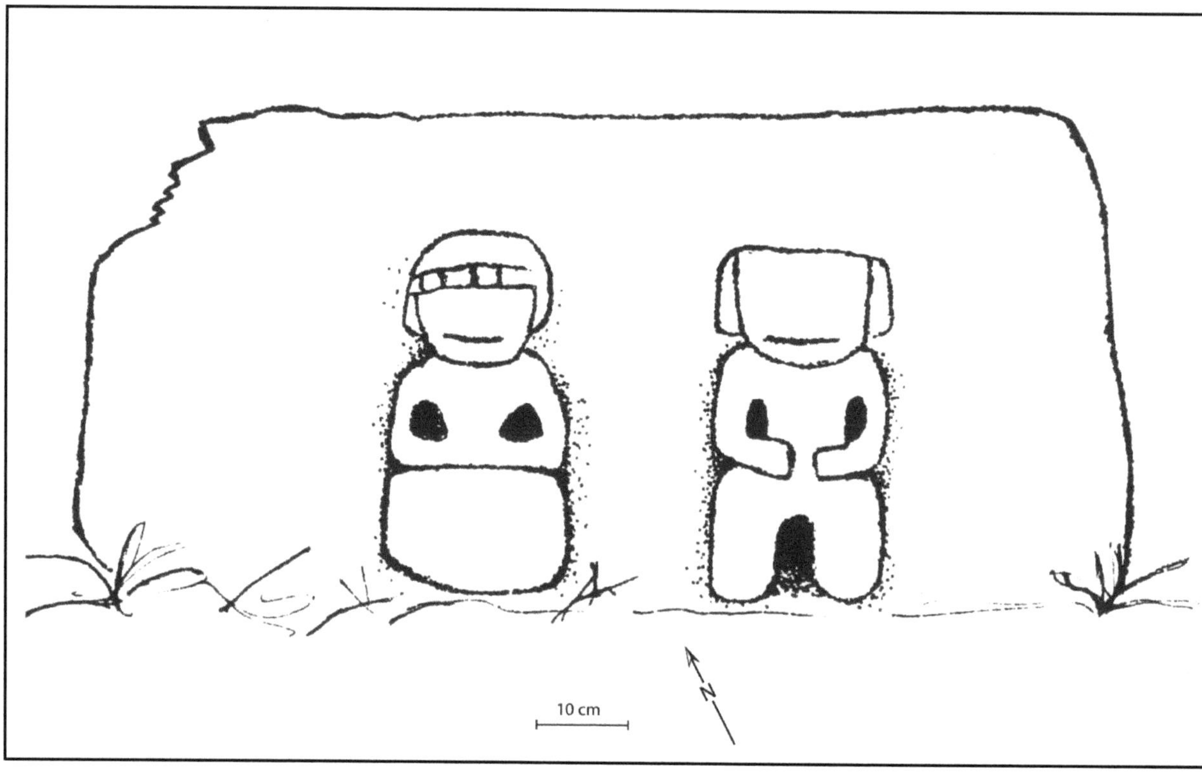

Figure 19. Hakatao, Ua Pou. Worked slab with figures in bas-relief (361hak 1).

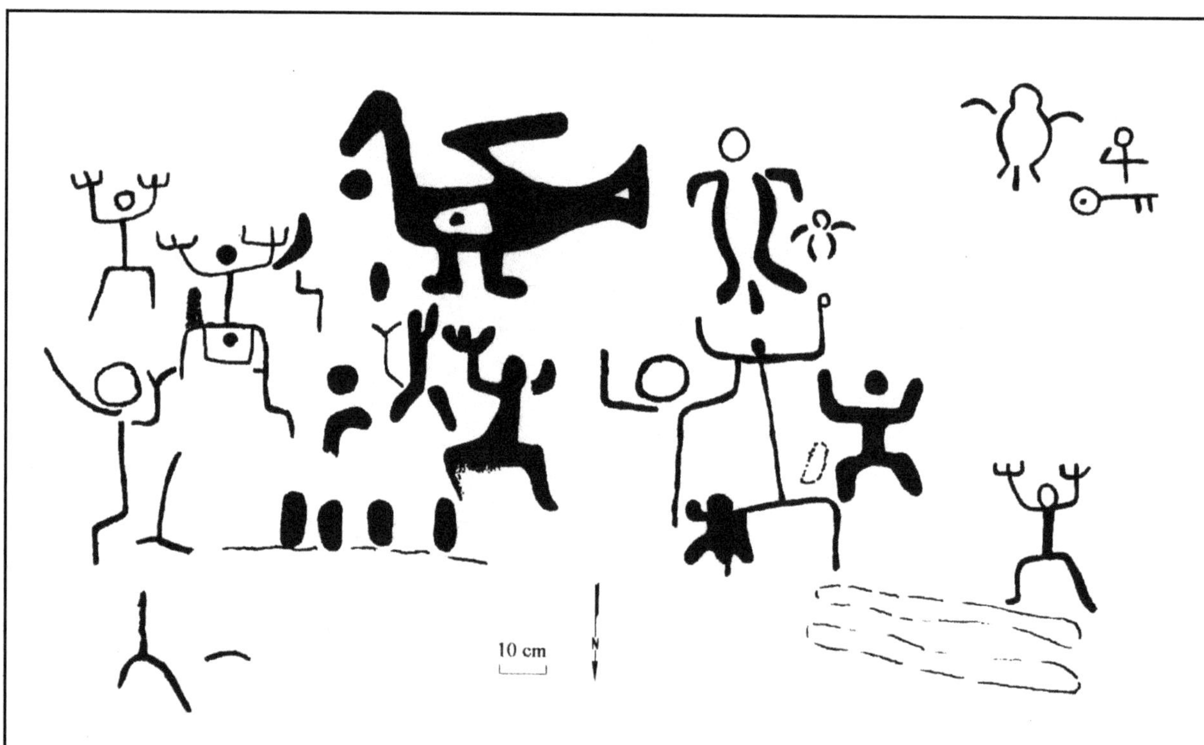

Figure 20. Hakahau, Ua Pou. Incised figures placed on the wall of a rectangular pit (361hak 1).

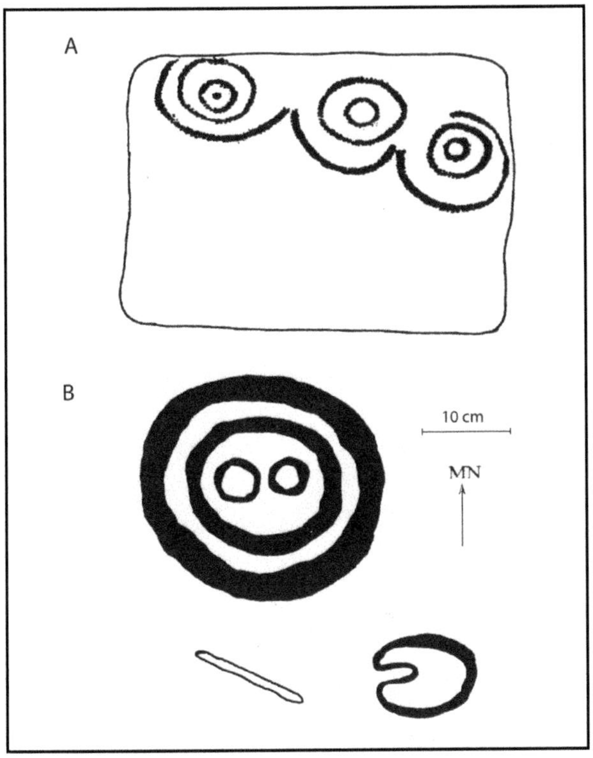

Figure 21 a–b. Hiva Oa; A) Taa Oa Valley (321taa 2); B) Puamau Valley (322pua 1).

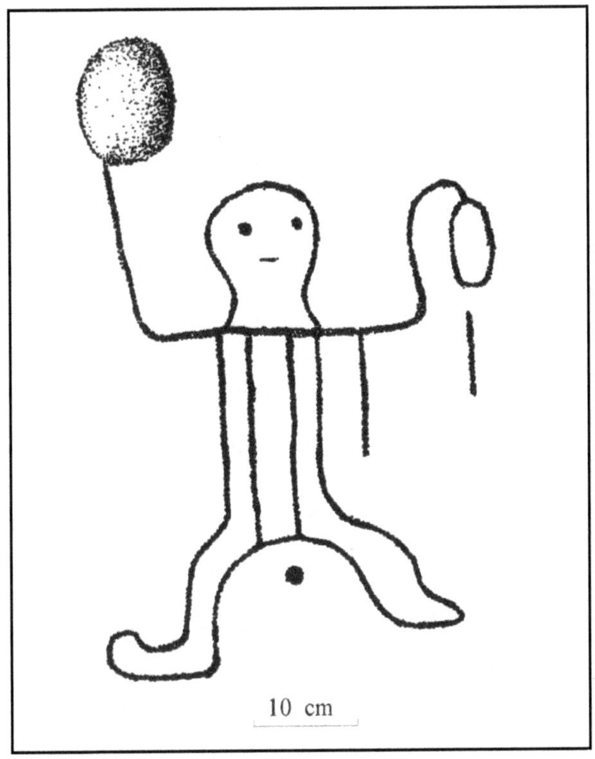

Figure 22. Hiva Oa. Female figure facing up on a flat slab. Located on a ridge between Puamau and Eiaone valleys (322vea 5).

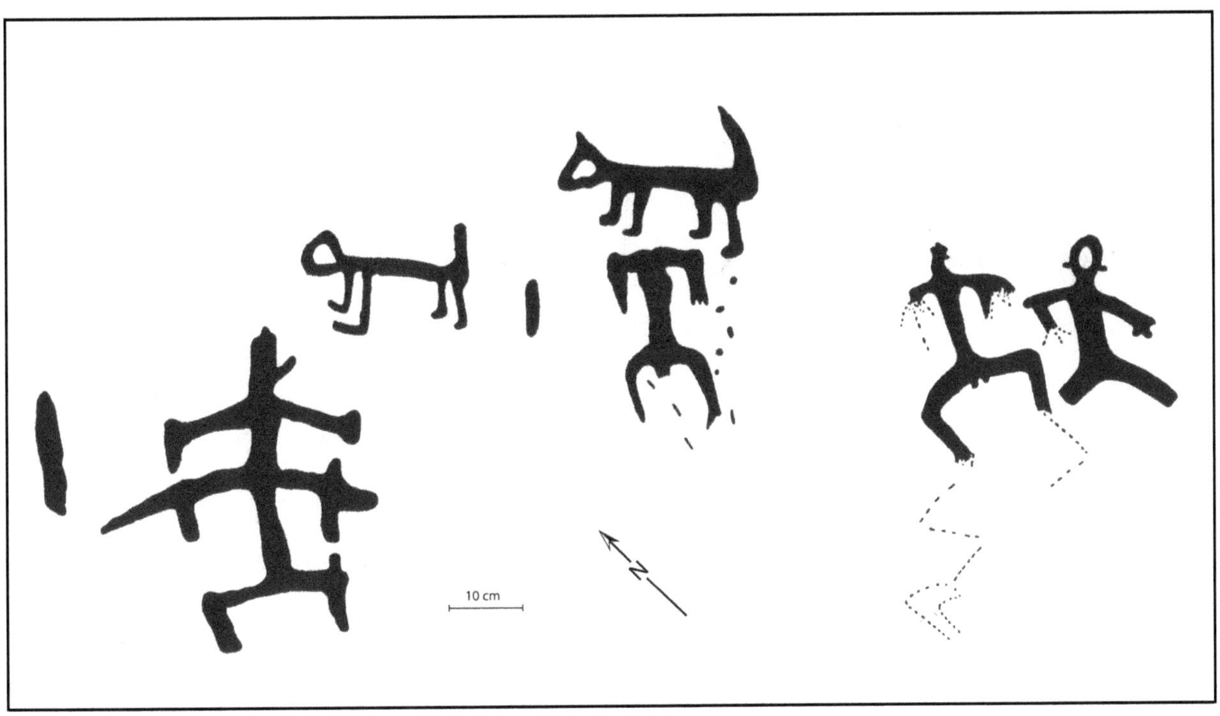

Figure 23. Eiaone Valley, Hiva Oa. Part of a pictograph panel (322eia 77/7).

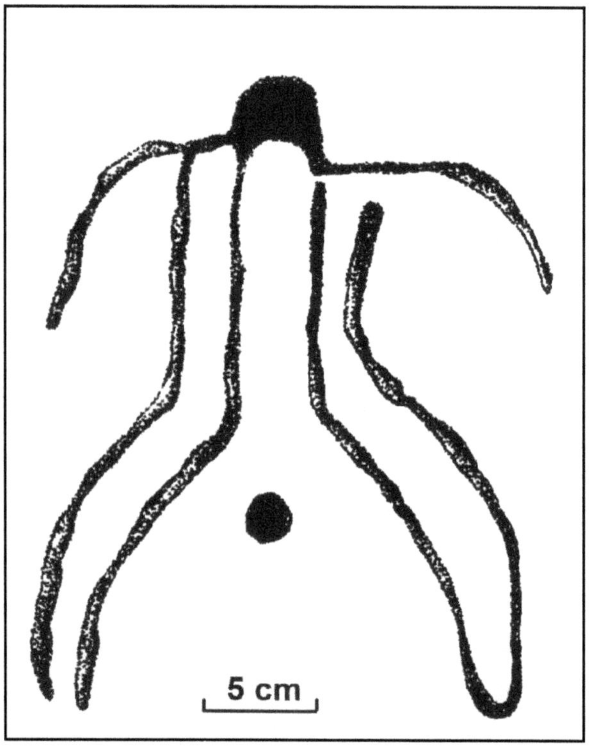

Figure 24. Omoa, Fatu Hiva. Female figure similar to a figure in Hatiheu Valley (310omo 1).

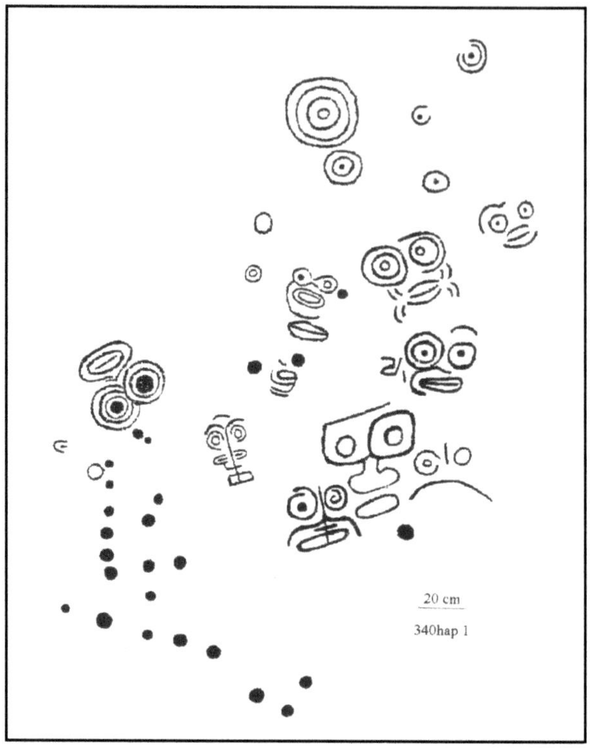

Figure 25. Hapatoni Beach, Tahuata. A megalithic boulder with anthropomorphic faces and vertically placed cupules (340hap 1).

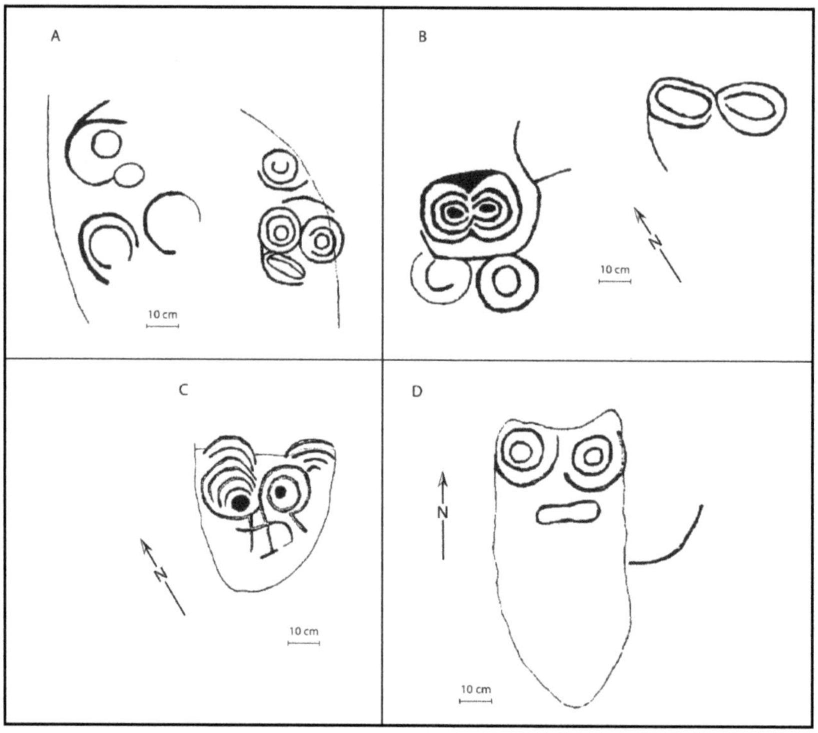

Figure 26a–d. Vaitahu and Hanamiai valleys, Tahuata. Modified boulders with anthropomorphic figures (340vai 1 and 2; 340han 1 and 3).

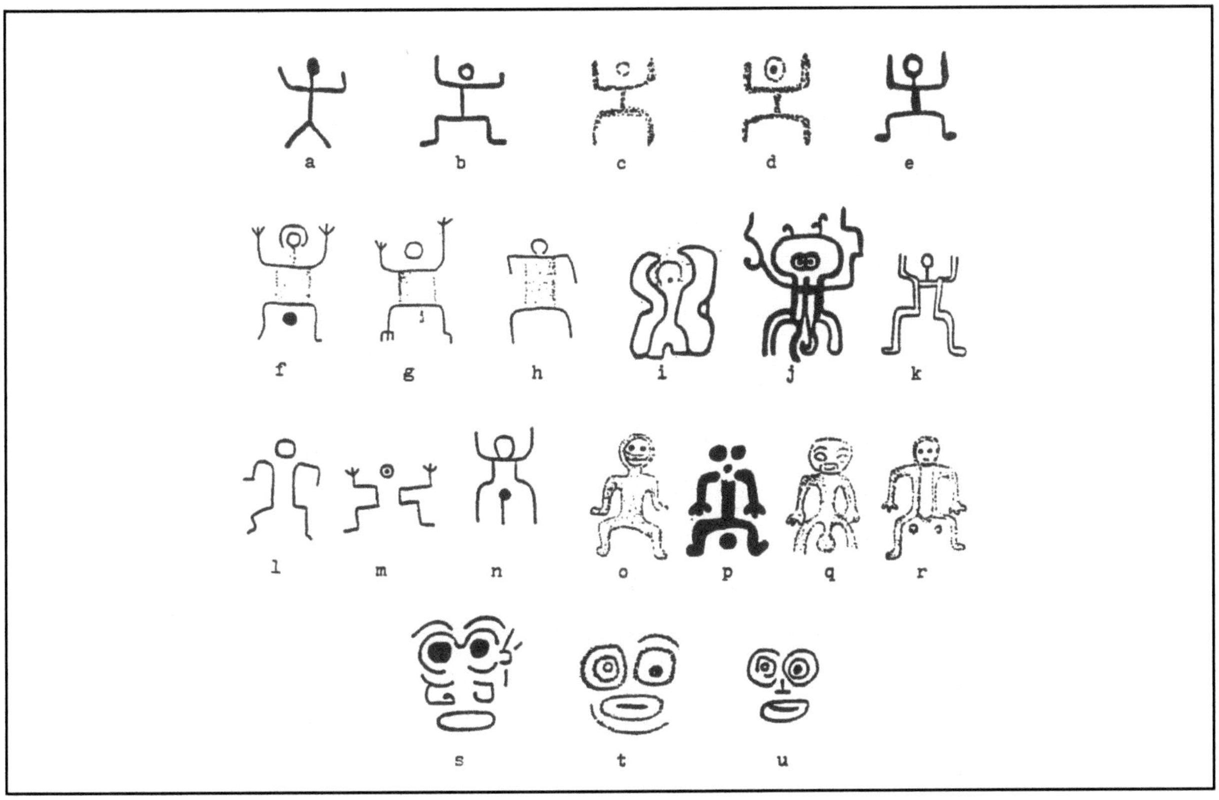

Figure 27. Various types of anthropomorphic figures and faces.

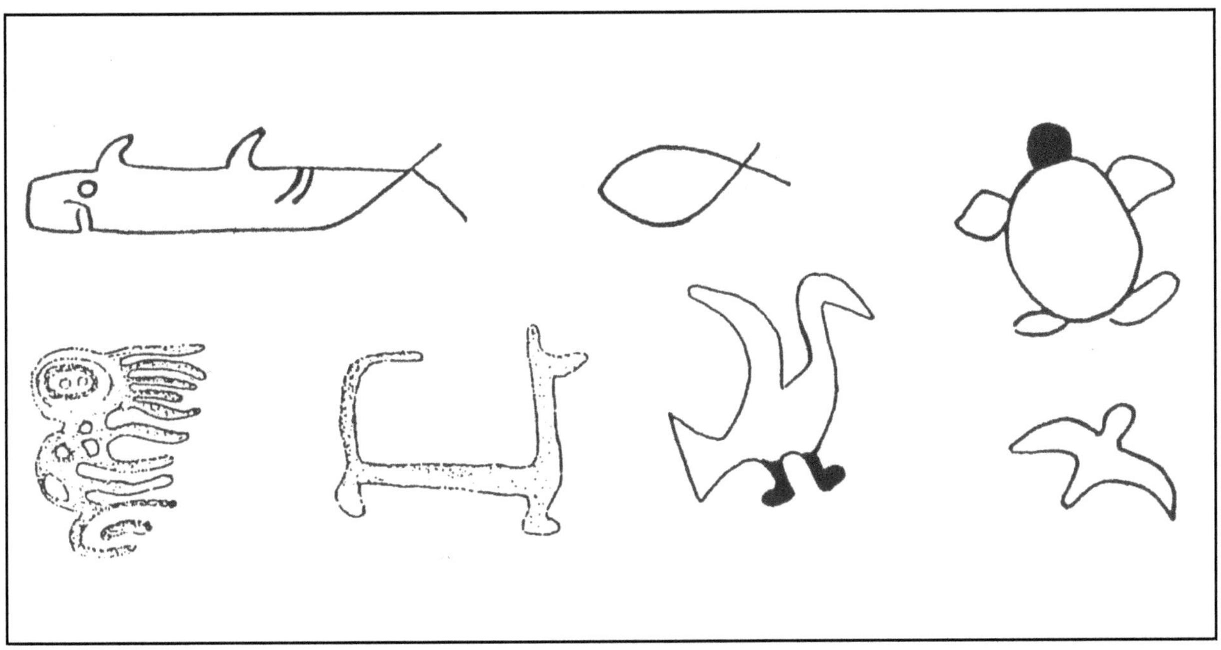

Figure 28. Various zoomorphic figures.

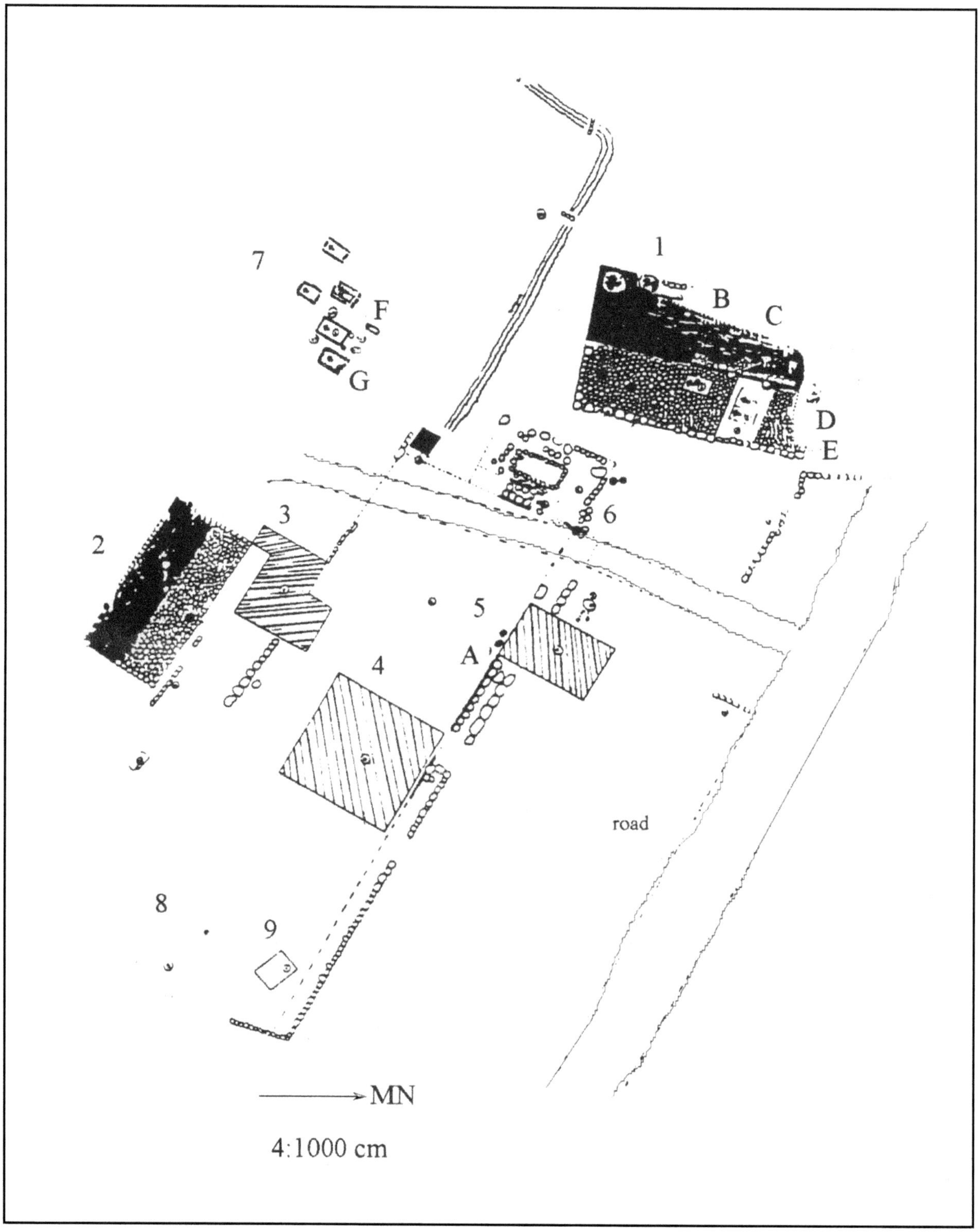

Figure 29. Tohua Pehe Kua. Map drawn by E. Edwards in 1985. The numbers refer to the structures; the capital letters refer to sculptures and petroglyphs.

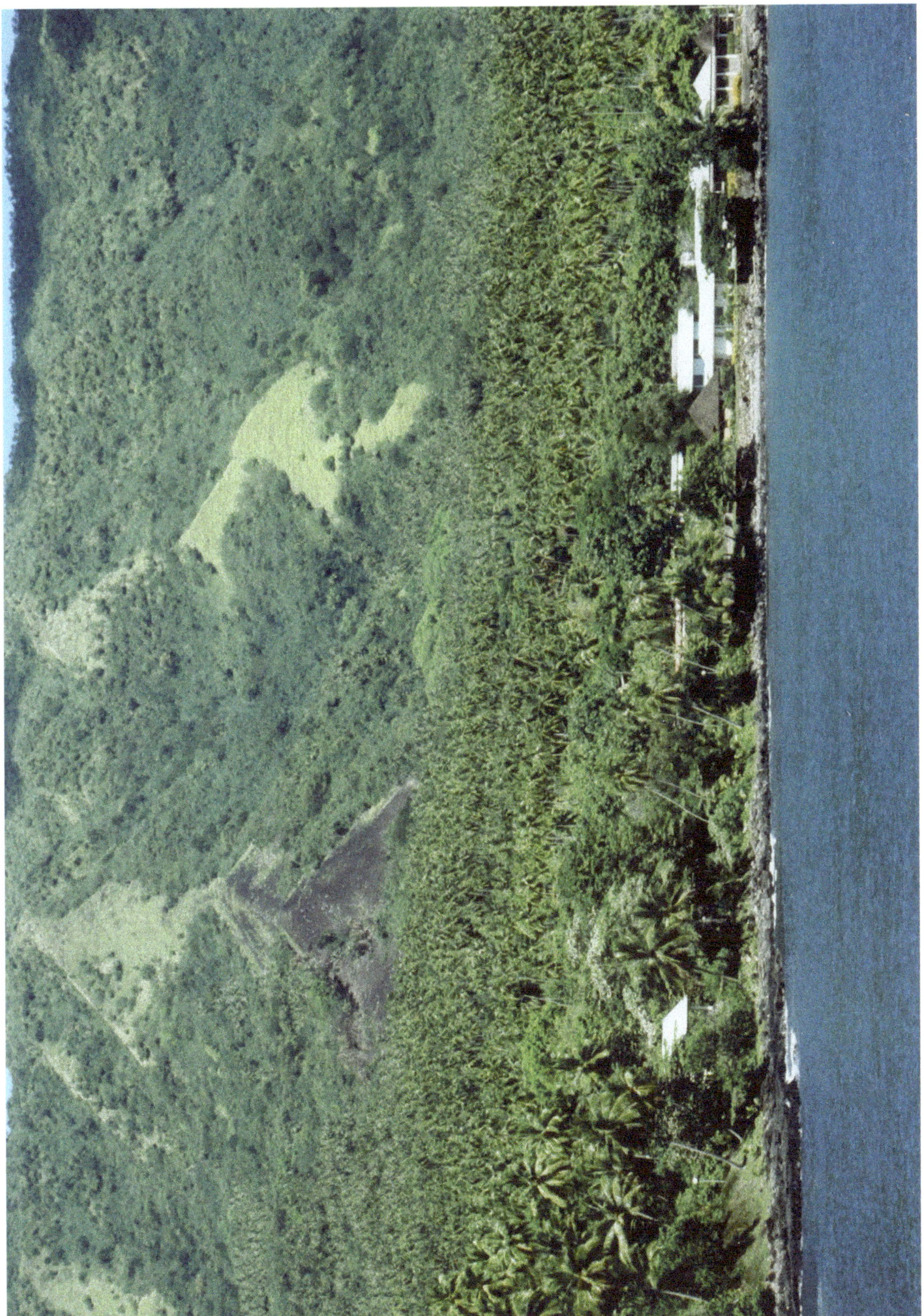

Figure 30. View of Hatiheu Valley from the bay. Note the burnt area, Te Ivi Maikuku, on the ridge to the left. Numerous petroglyphs were discovered in the area after the fire.

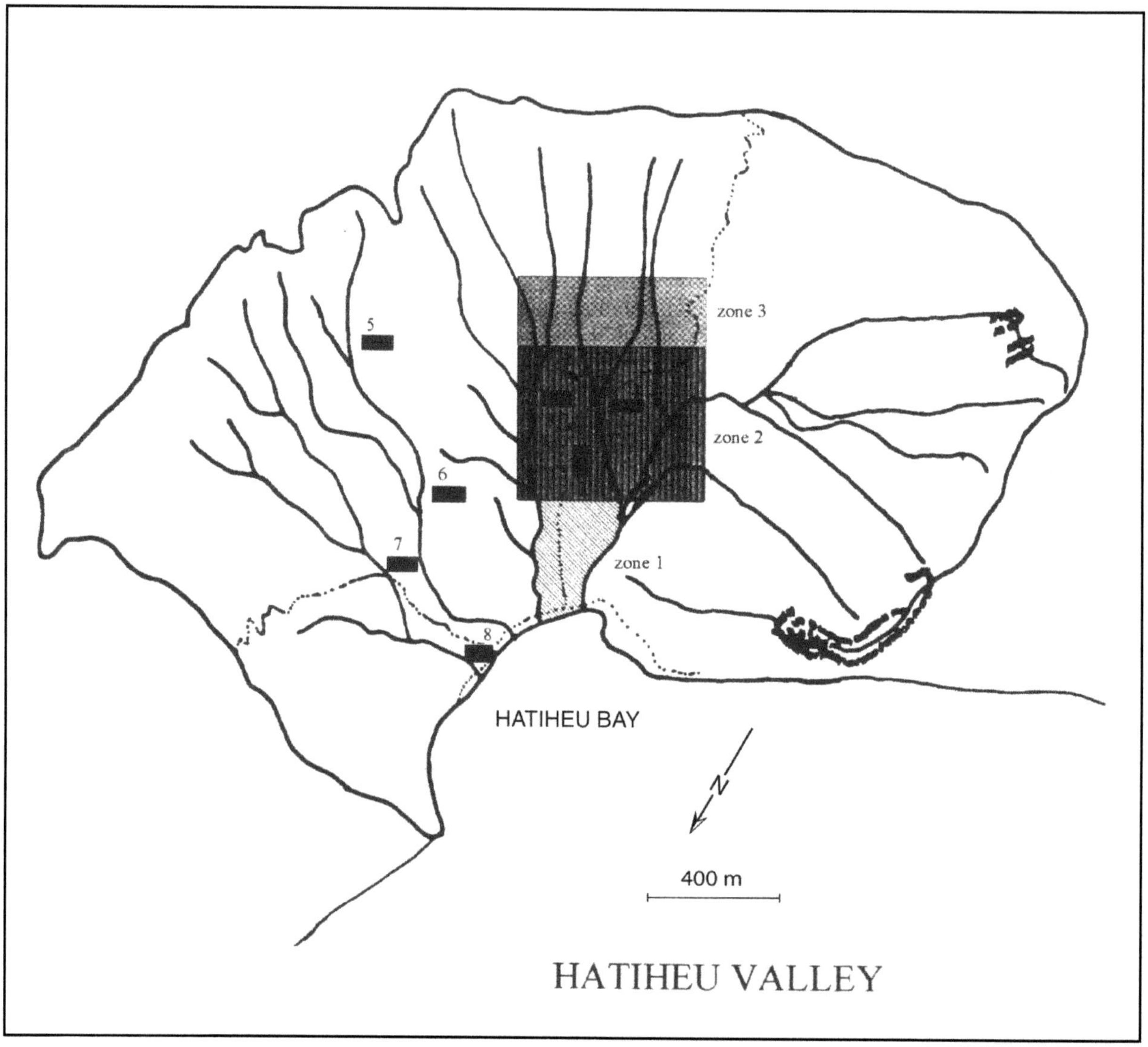

Figure 31. Hatiheu Valley. Research area showing three defined settlement zones.

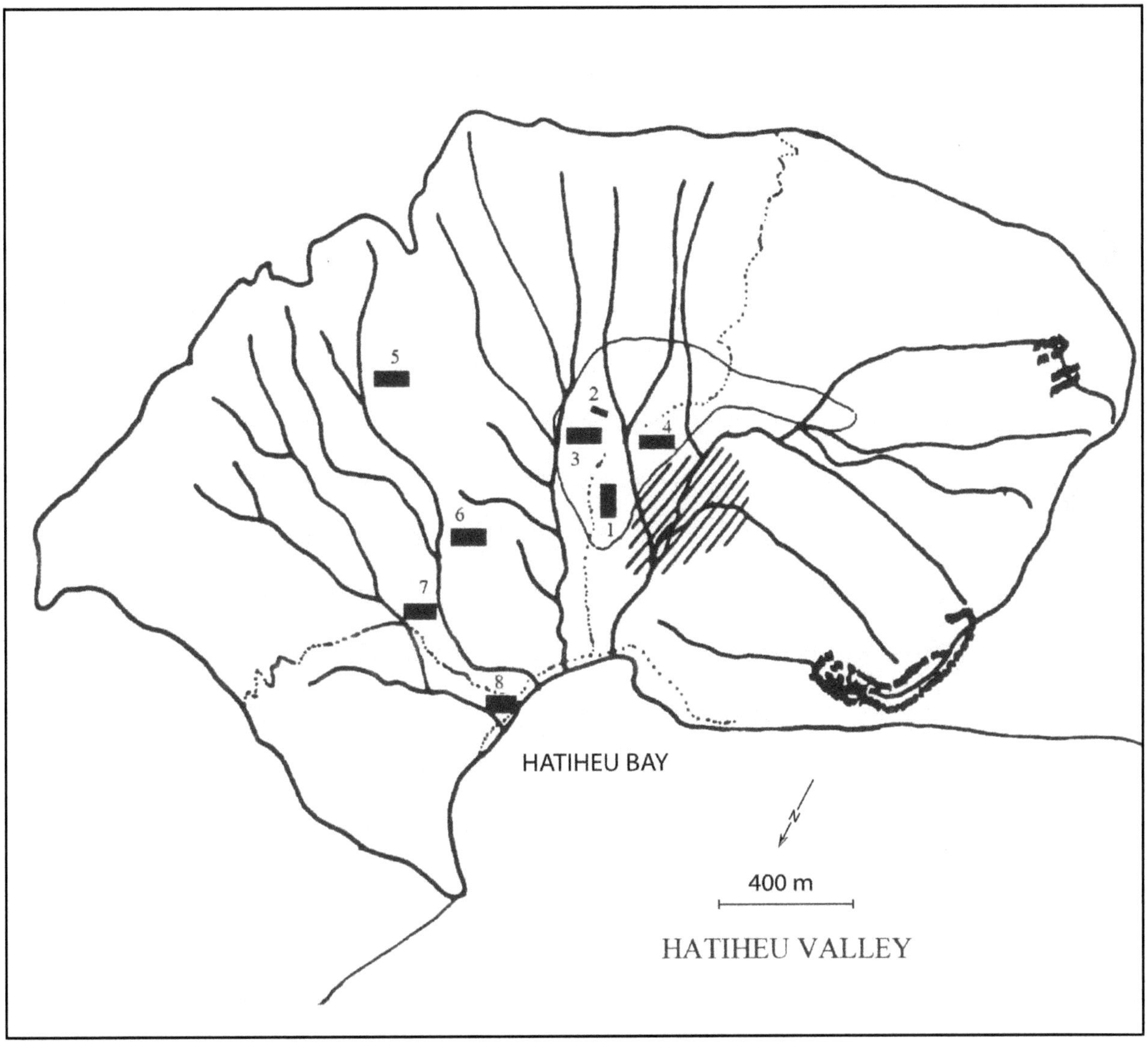

Figure 32. Locations of Tohua Hikokua (1); Kamuihei II (2); Kamuihei I (3); Tahakia (4); Maikuku (5); Paahaua (6); Naniuhi (7); and Pahumano (8). The solid line encloses the chiefly area, while the striped section indicates the agricultural sector.

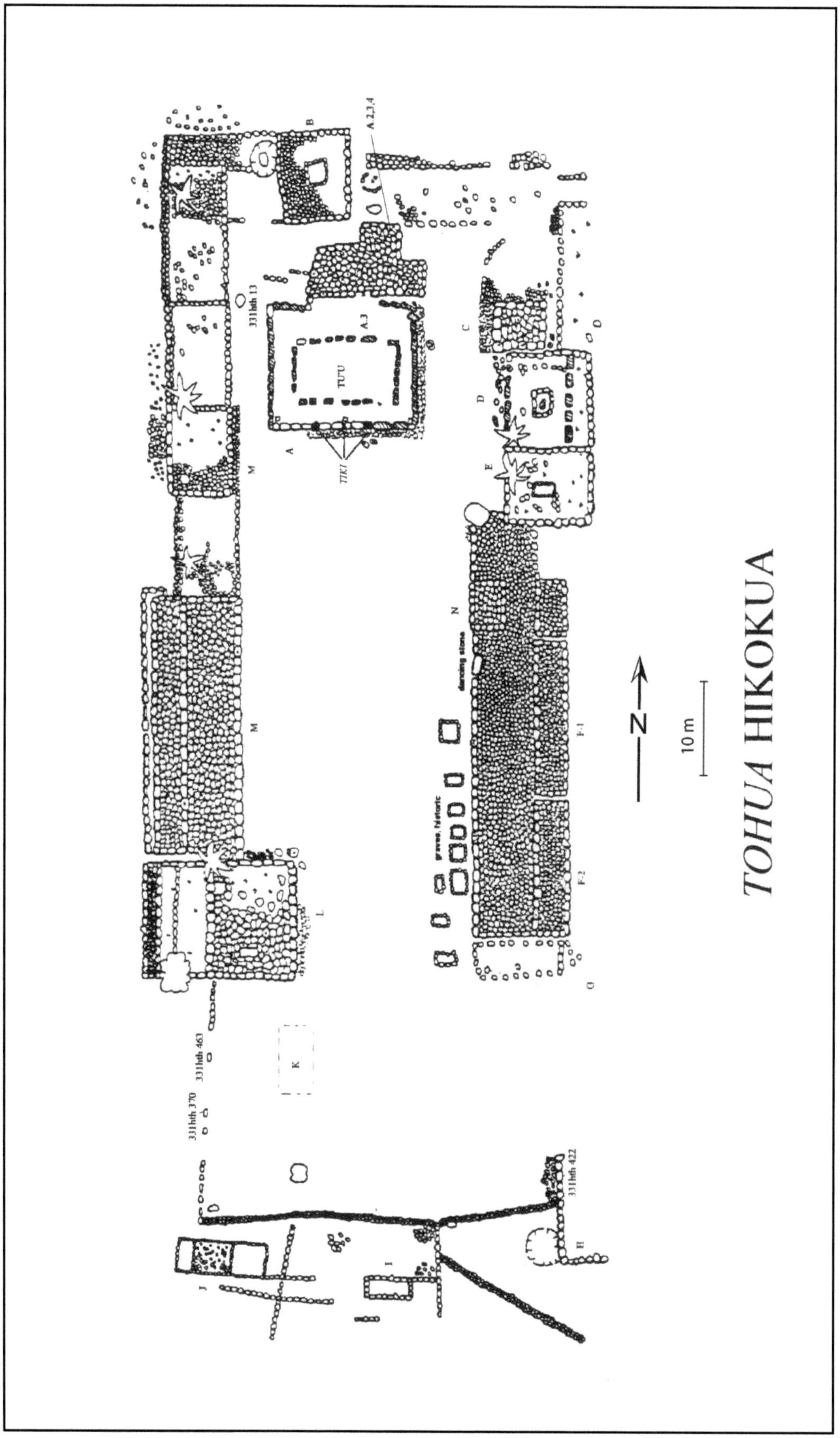

Figure 33. Tohua Hikokua.

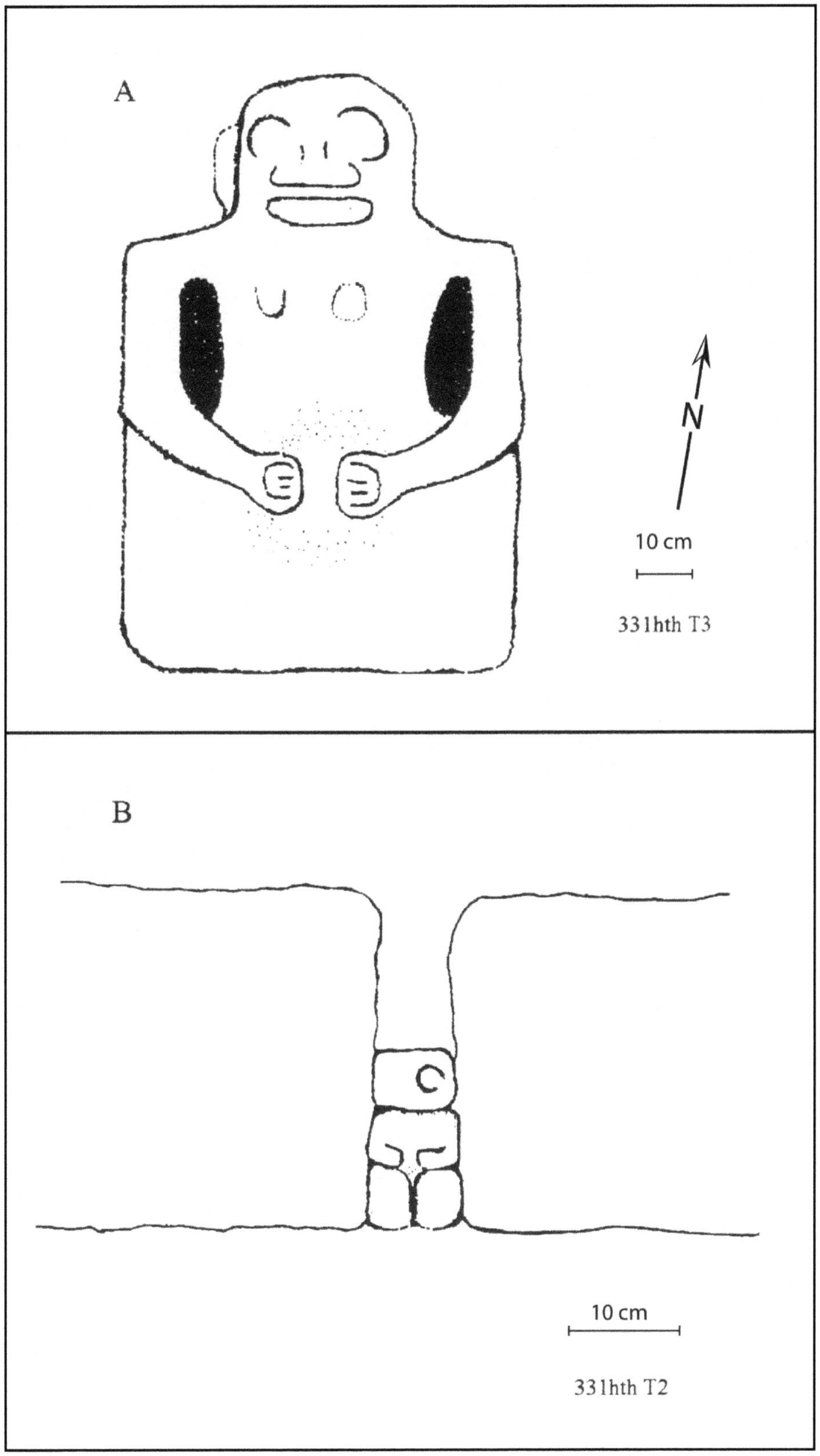

Figure 34 a–b. Tohua Hikokua. Two tiki placed in the facing wall of a ritual structure (331hthT3 and T2).

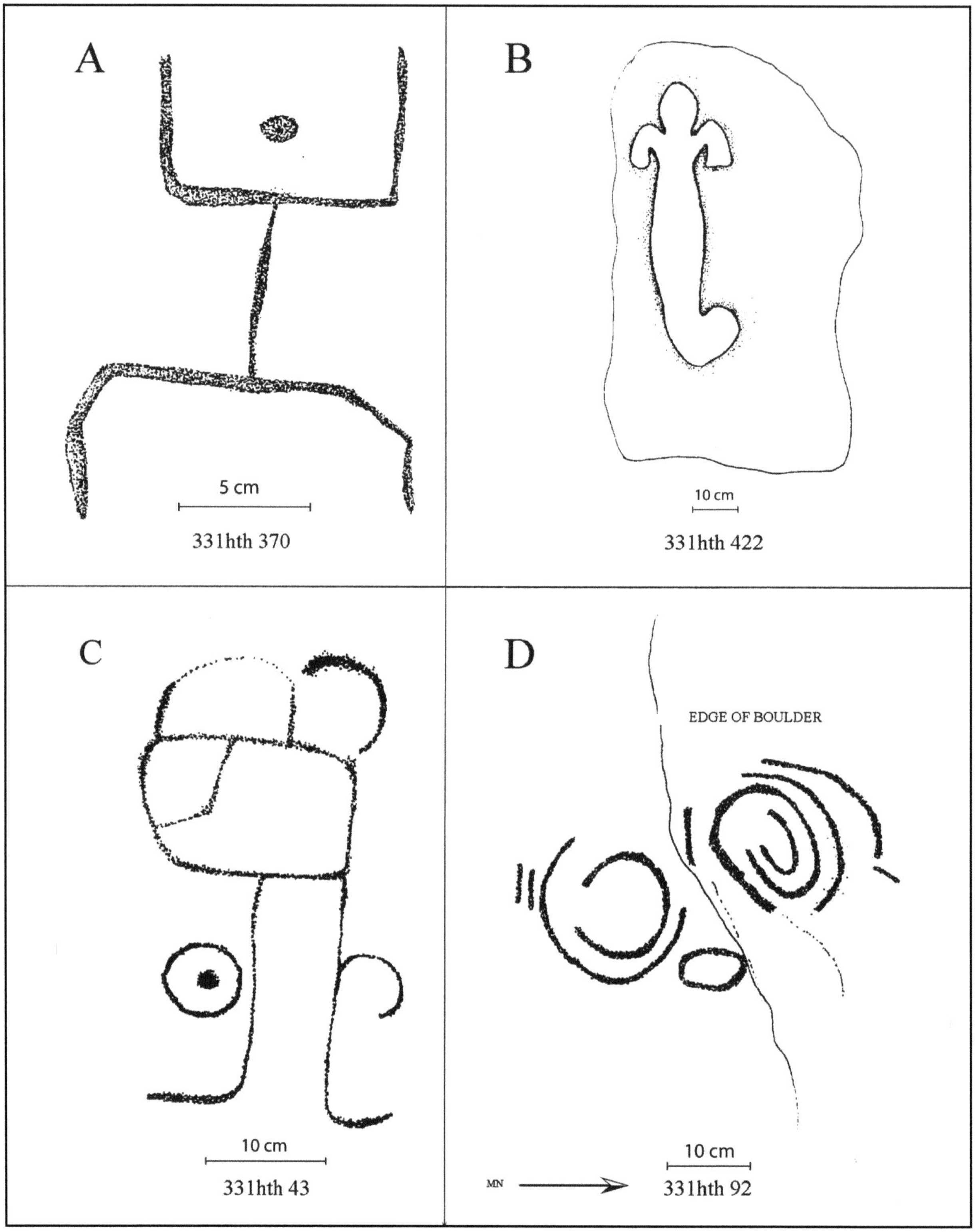

Figure 35 a–d. Tohua Hikokua: A) an anthropomorphic stick figure (331hth 370); B) a lizard image in bas-relief (331hth 422); C) a geometric figure (331hth 43); and D) an anthropomorphic face (331hth 92).

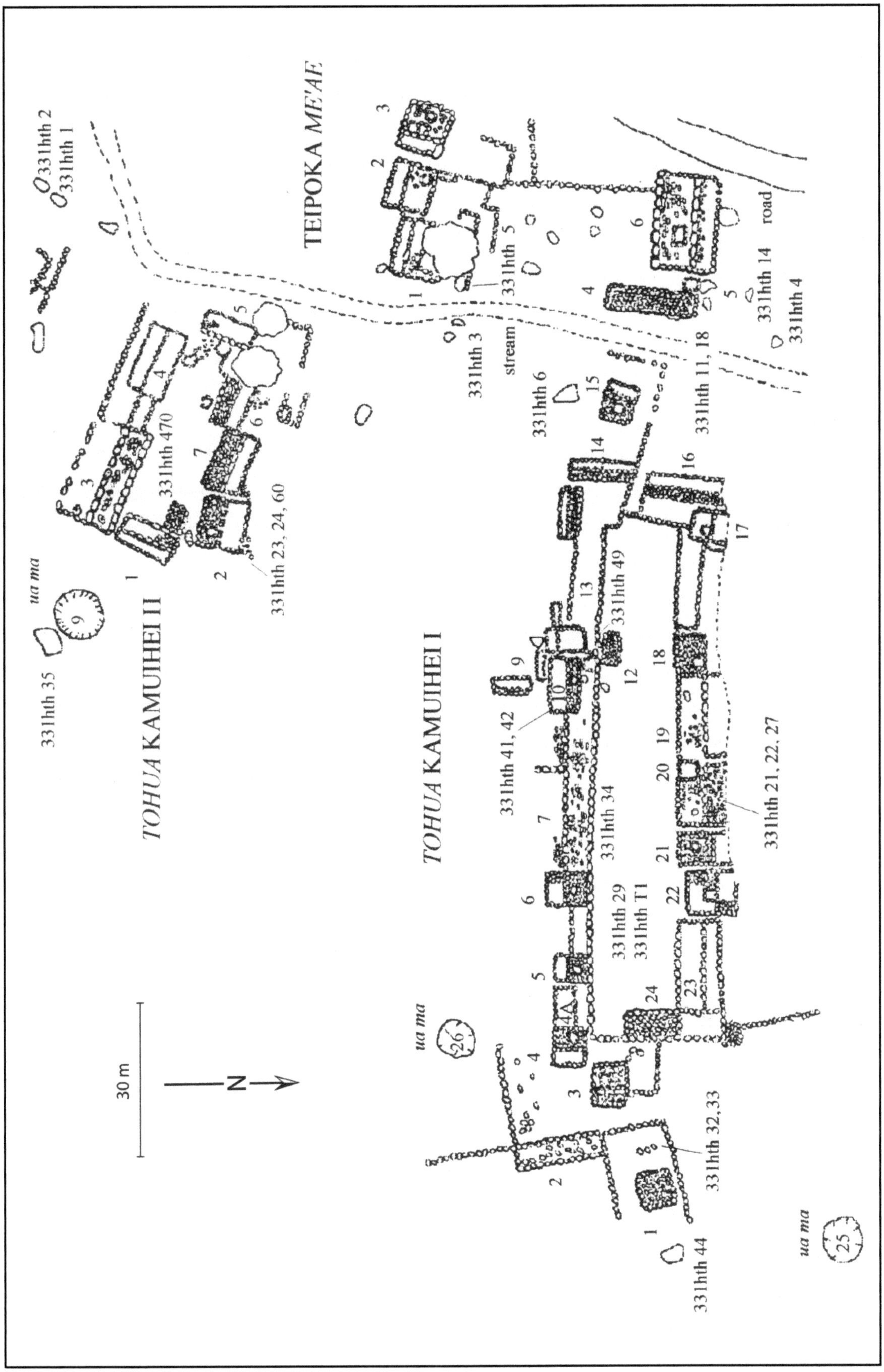

Figure 36. Tohua Kamuihei I, II, and the Te Iipoka area (map drawn by E. Edwards in 1986).

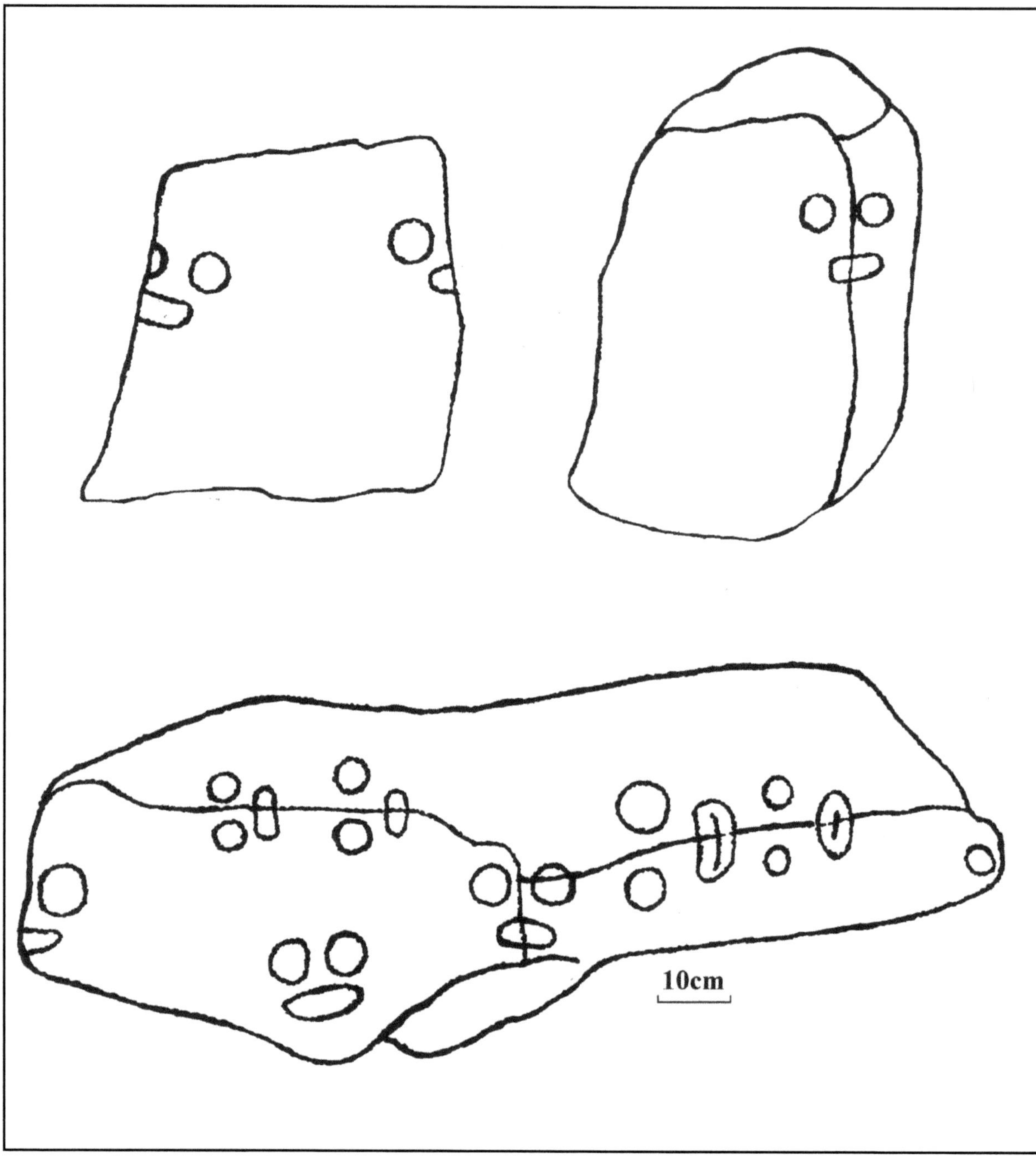

Figure 37. Tohua Kamuihei I. Two upright boulders with a total of eleven petroglyph faces placed on a *meae* (331hth 33/1 and 33/2; 331hth 32).

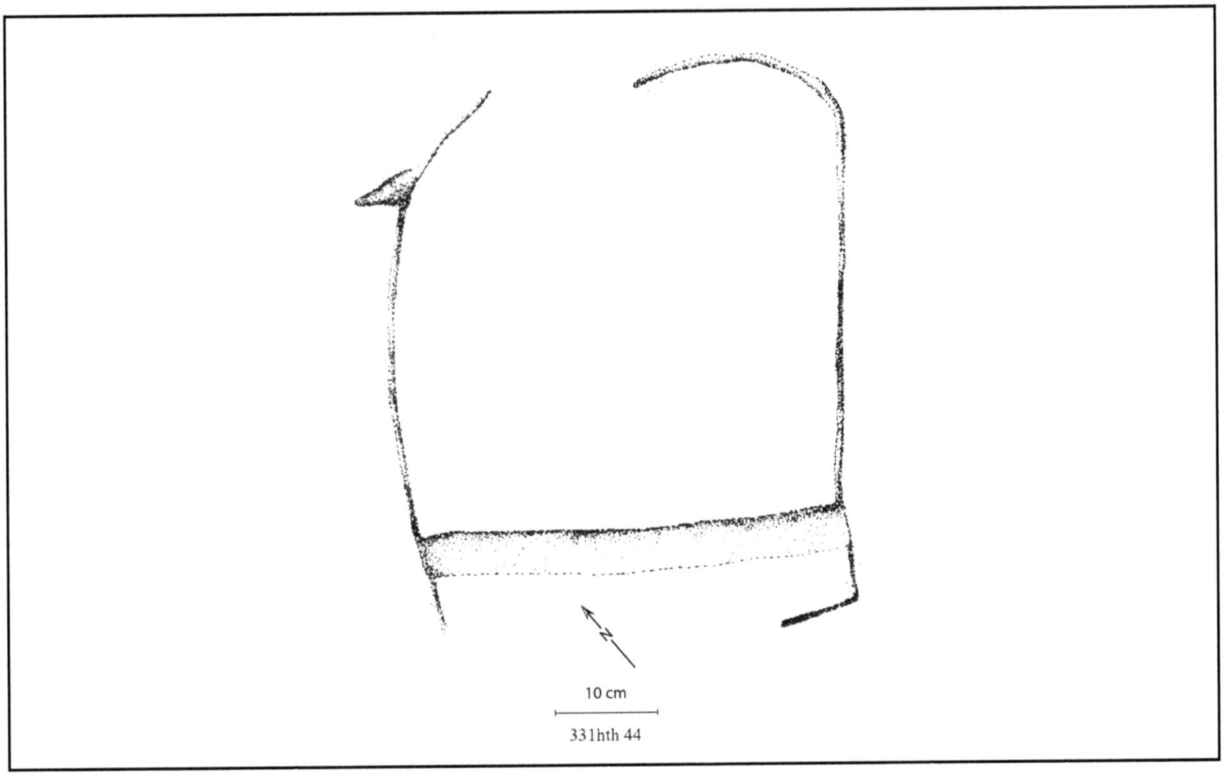

Figure 38. Tohua Kamuihei I. A dog petroglyph located near the *meae* (331hth 44).

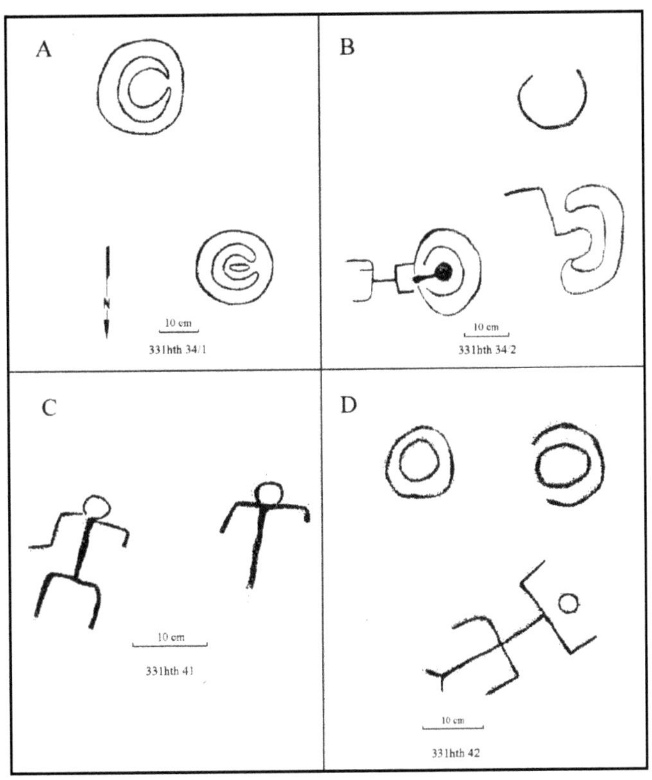

Figure 39 a–d. Tohua Kamuihei I. Petroglyphs a and b are located on the warrior's platform, while images c and d are found on the chief's residential unit (331hth 34/1 and 34/2; 331hth 41 and 42).

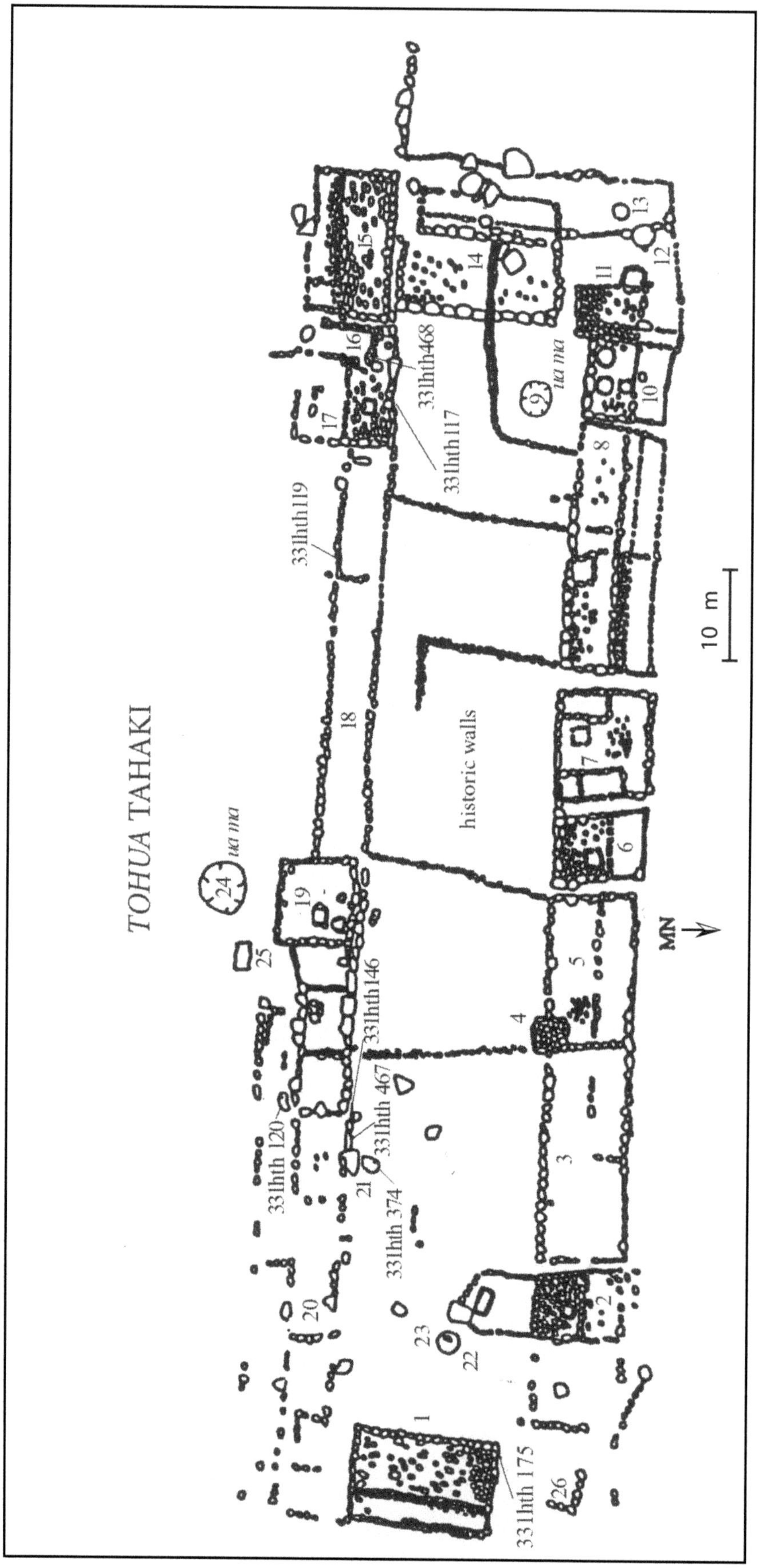

Figure 40. Tohua Tahakia (map courtesy Pierre Ottino, 1995).

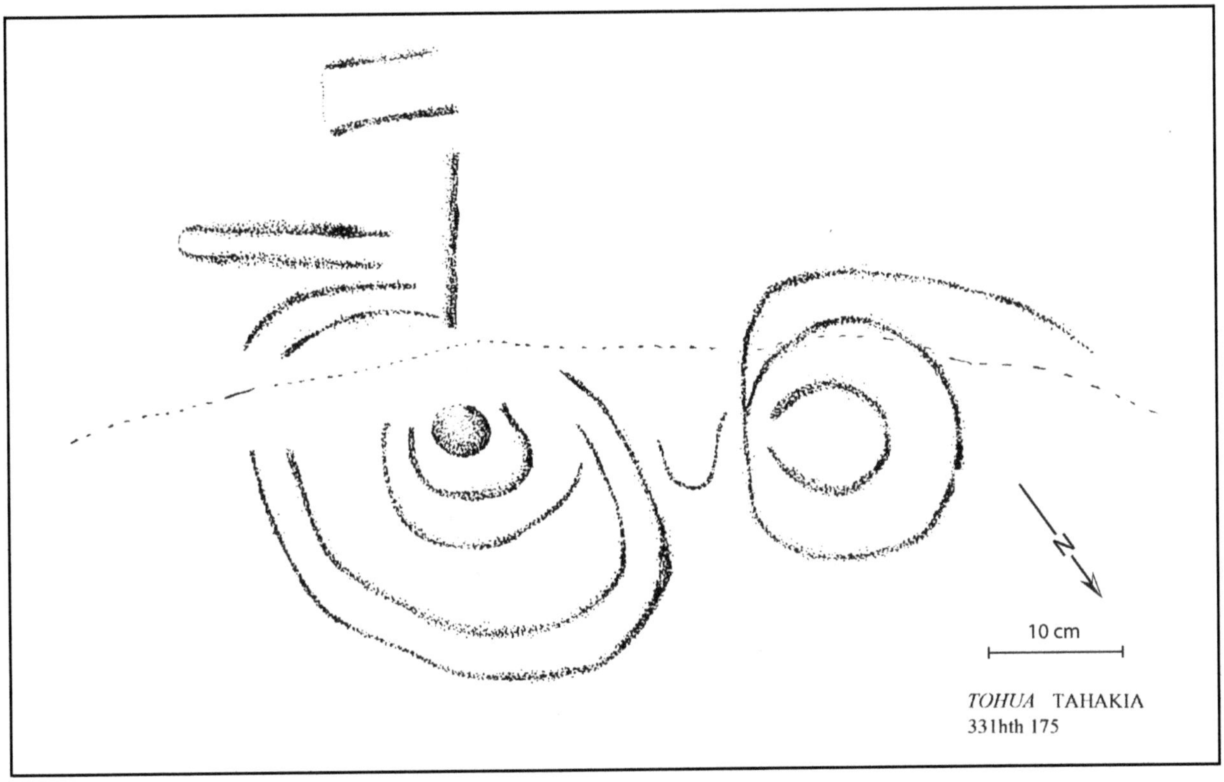

Figure 41. Tohua Tahakia. Anthropomorphic face placed on a megalithic platform (331hth 175).

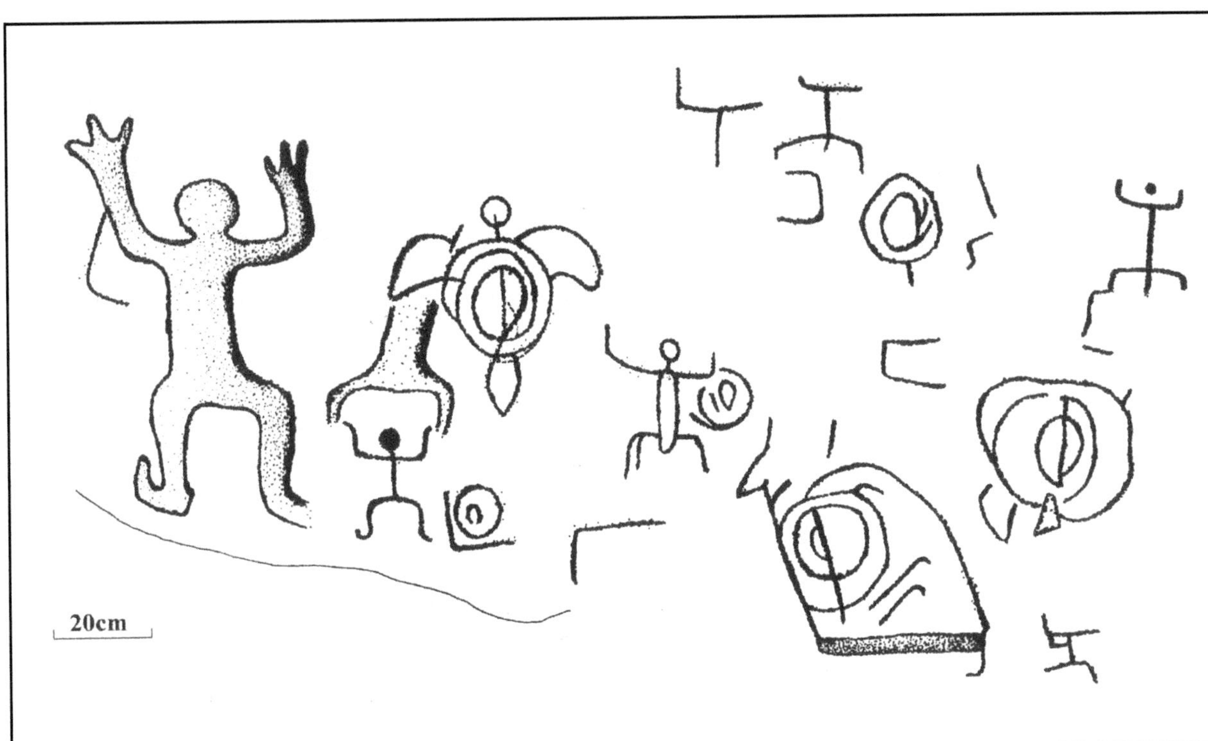

Figure 42. Tohua Tahakia. A large outcrop with some overlapping petroglyphs (331hth 389).

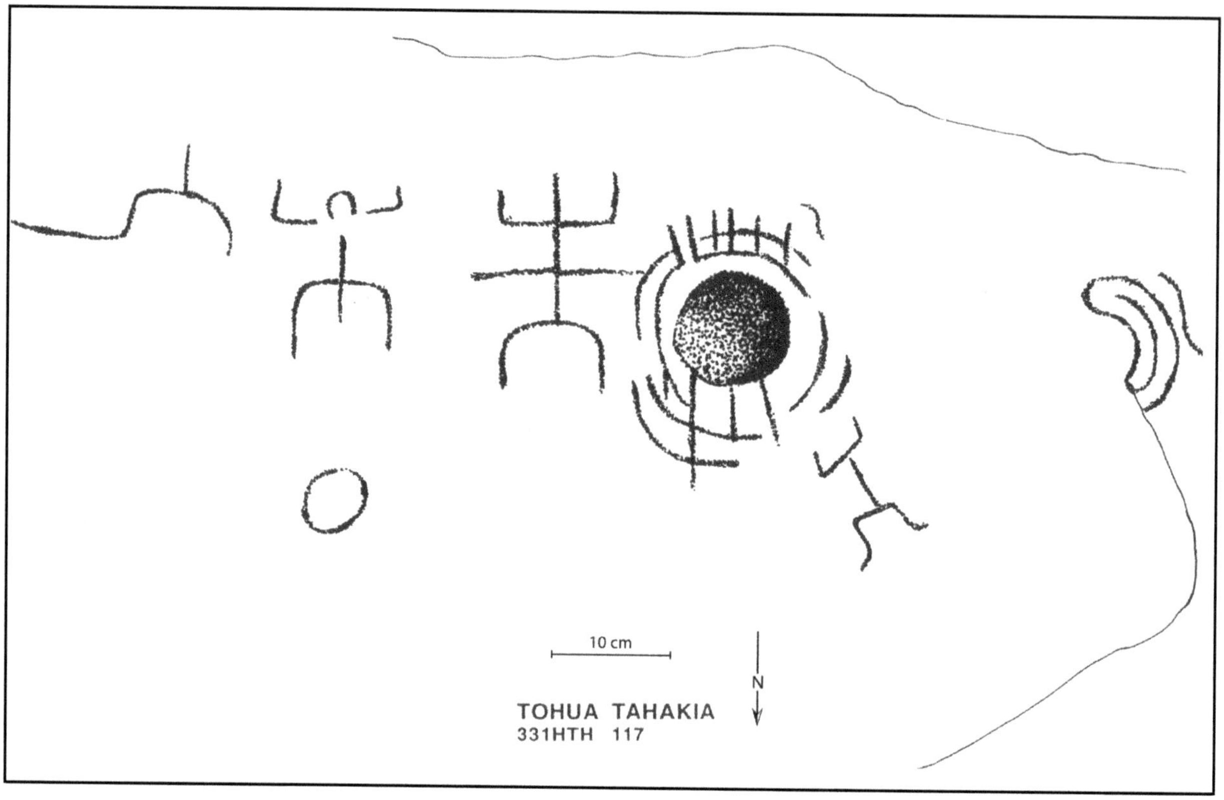

Figure 43. Tohua Tahakia. Vertically placed anthropomorphic and geometric images (331hth 117/1).

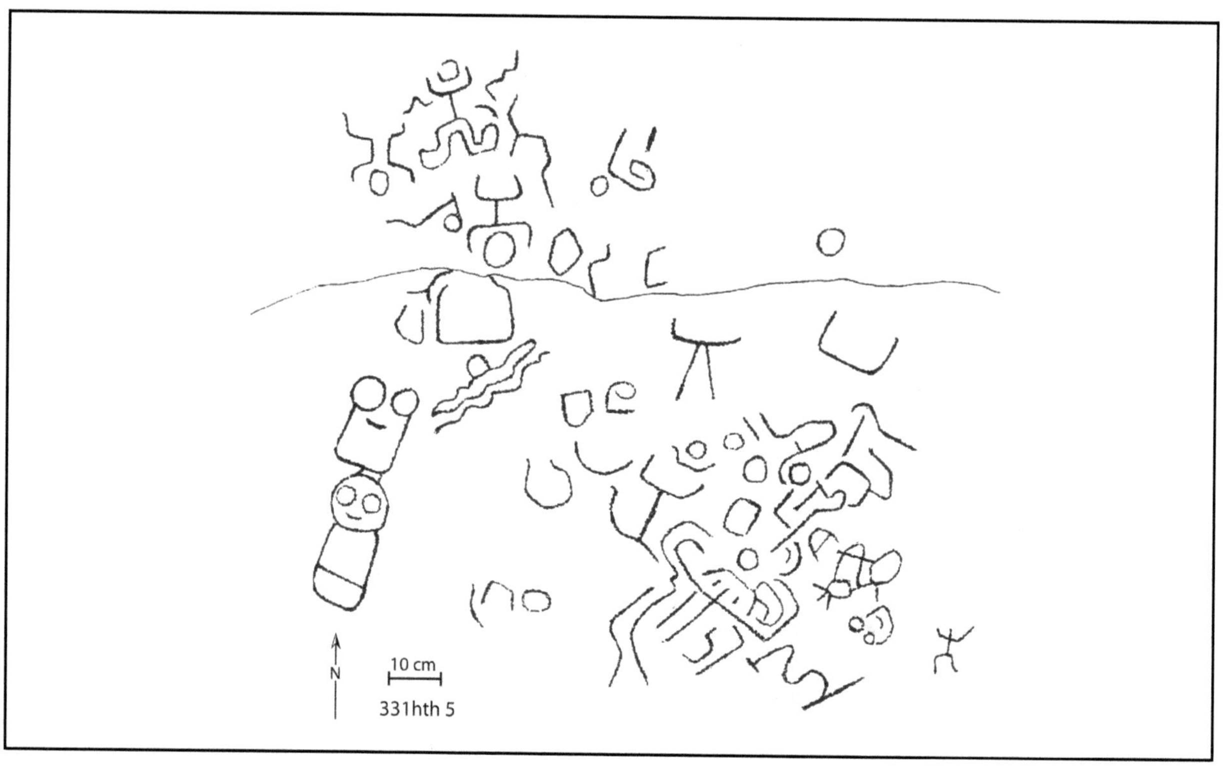

Figure 44. Kamuihei-Te Iipoka area. Petroglyph boulder is associated with priest's platform and a large sacred banyan tree (331hth 50).

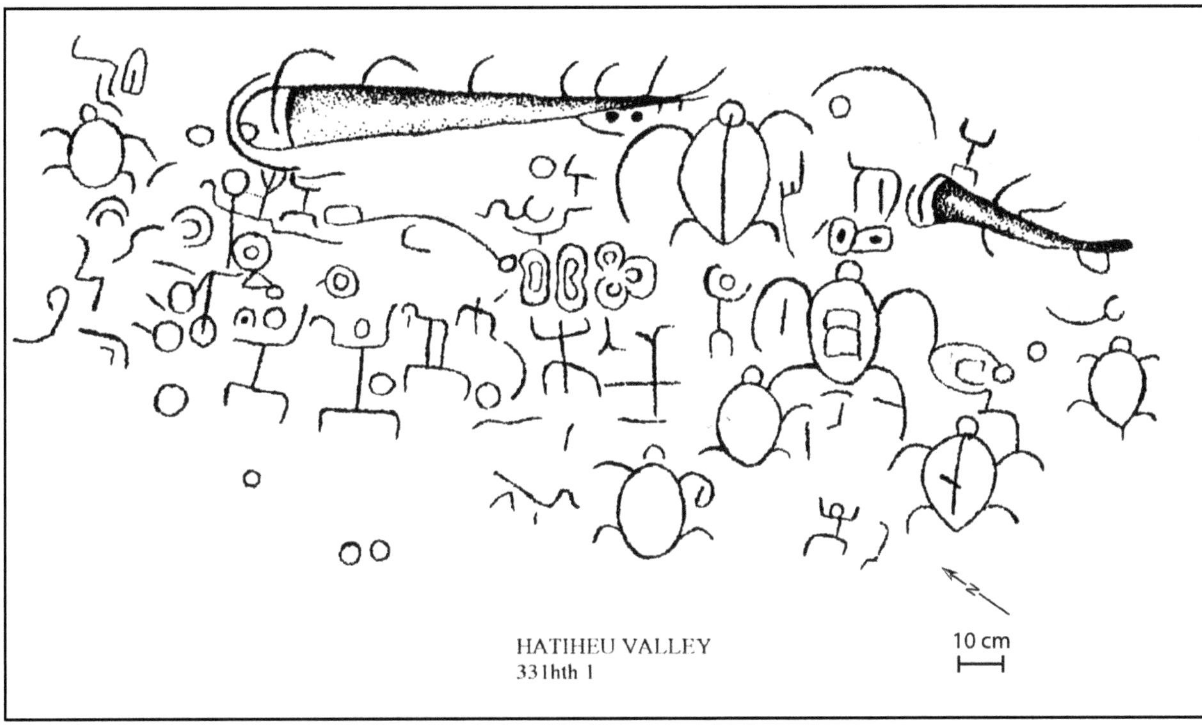

Figure 45. Kamuihei-Te Iipoka area. Panel with anthropomorphs, turtles, and sea creatures. Other panels depict a hammerhead and turtles (331hth 1 1/1).

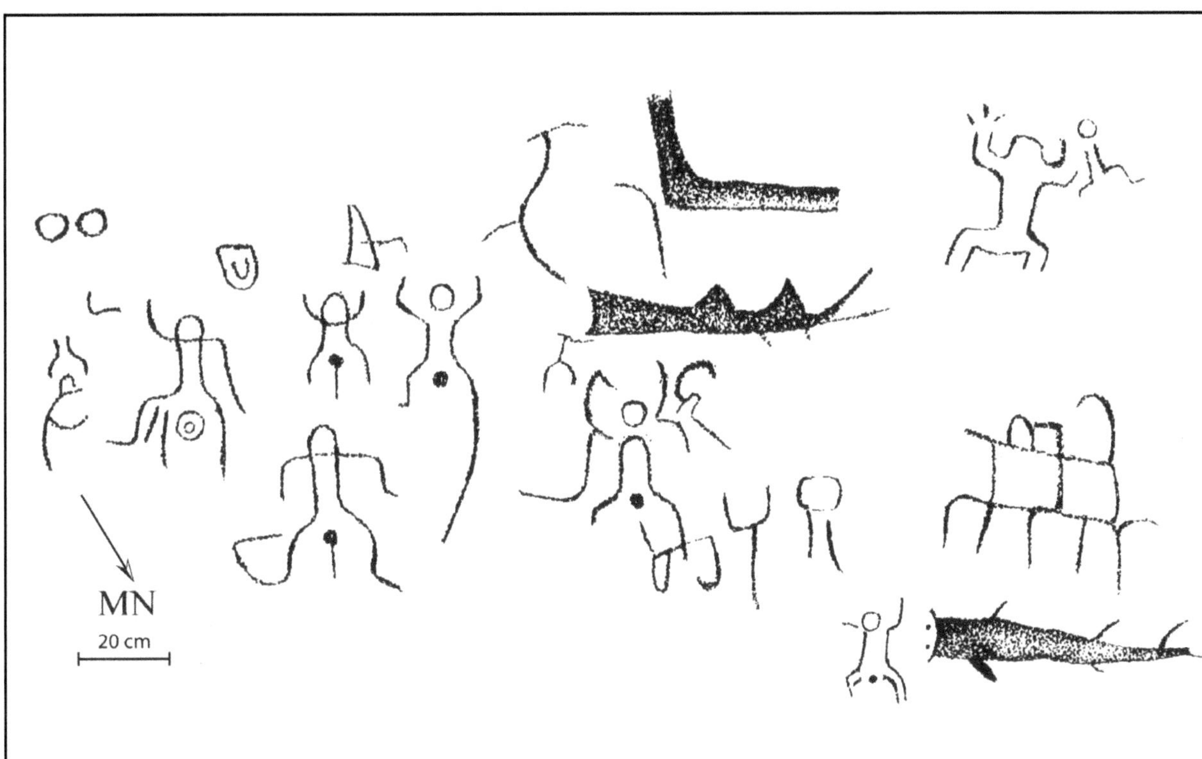

Figure 46. Kamuihei-Te Iipona area. Panel depicts several female figures and sea creatures (331hth 2/1).

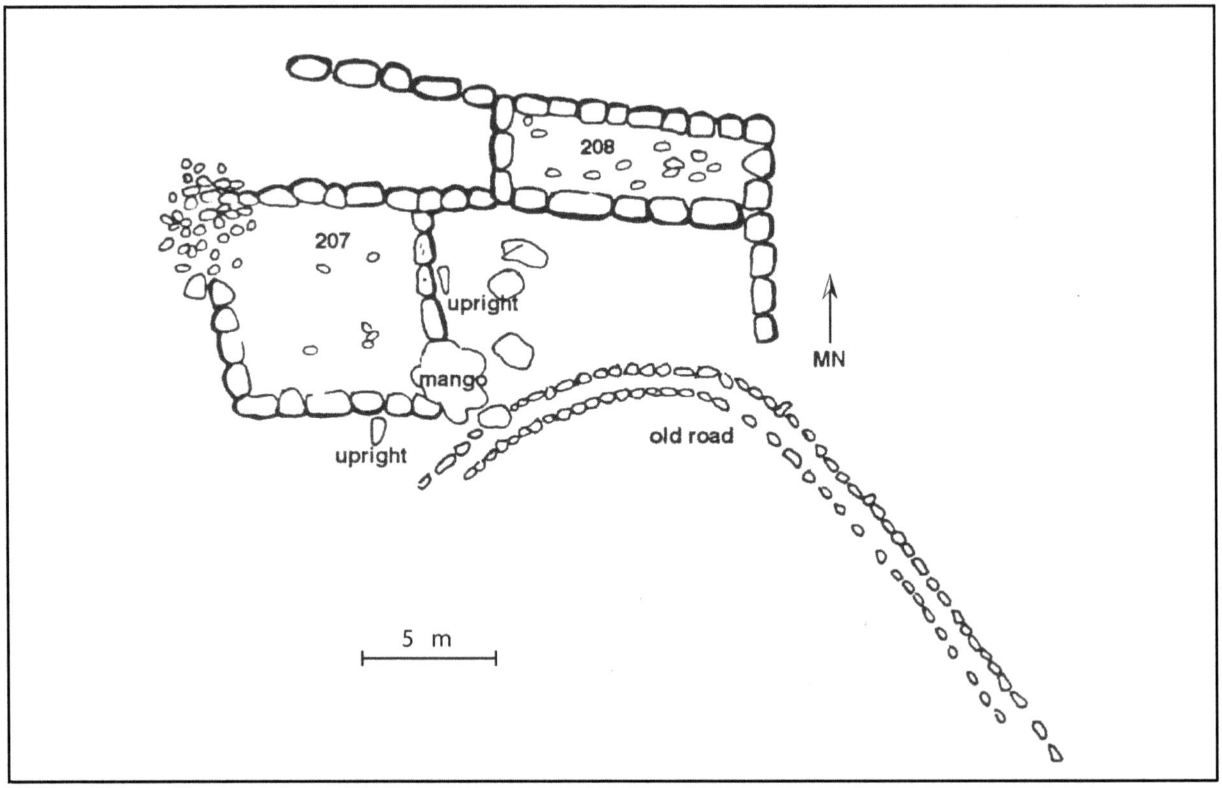

Figure 47. A *meae* complex.

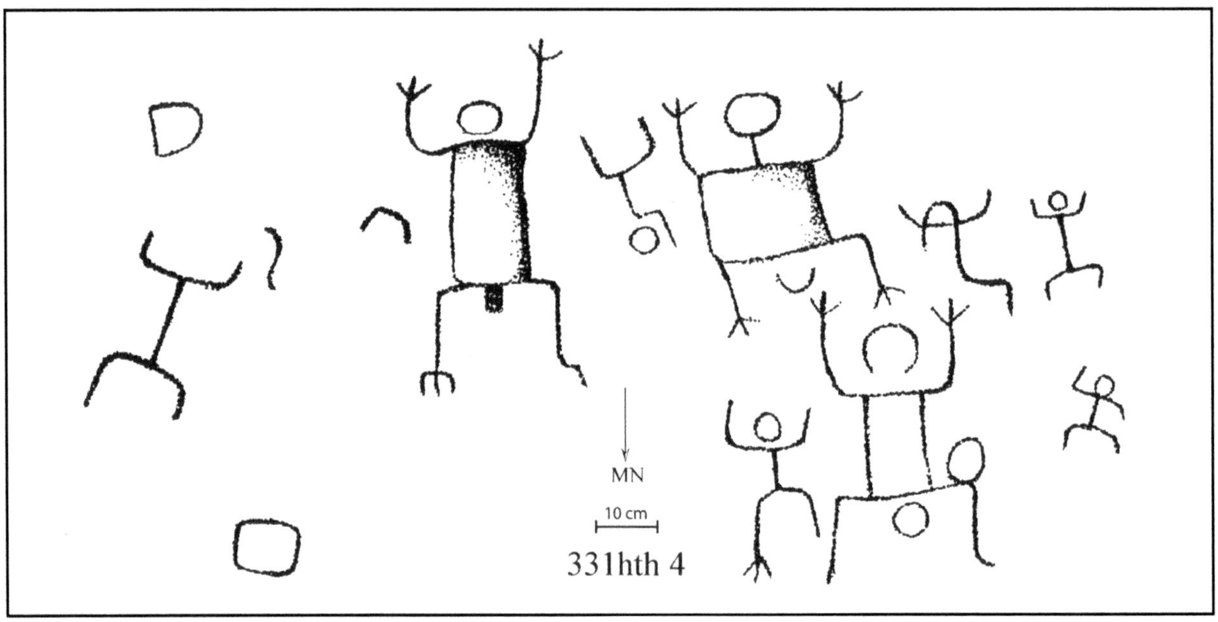

Figure 48. Hikokua-Kamuihei-Te Iipoka-Tahakia area. Various types of anthropomorphs (331hth 4).

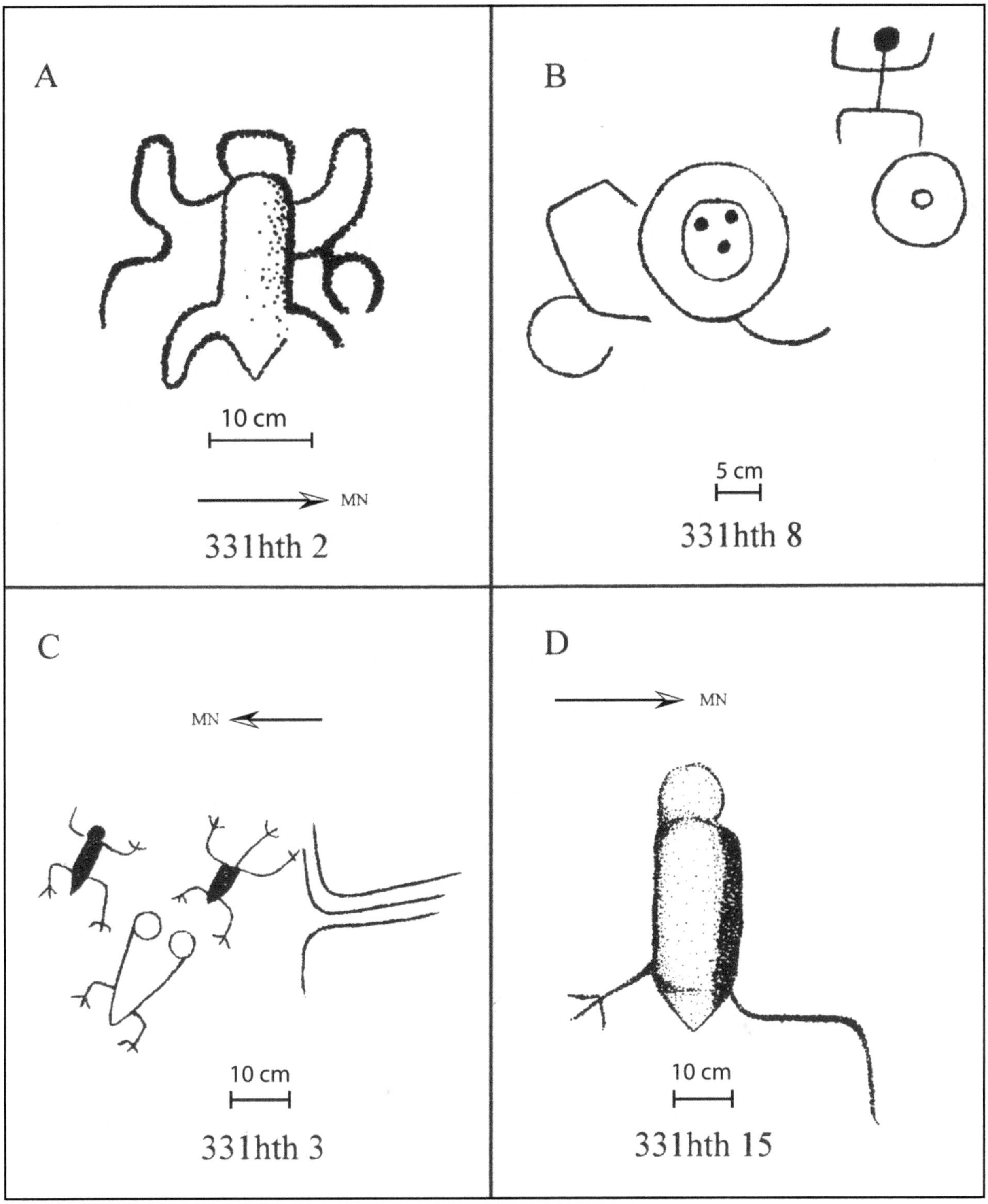

Figure 49. Kamuihei-Te Iipoka area. Various unidentified exotic figures (331hth 2, 3, 8, and 15).

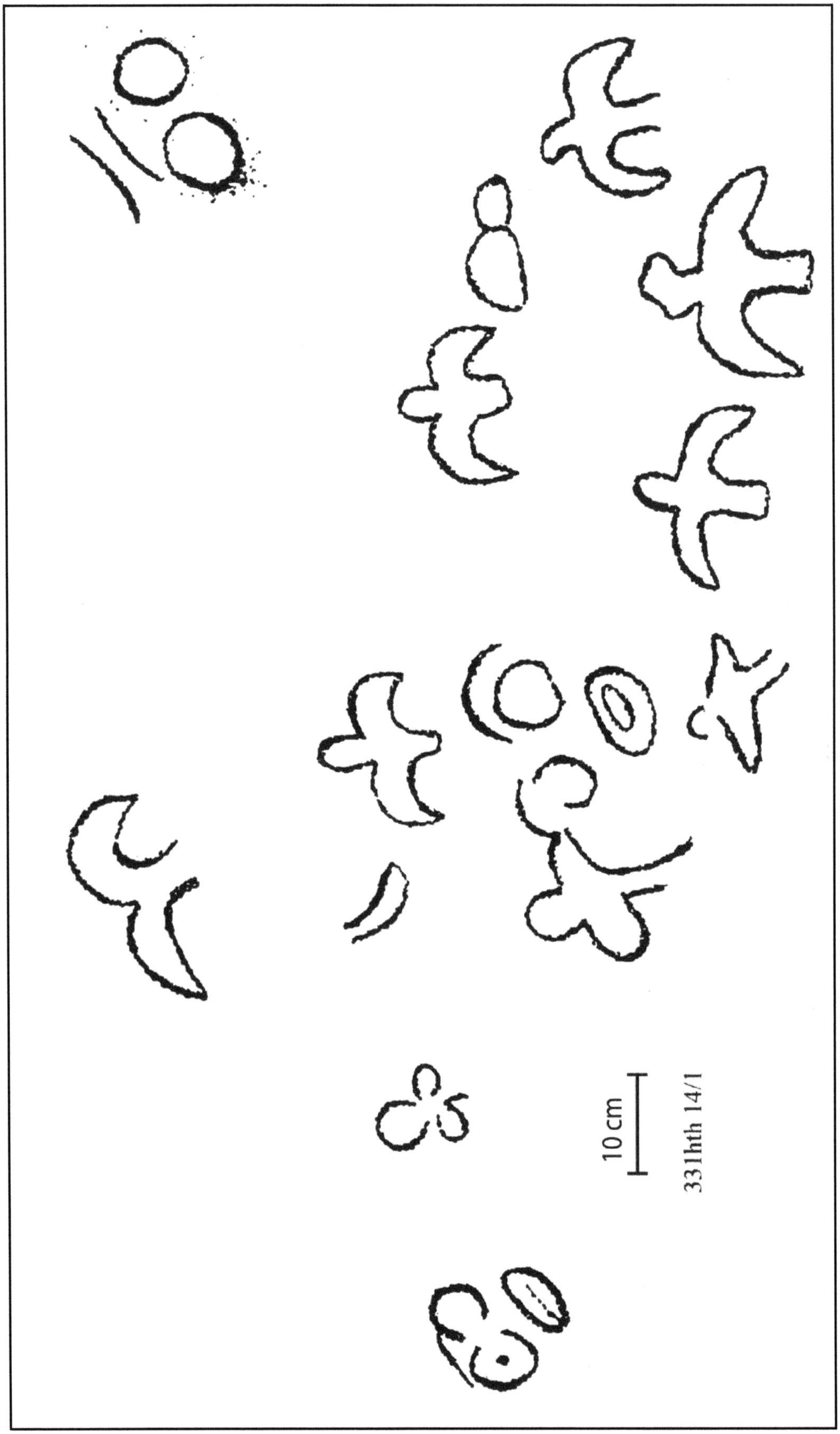

Figure 50. Kamuihei-Te Iipoka area. Bird images seen from below or above and three faces (331hth 14/1).

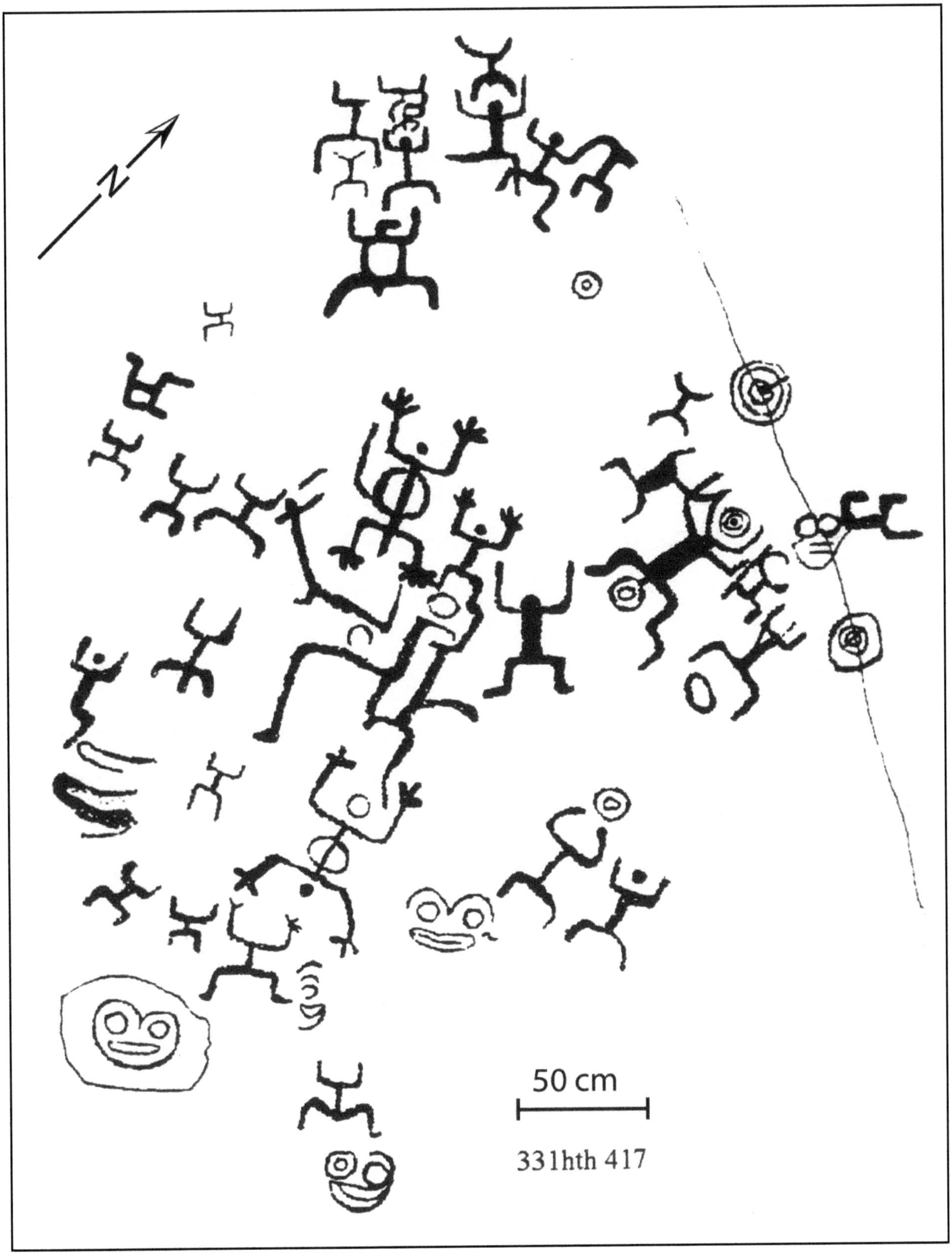

Figure 51. Ototemoui. Petroglyph boulder (331hth 417). The encircled face is a piece of stone that flaked off. It is presently housed in the Hatiheu Village Museum.

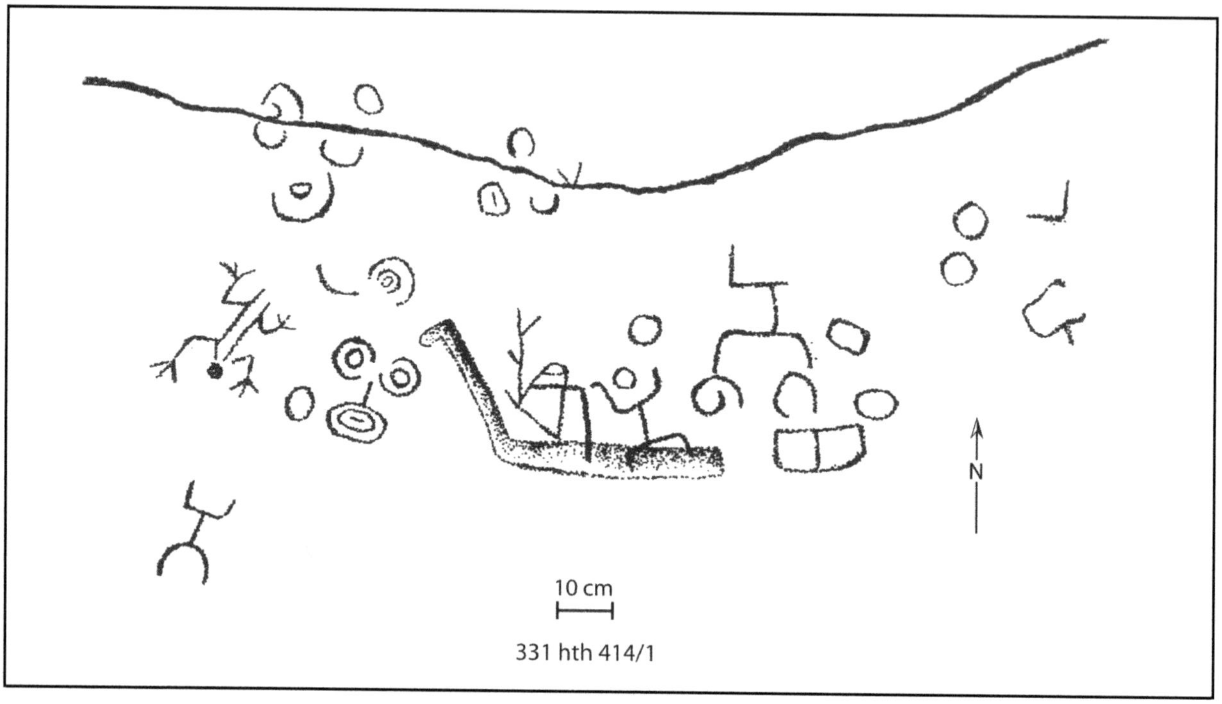

Figure 52. Ototemoui. An anthropomorph is superimposed on a legless dog figure (331hth 414/1).

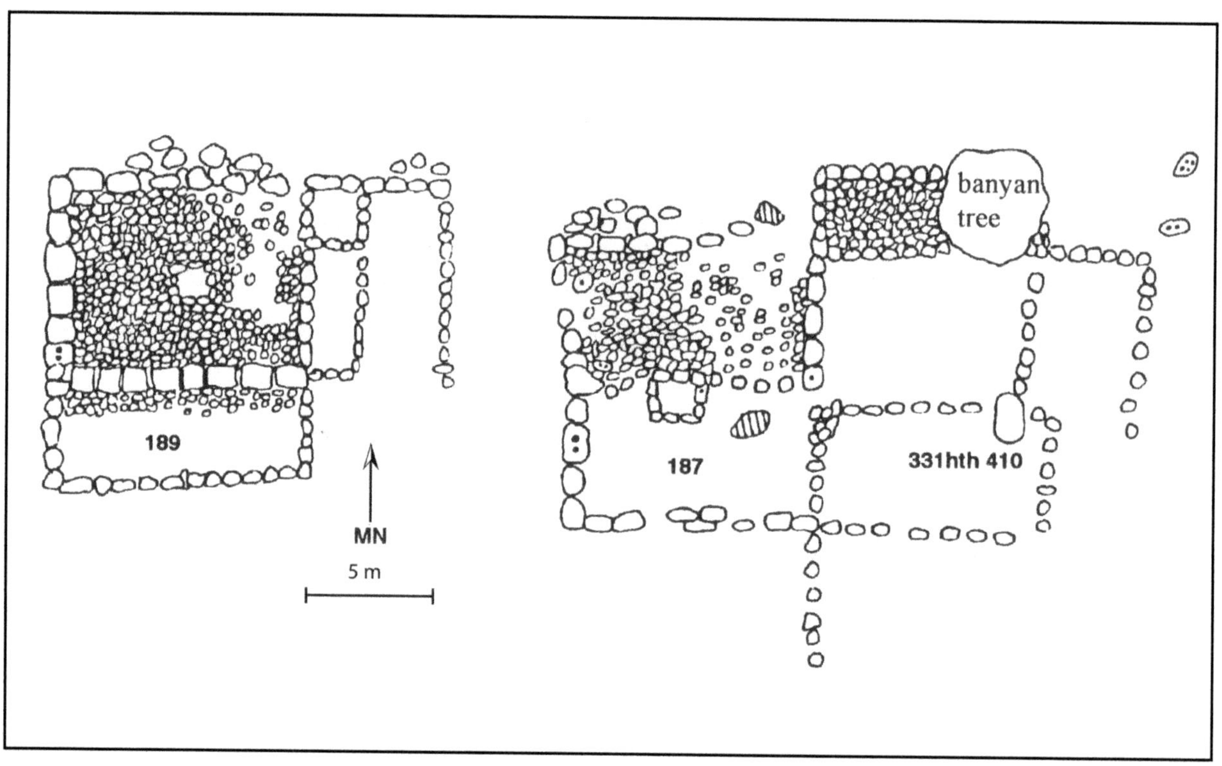

Figure 53. A *meae* complex (Structures 187 and 189) with the location of petroglyph boulder, 331hth 410, marked.

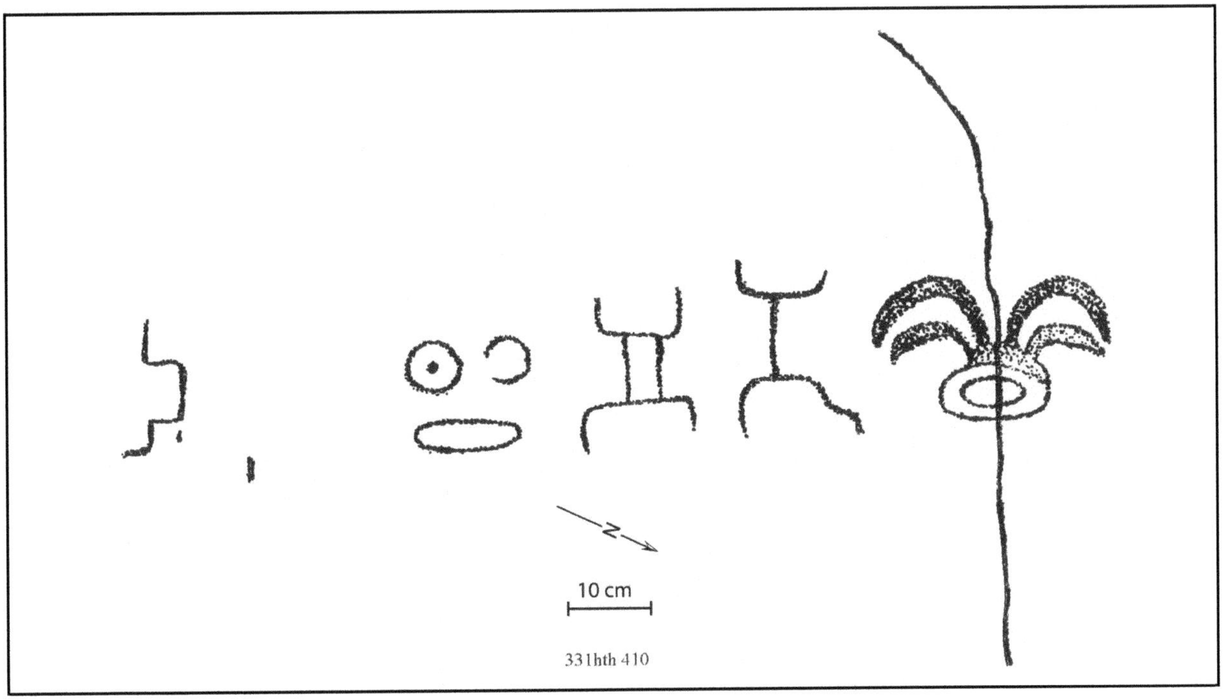

Figure 54. A *meae* complex with petroglyph boulder (331hth 410).

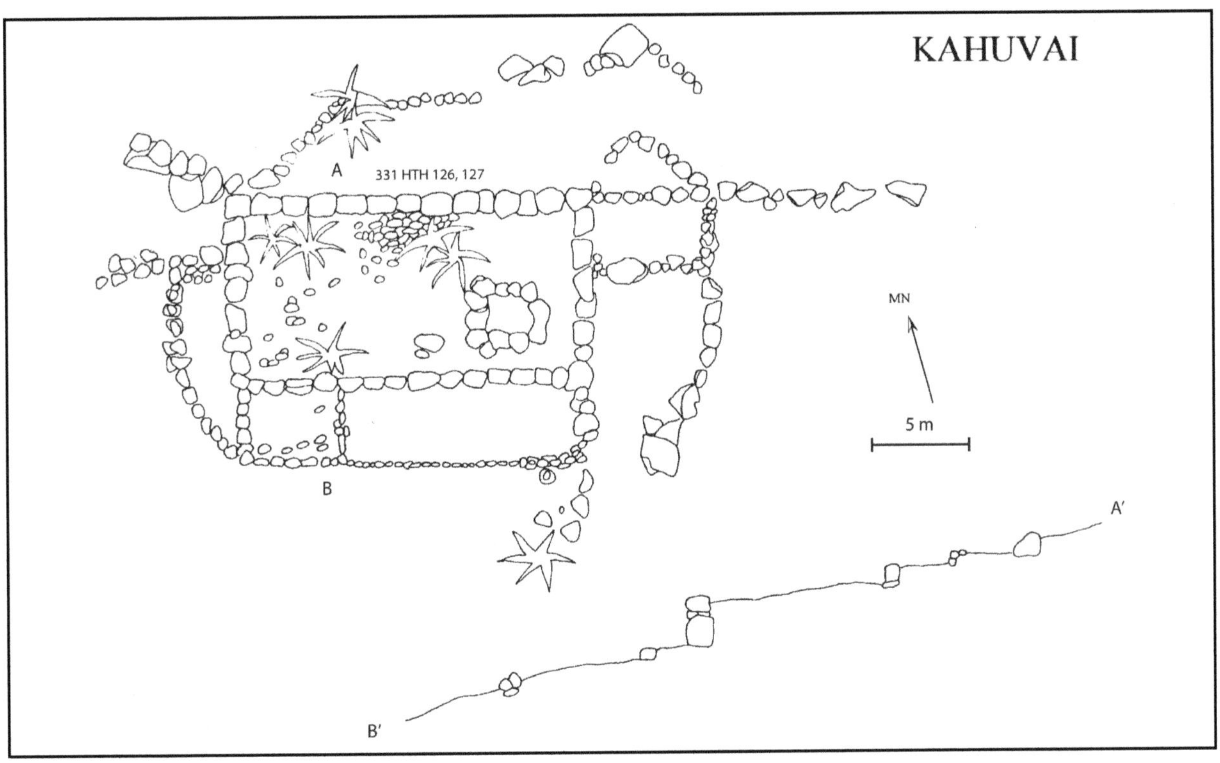

Figure 55. Kahuvai. A megalithic house platform. Two boulders with anthropomorphs are placed in the facing wall (331hth 126, 127).

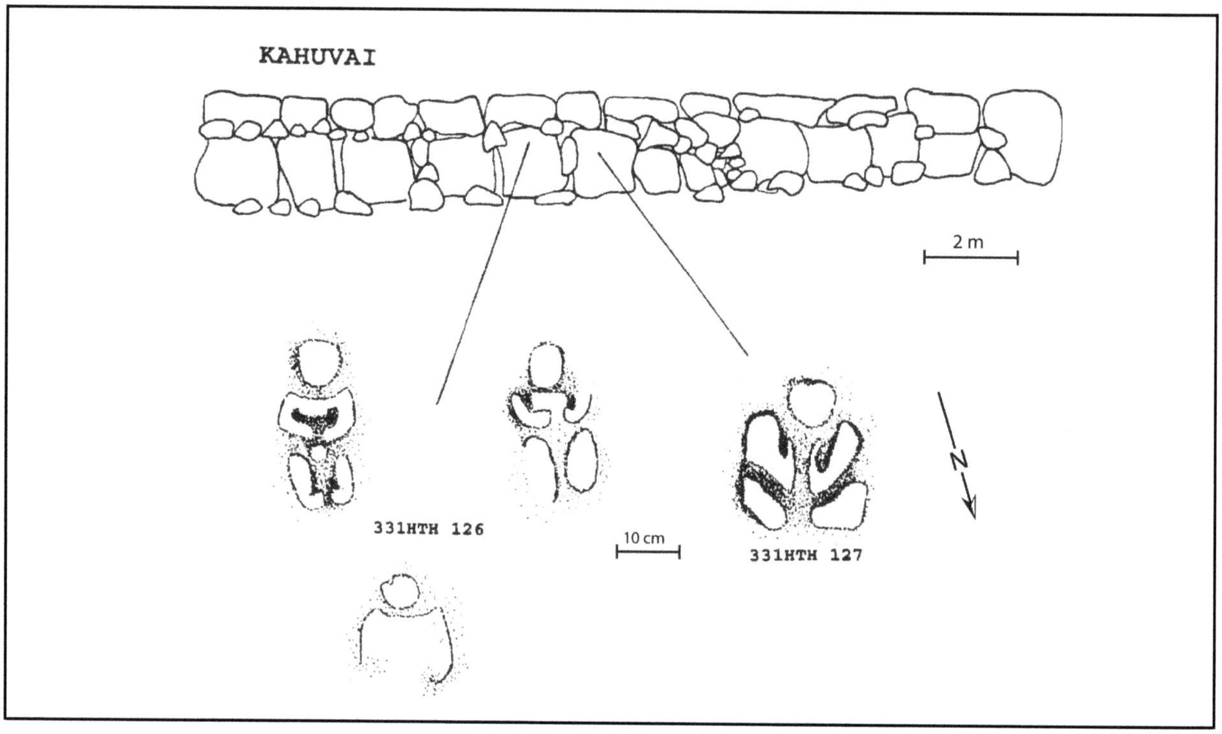

Figure 56. Kahuvai. Facing wall of Structure 221 with anthropomorphs in bas-relief. Boulder 331hth 126 has three figures, one is unfinished. Boulder 331hth 127 has one figure.

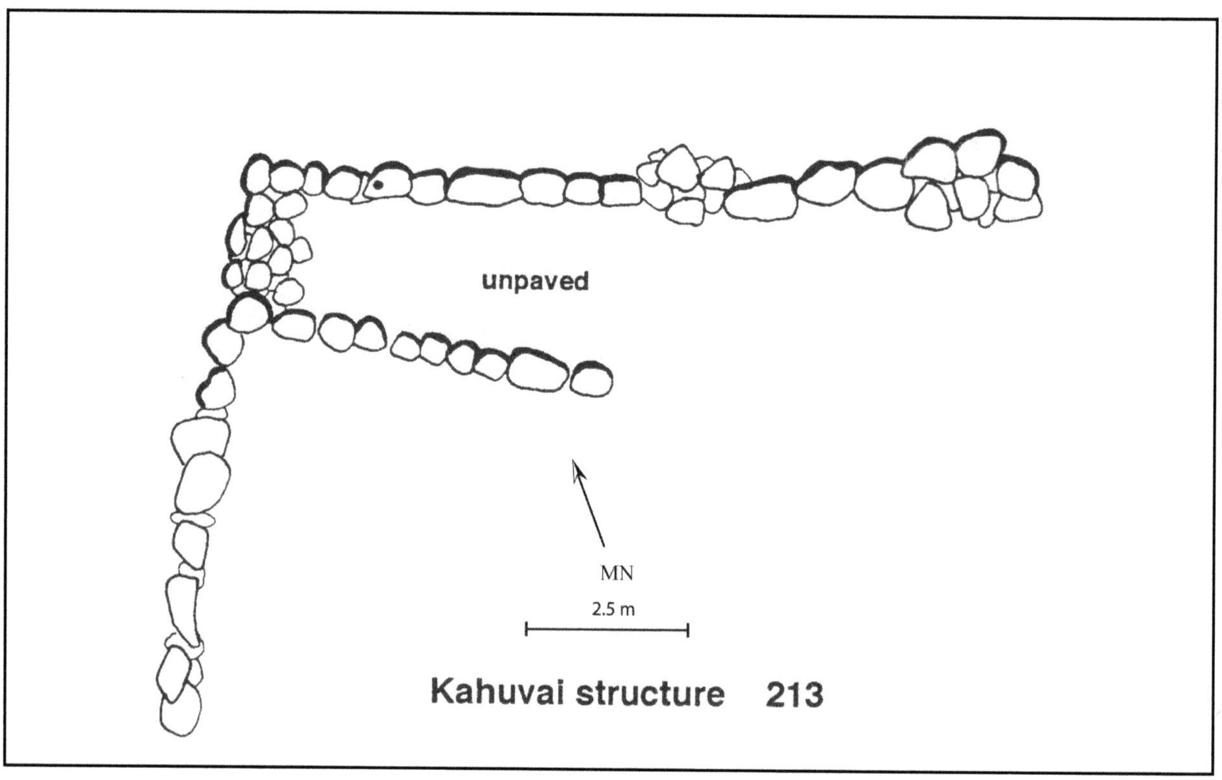

Figure 57. Kahuvai, An agricultural terrace (?), Structure 213.

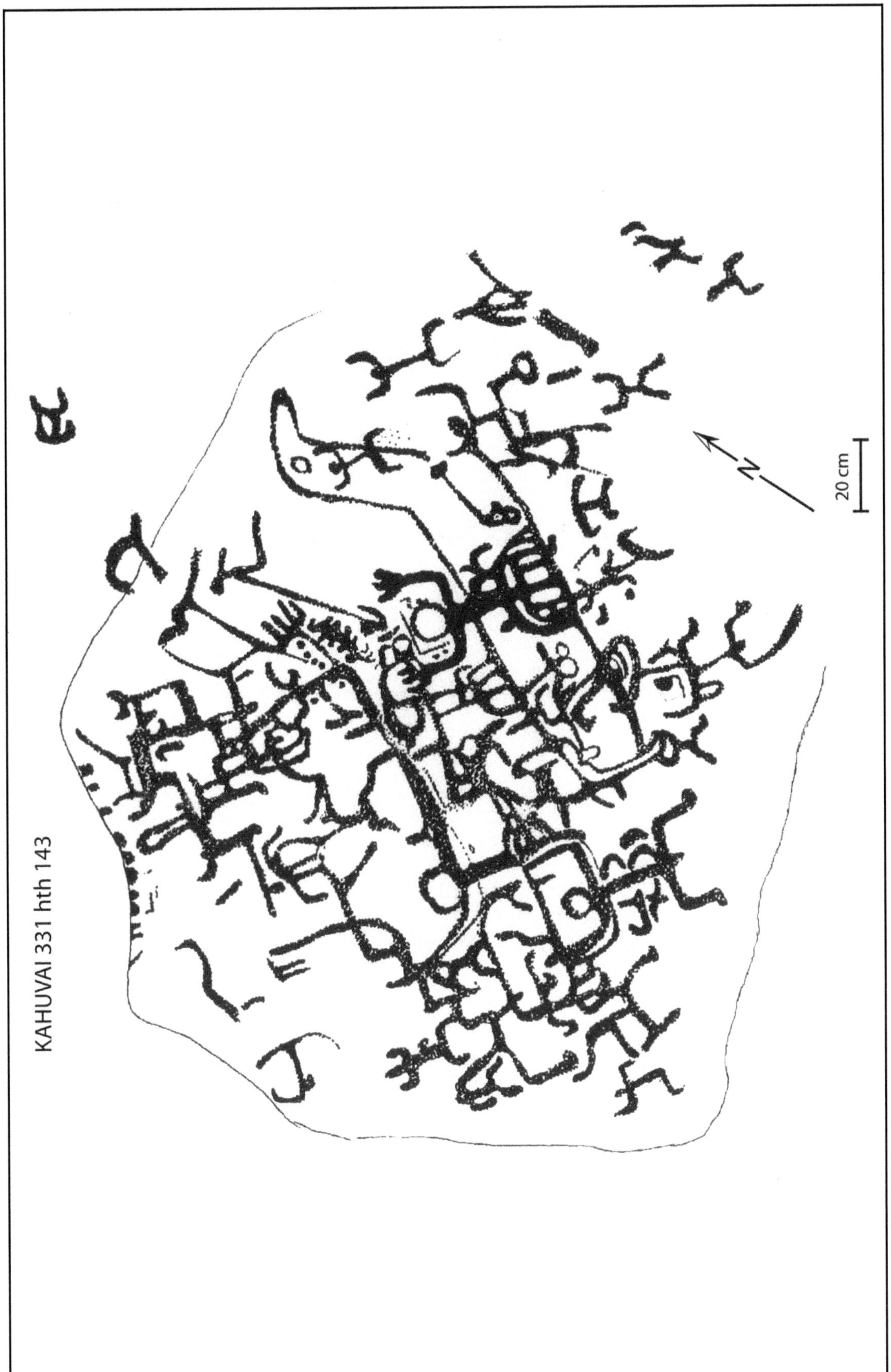

Figure 58. Kahuvai. A megalithic boulder with several anthropomorphic figures overlapping two dog figures (331hth 143).

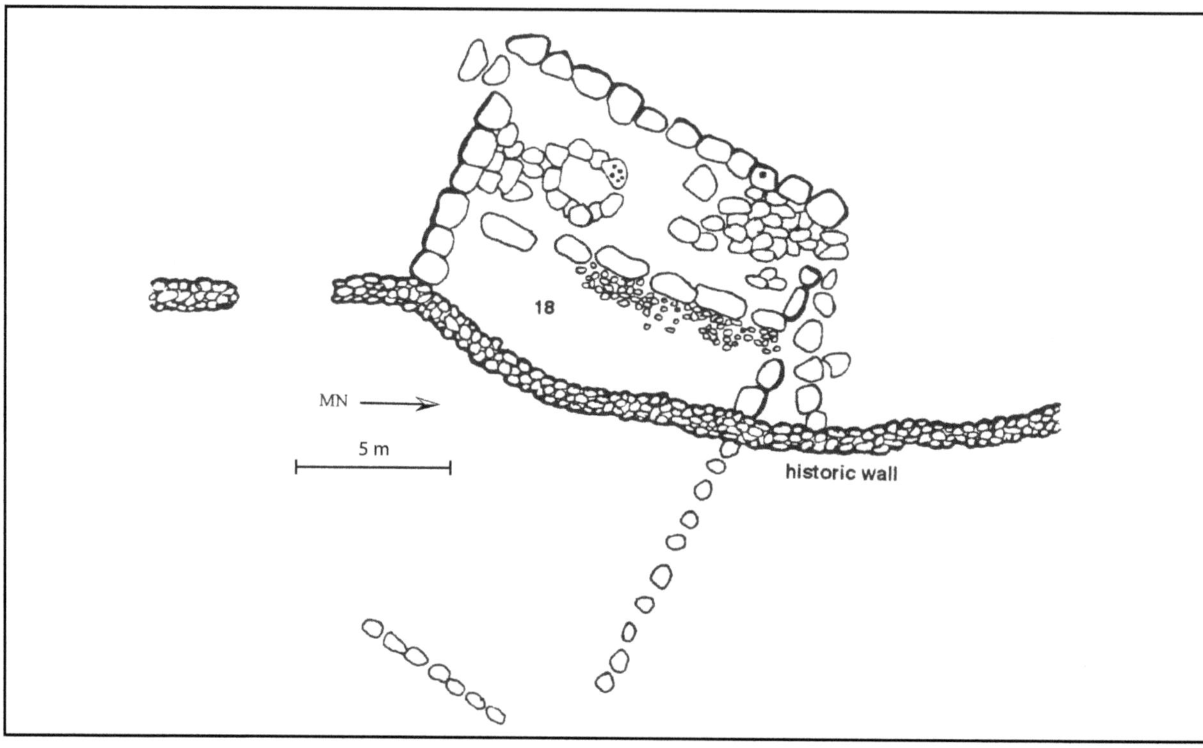

Figure 59. A historic wall constructed over a traditional sleeping house (Structure 18).

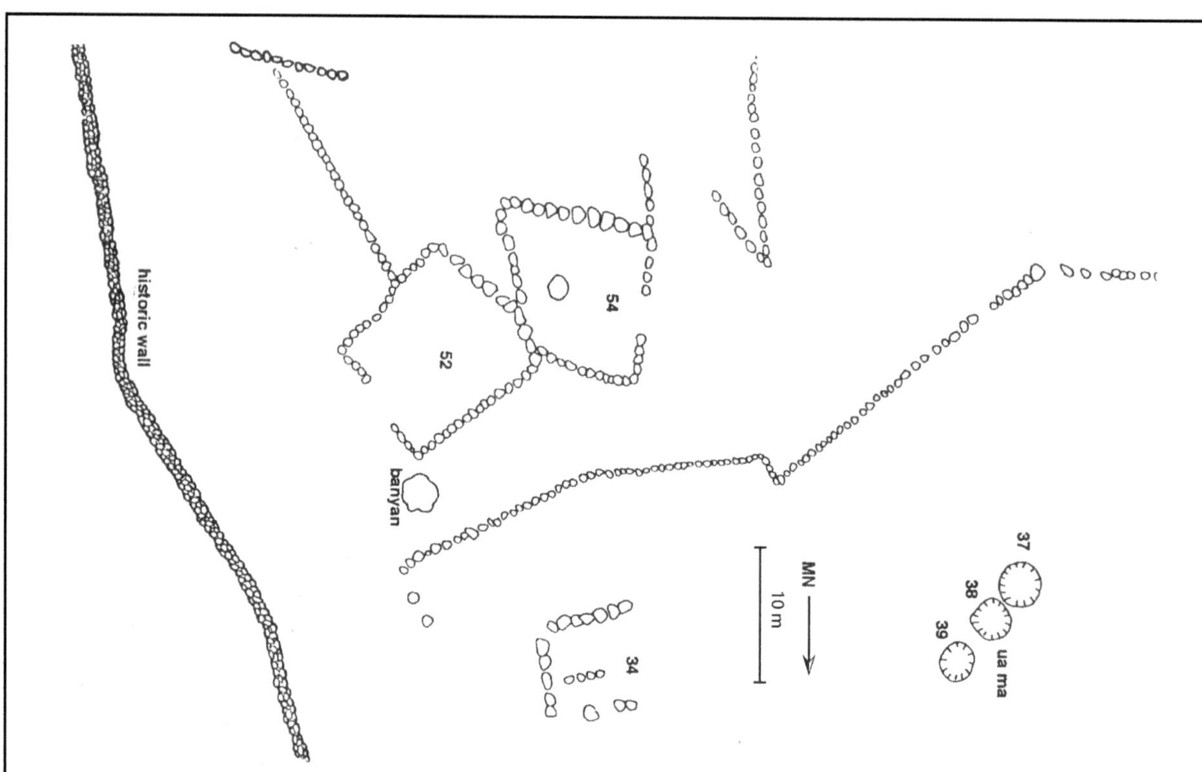

Figure 60. A section of the walled area. Note the three breadfruit silos to the right (Structures 37, 38, and 39).

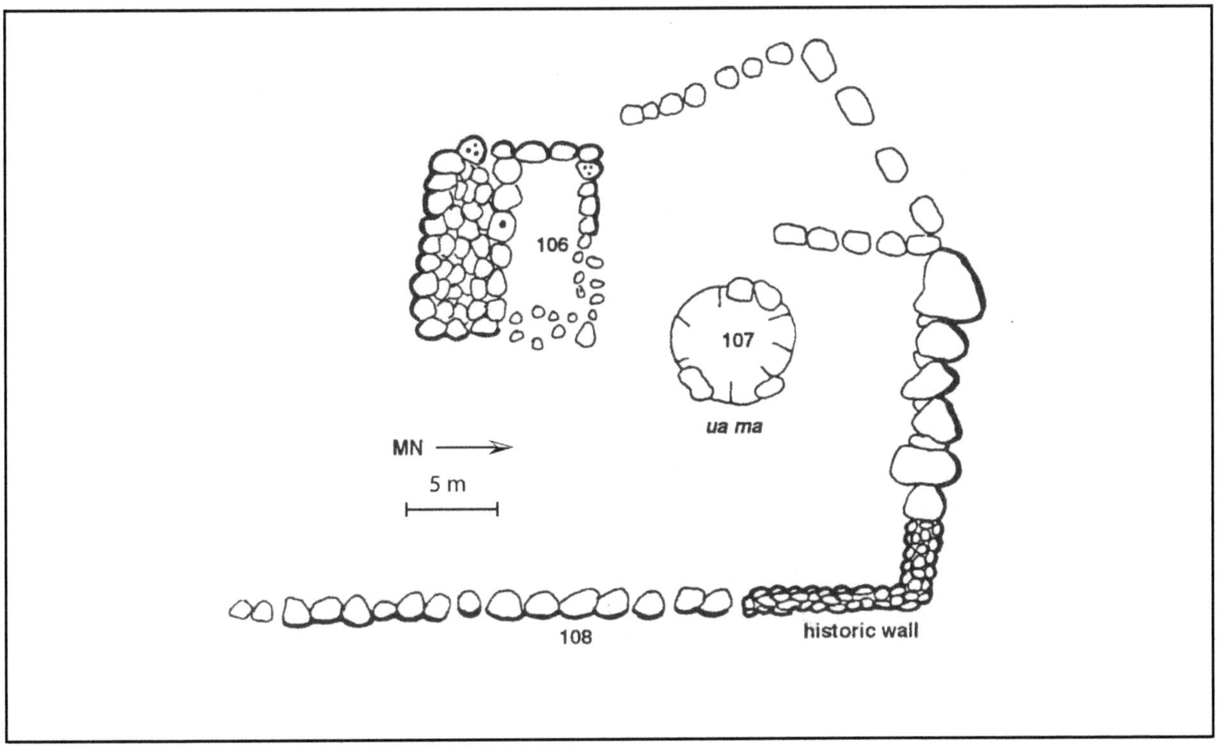

Figure 61. A small residential unit with a breadfruit silo located in the agricultural area.

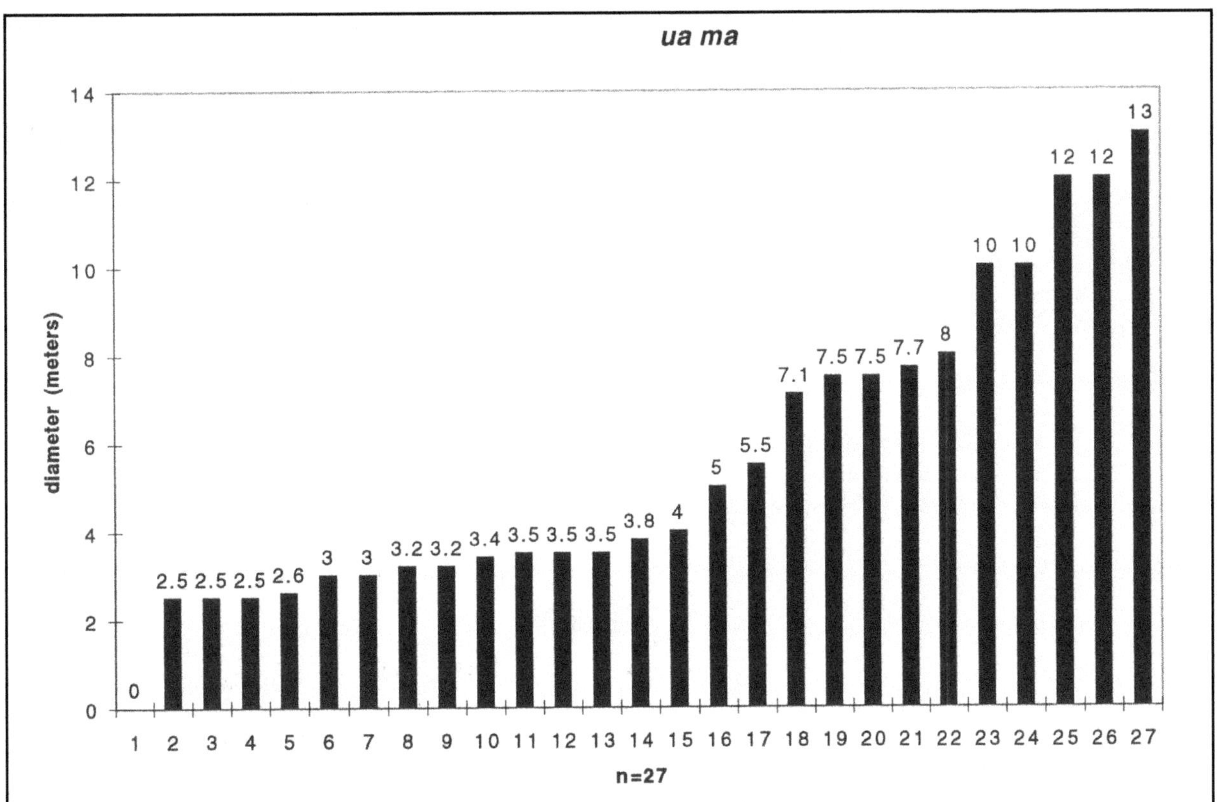

Figure 62. The diameters (in meters) of the breadfruit silos in the western section of Hatiheu Valley.

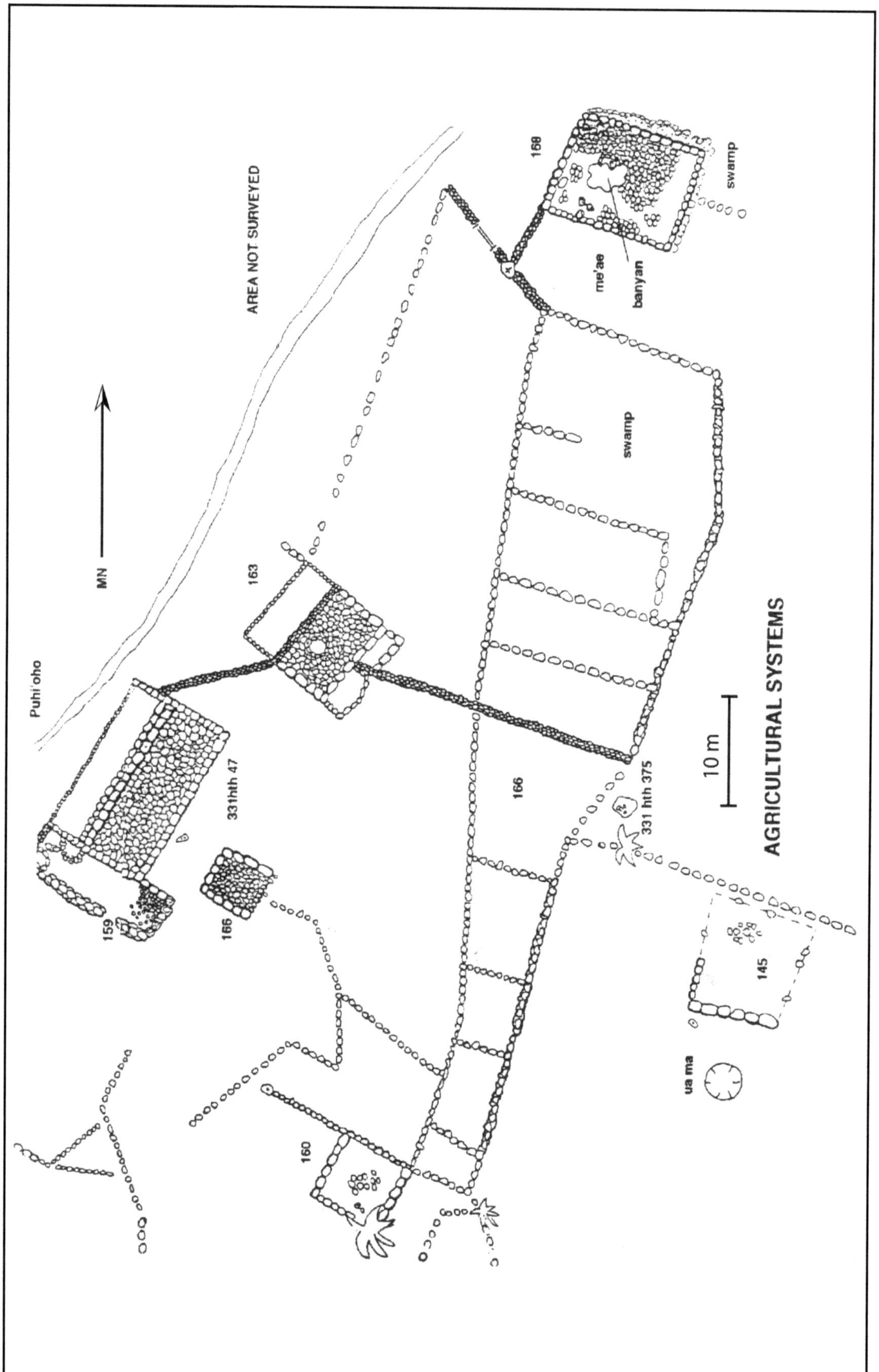

Figure 63. Abandoned irrigated taro fields. Petroglyph boulders 331hth 47 and 375 are marked.

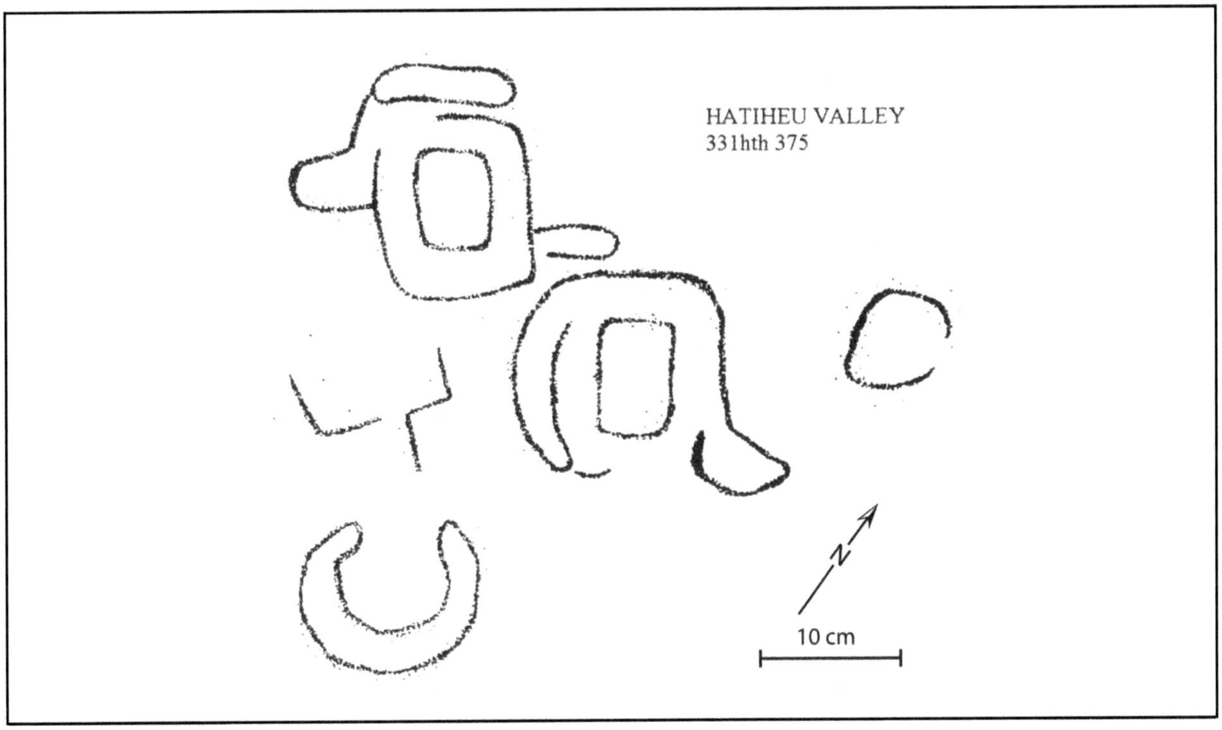

Figure 64. The petroglyphs are related to tattoo motifs.

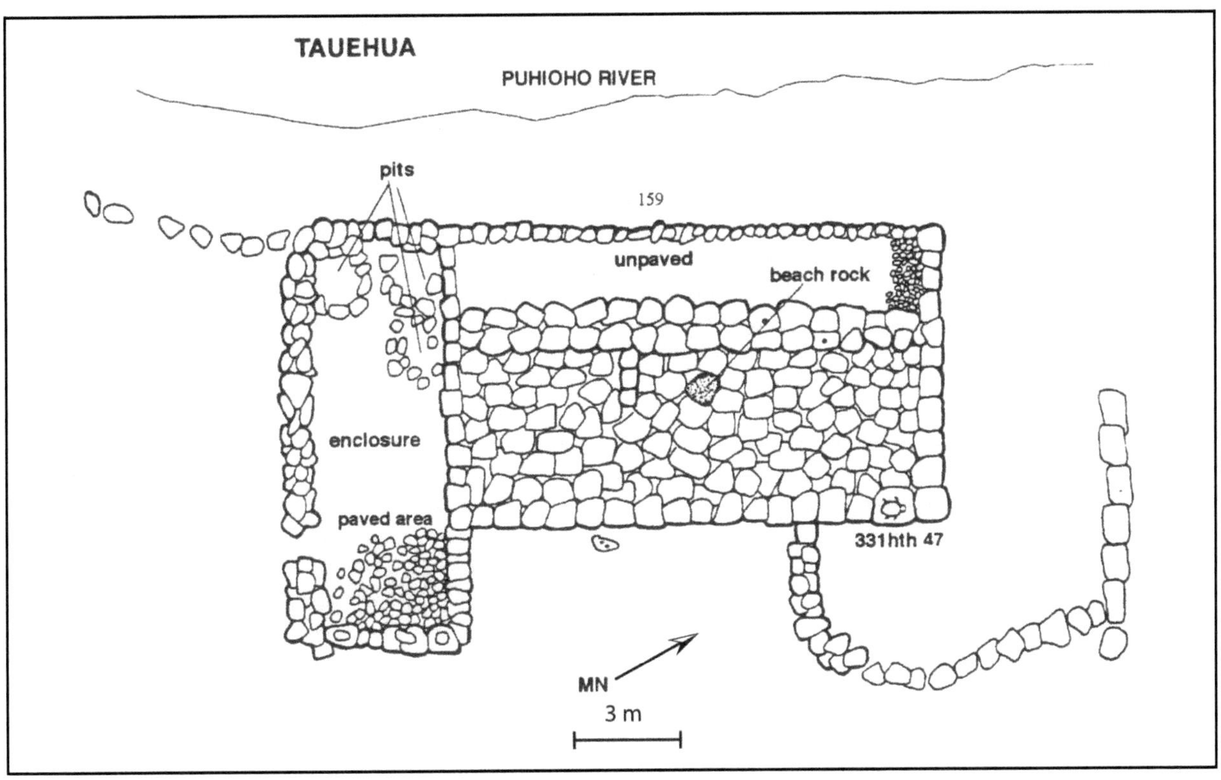

Figure 65. Structure 159 with the location of a bas-relief turtle depicted on top the northeast-facing wall (331hth 47).

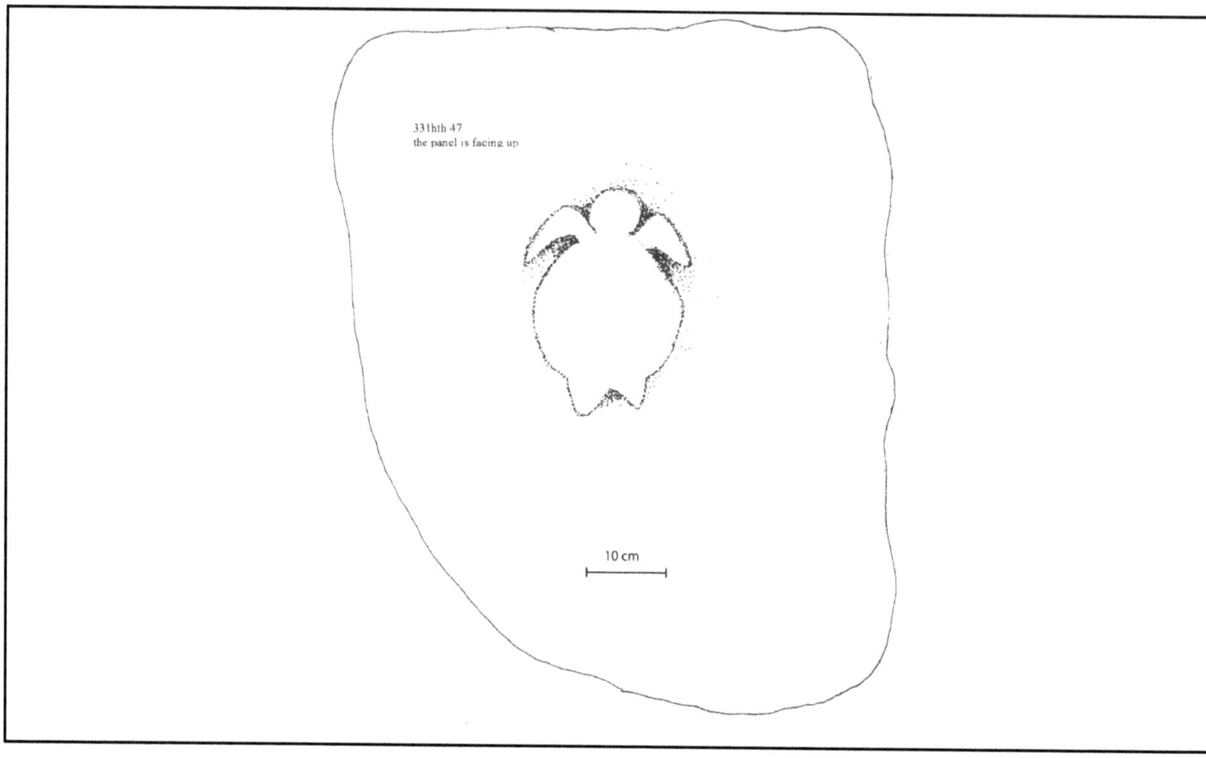

Figure 66. Bas-relief turtle placed on Structure 159 (331hth 47).

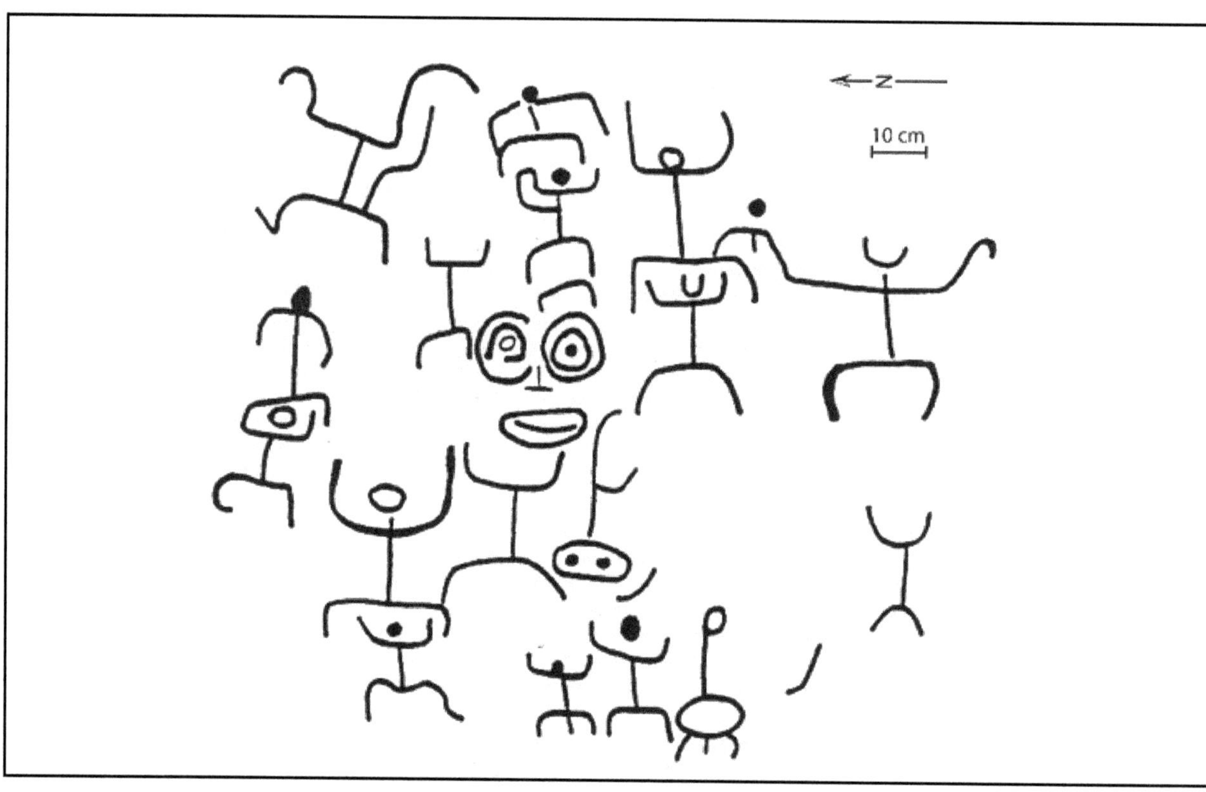

Figure 67. Niuamapu area, Hatiheu Valley. Anthropomorphic figures on a boulder located in the center of a Puhioho River (331hth 46/1).

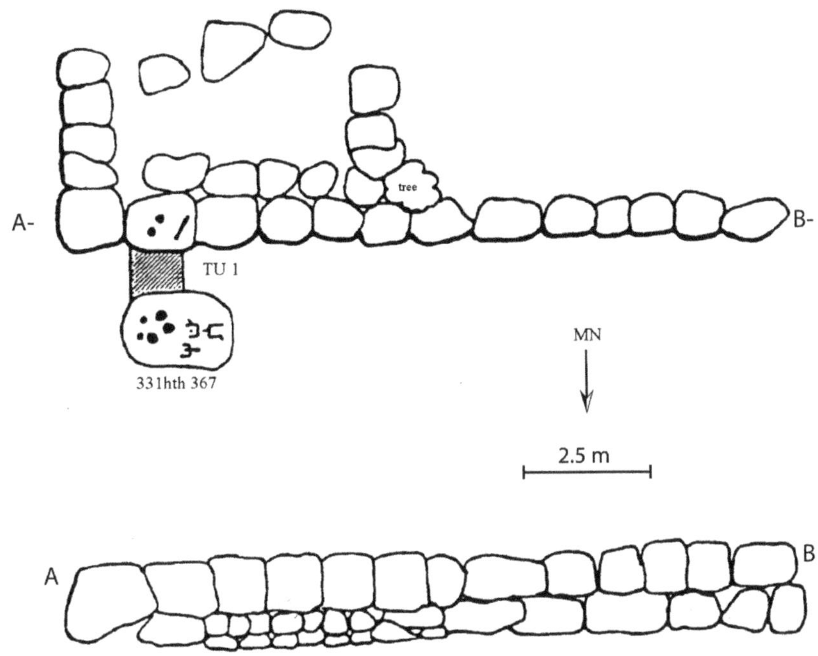

Figure 68. Structure 200 showing Test Unit 1 and petroglyph boulder 331hth 367.

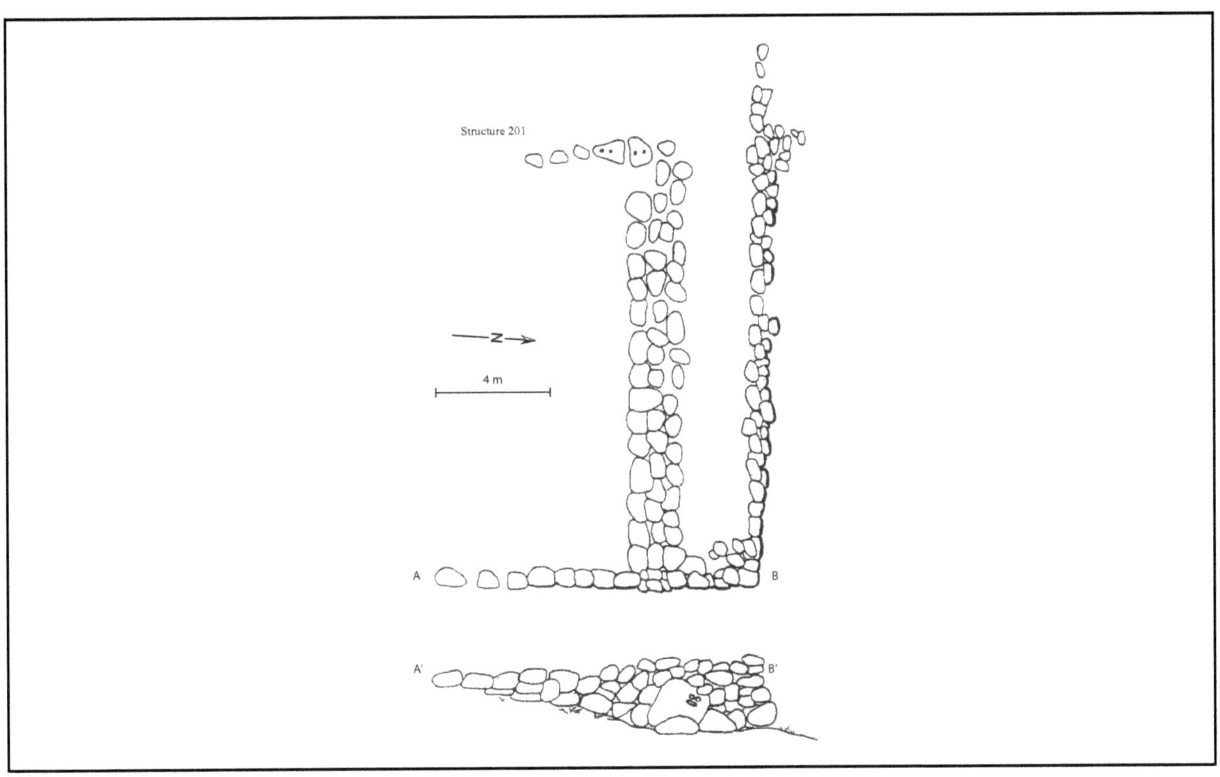

Figure 69. Structure 201 has been constructed around a petroglyph boulder (331hth 390).

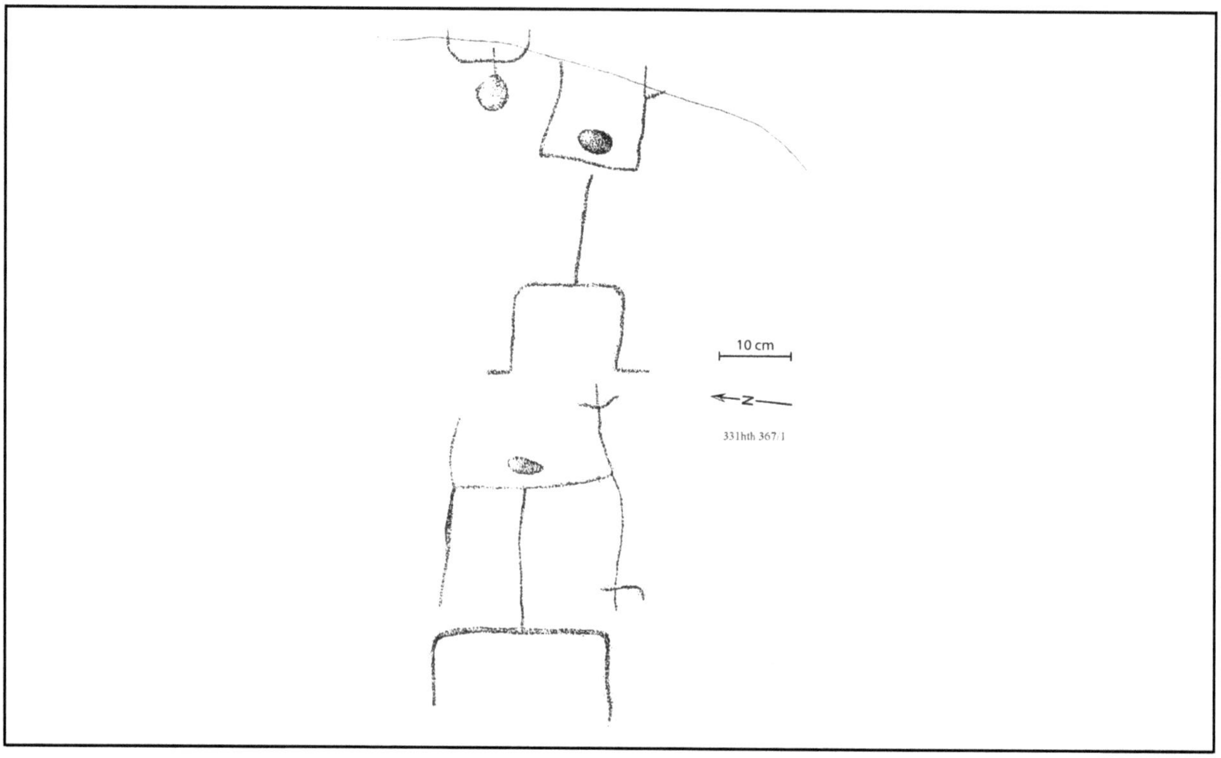

Figure 70. Petroglyphs boulder with anthropomorphs (331hth 367/1).

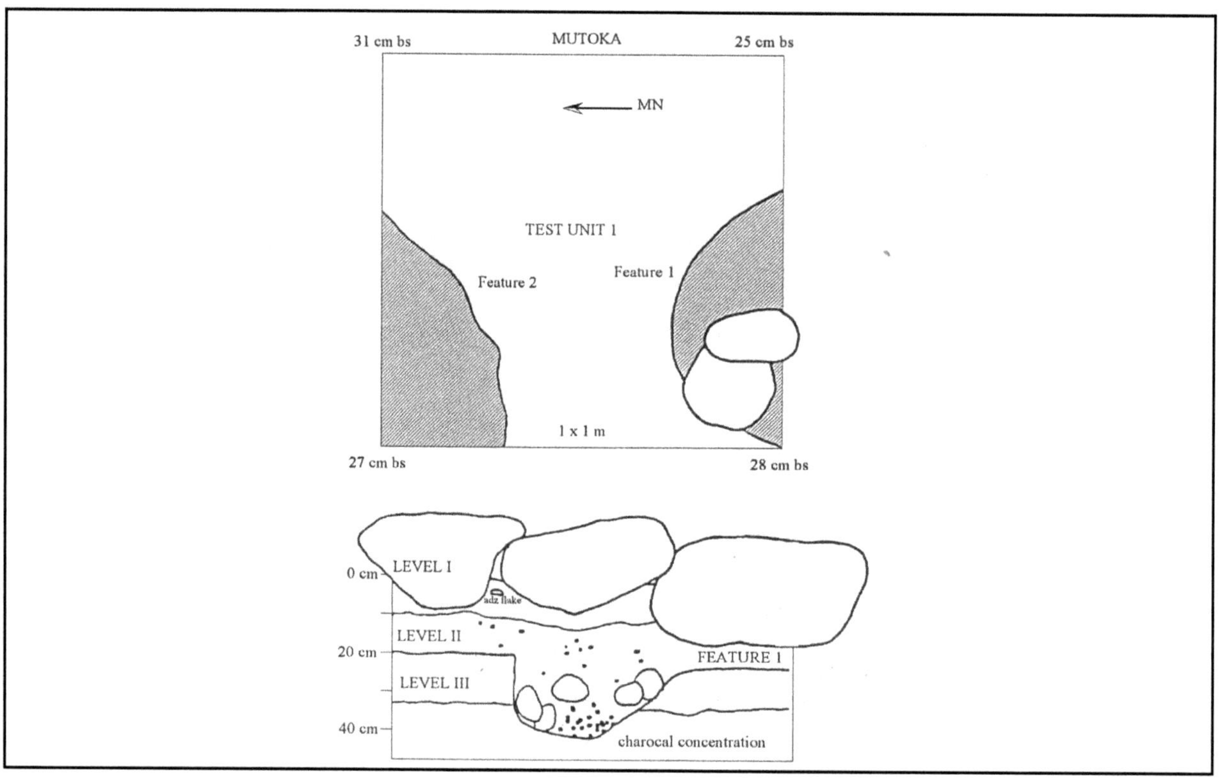

Figure 71. Test Unit 1 showing Features 1 and 2 and a stratigraphic section of the north wall of Structure 200.

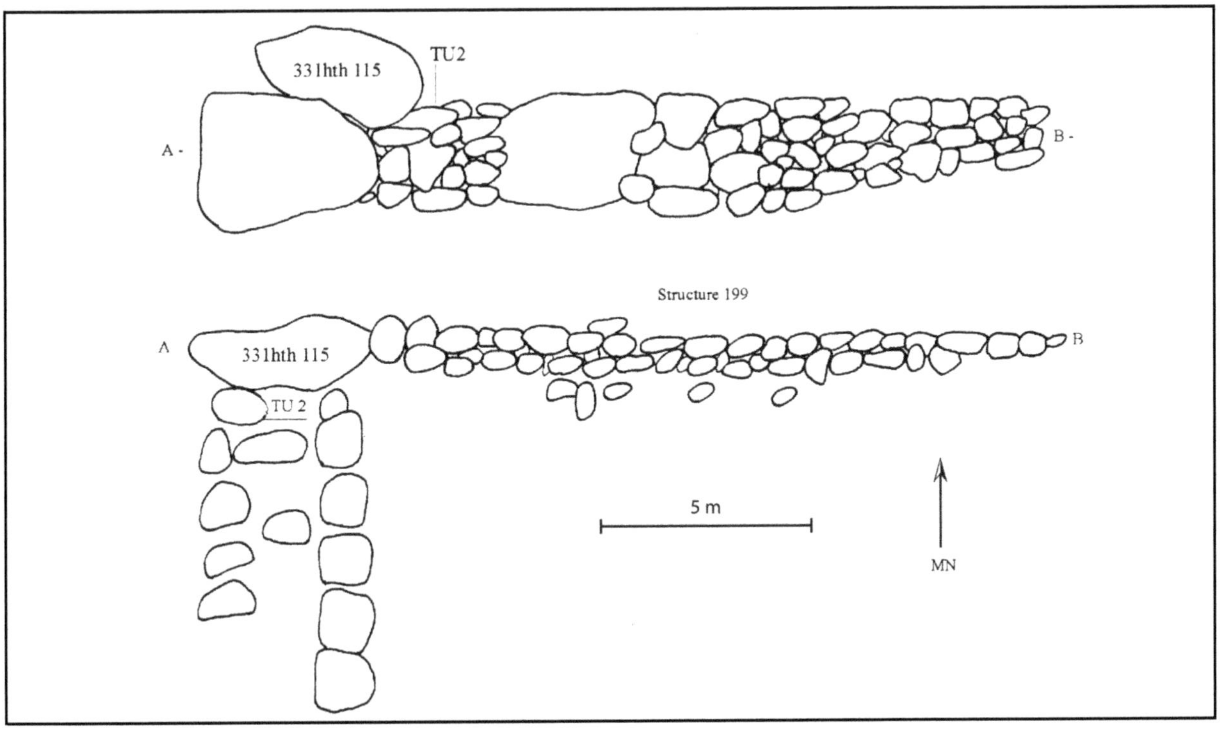

Figure 72. Kamuihei-Te Iipoka area showing Test Unit 2 and terrace (Structure 199) with petroglyph boulder (331hth 115).

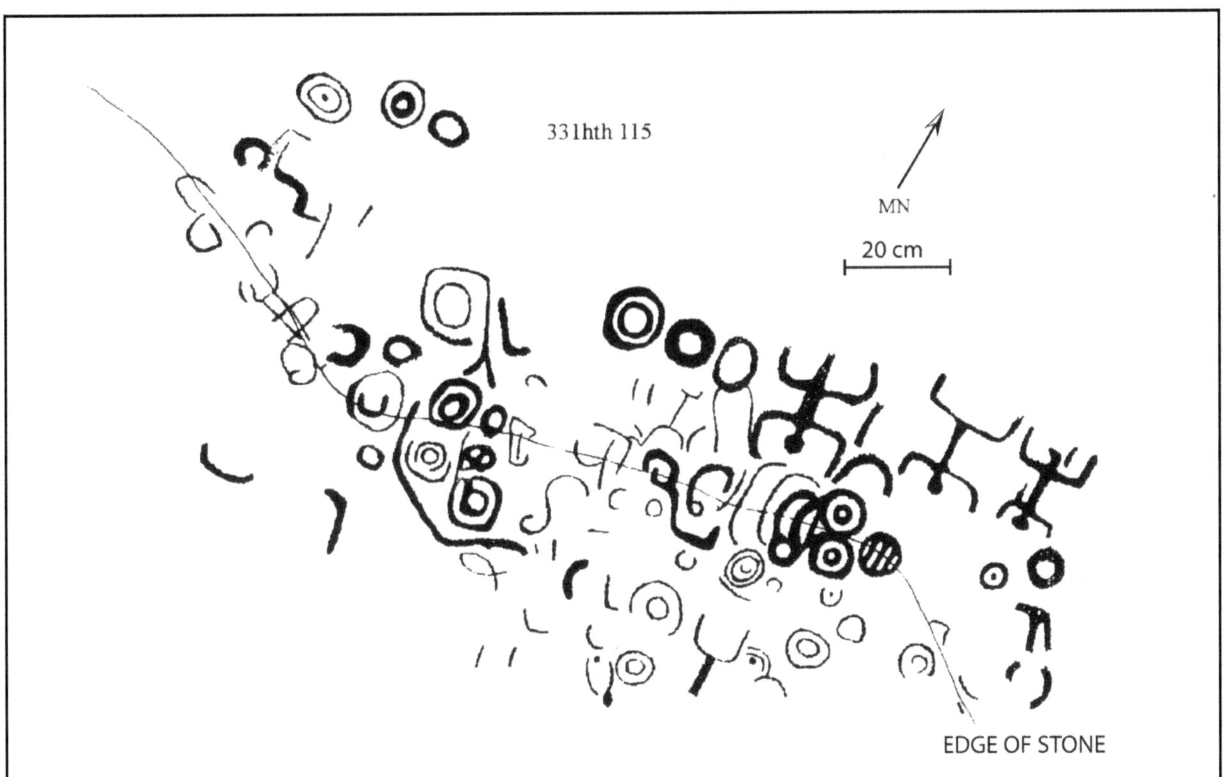

Figure 73. Kamuihei- Te Iipoka area. Petroglyph boulder next to Test Unit 2 (331hth 115).

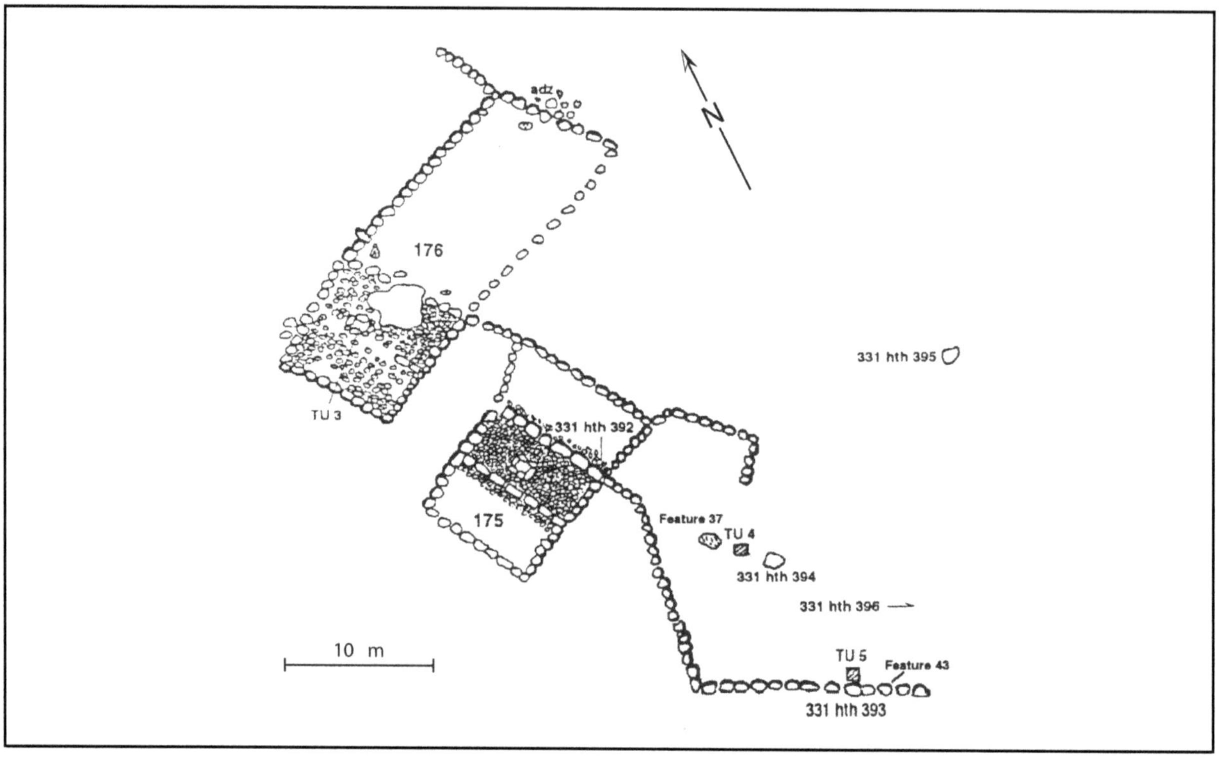

Figure 74. Ototemoui Ridge showing Structures 175 and 176, Test Units 3, 4, and 5 and various features. Locations of petroglyph boulders (331hth 392, 393, 394, 395, and 396) are marked.

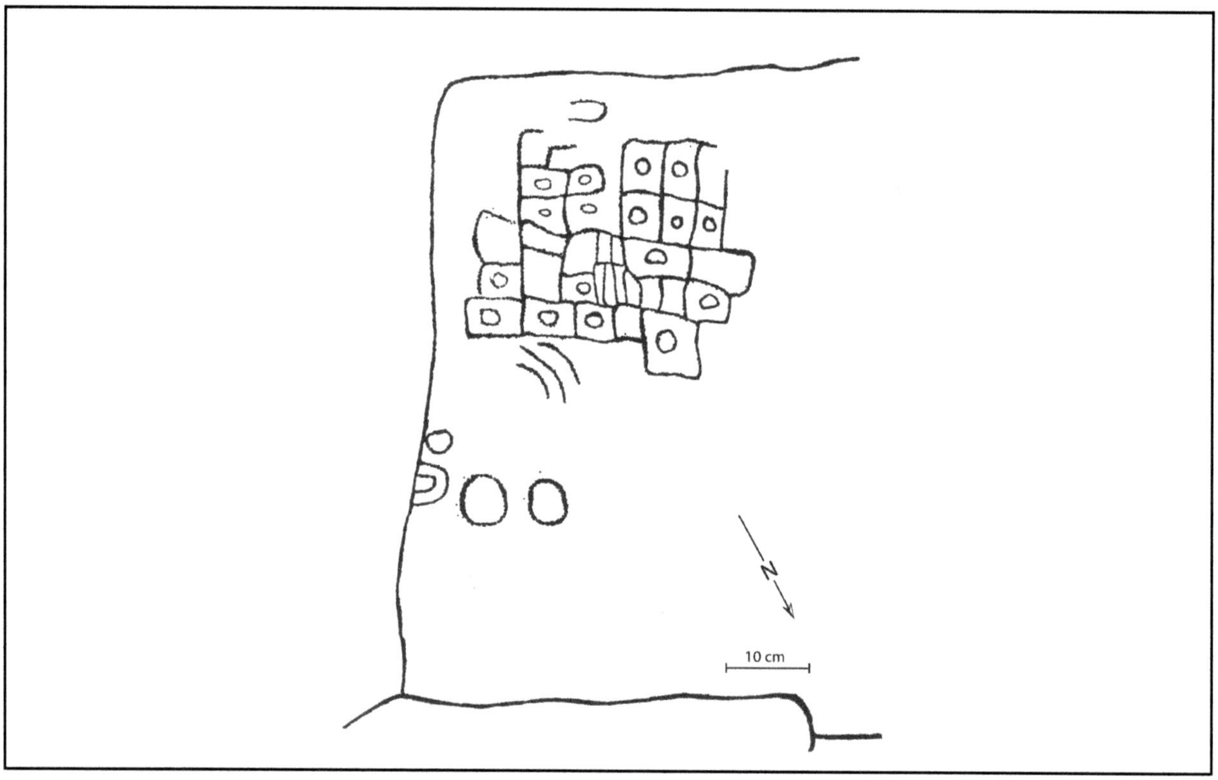

Figure 75. Ototemoui Ridge. Petroglyph boulder with faces and geometric figures (331hth 392).

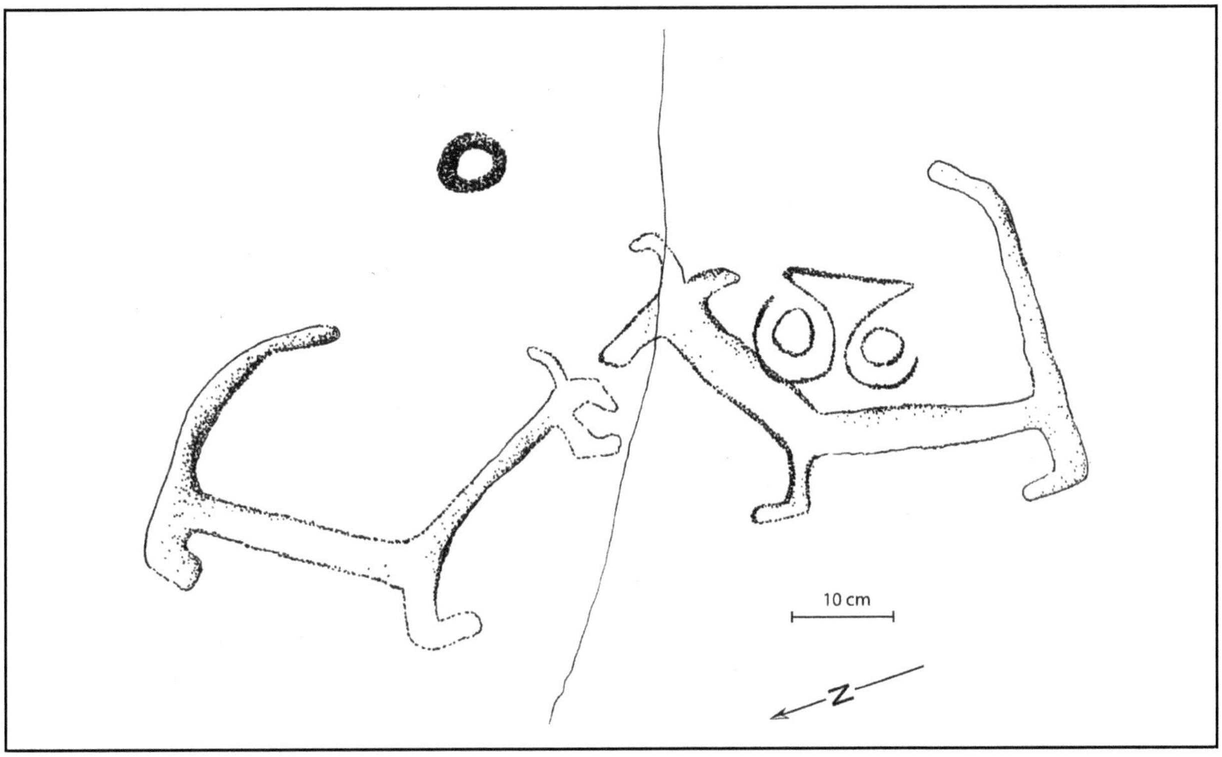

Figure 76. Ototemoui Ridge. Two dog petroglyphs and an anthropomorphic face wrapped around the edge of a boulder (331hth 394/1).

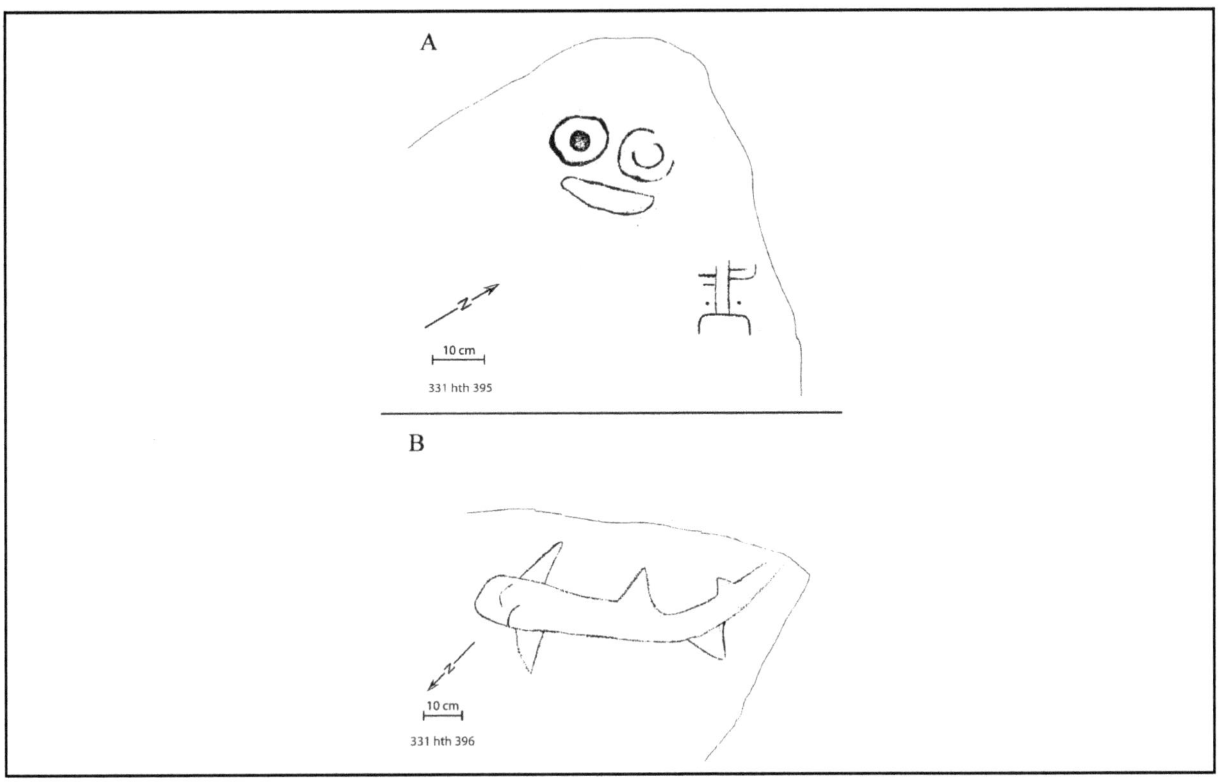

Figure 77 a–b. Ototemoui Ridge. A) a face figure and an unfinished anthropomorph (331hth 395); B) a shark petroglyph (331hth 396).

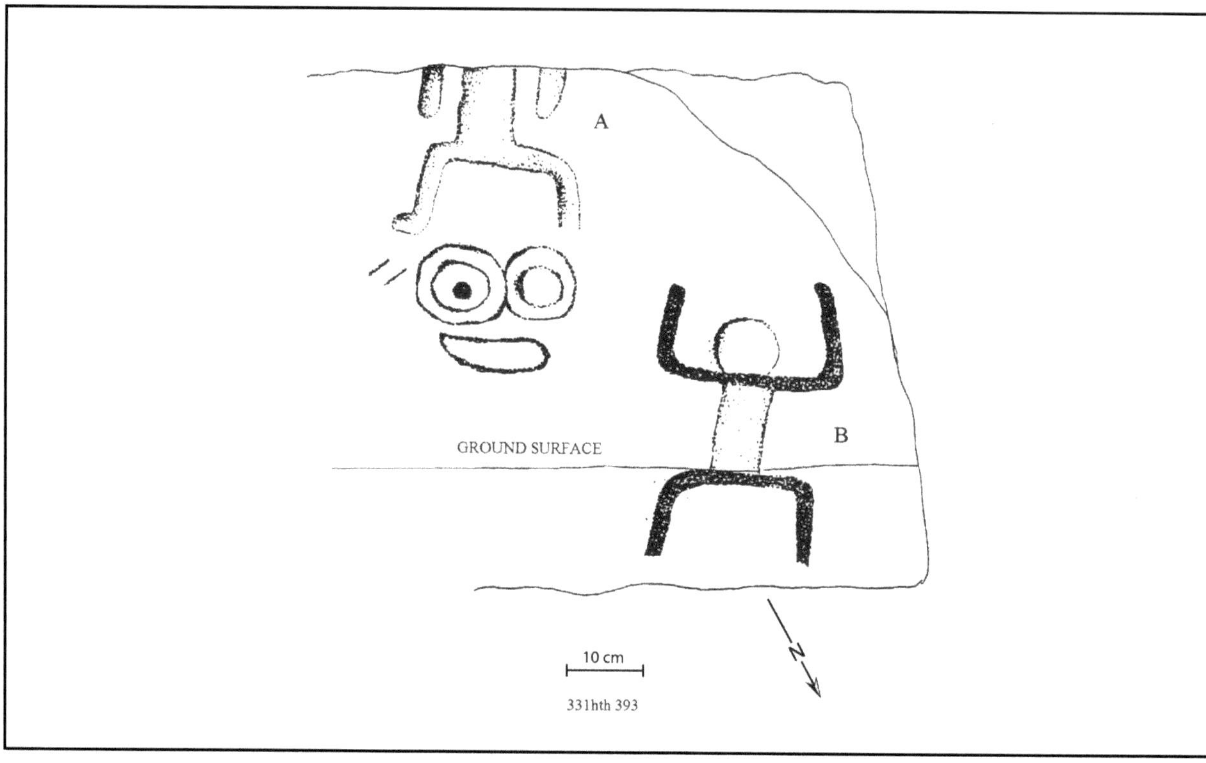

Figure 78. Ototemoui Ridge. A boulder with anthropomorphs is located near Test Unit 5 (331hth 393).

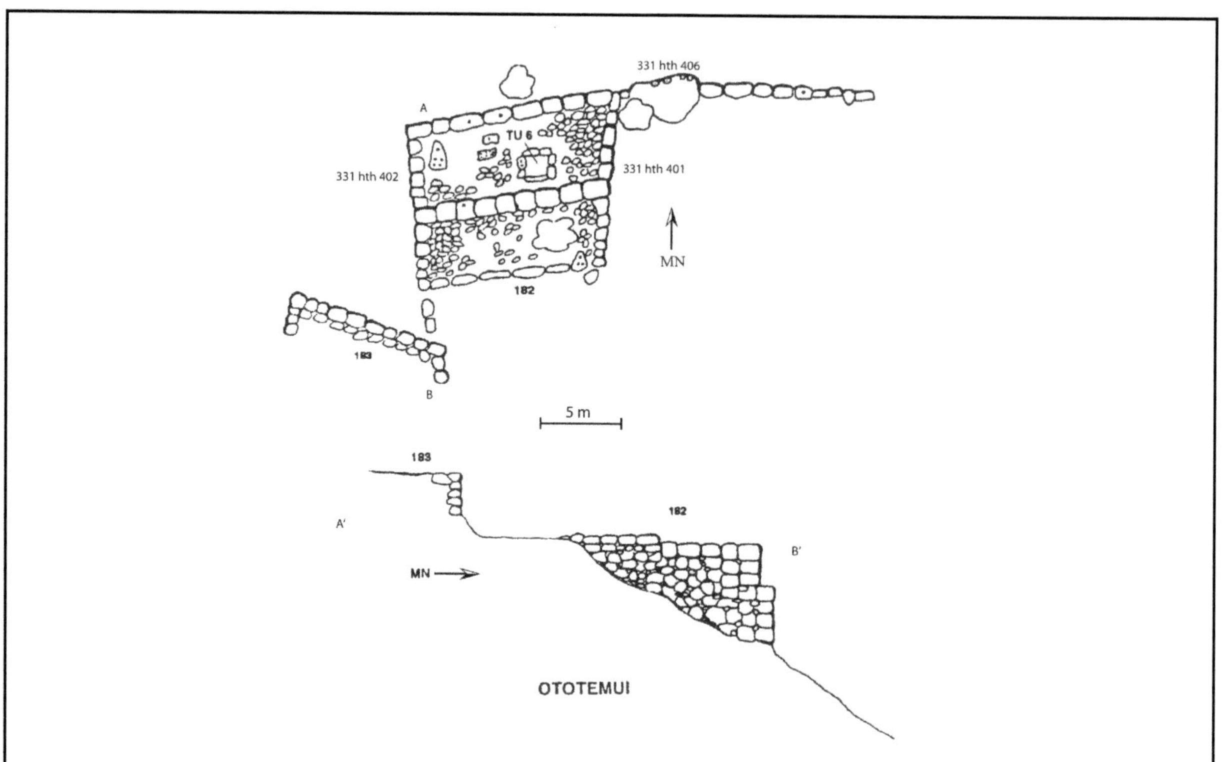

Figure 79. Ototemoui Ridge. Terrace (Structure 182) with Test Unit 6 and petroglyph boulders marked (331hth 401, 402, and 406).

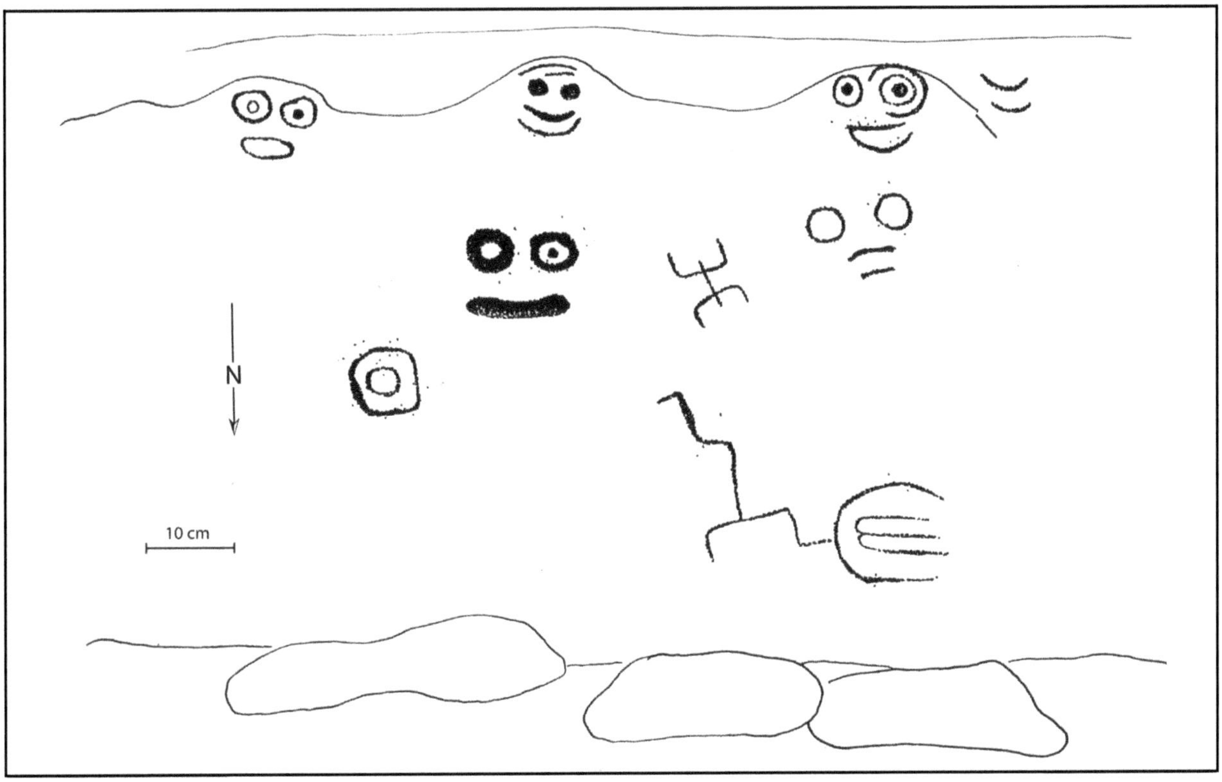

Figure 80. Ototemoui Ridge. Petroglyph boulder (331hth 401) is located directly above Test Unit 6.

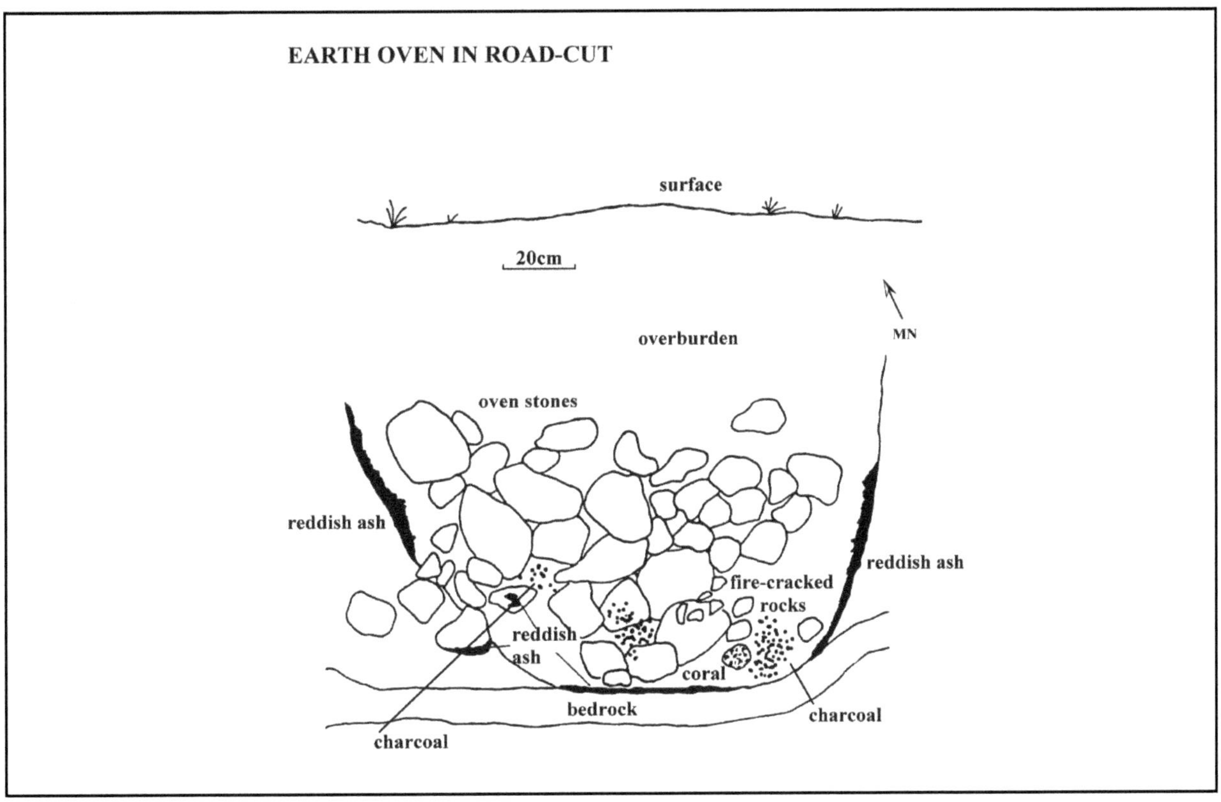

Figure 81. Earth oven located near Vaiuu River showing stratigraphic section.

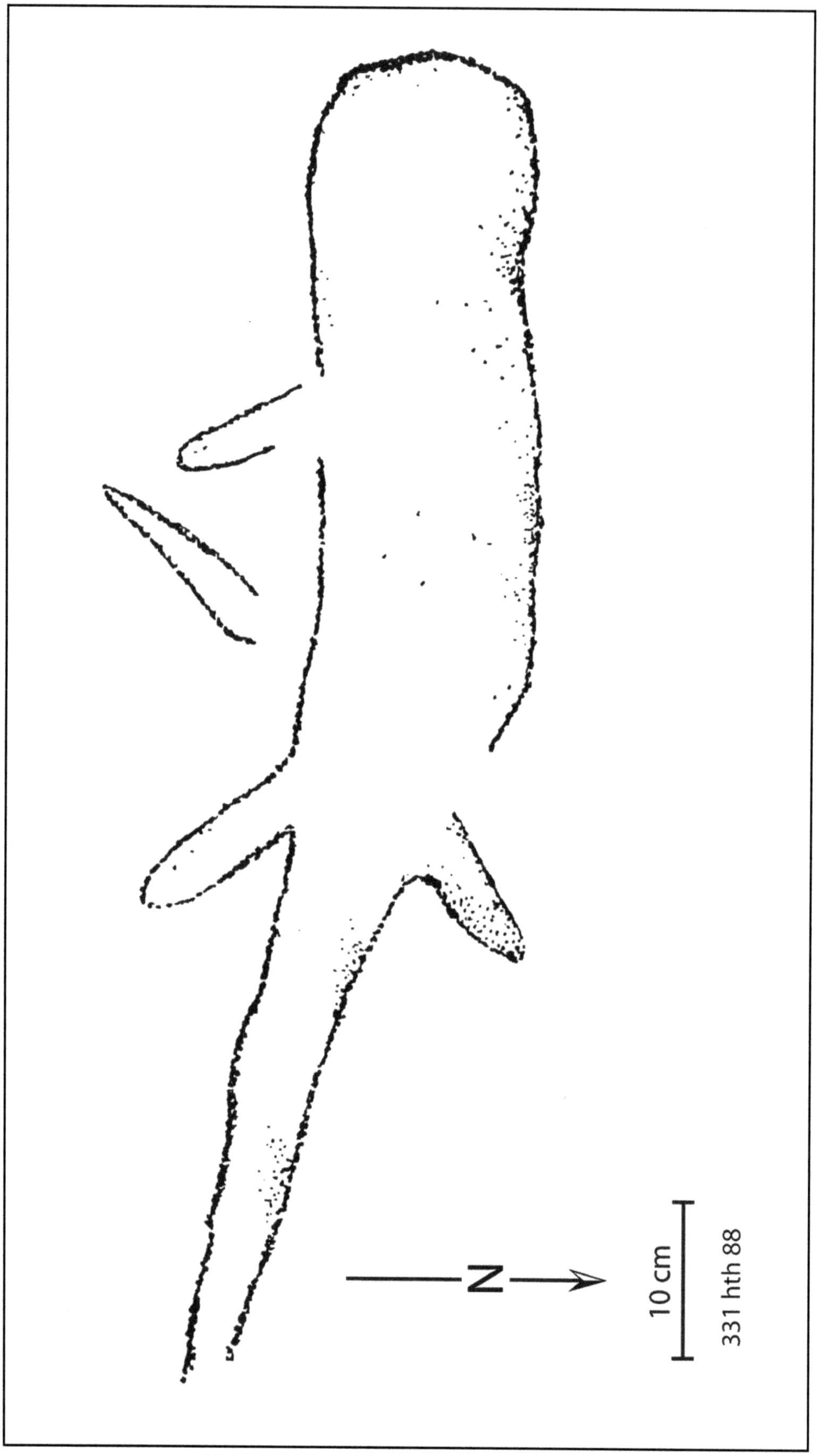

Figure 82. Petroglyph boulder is located near earth oven (331hth 88).

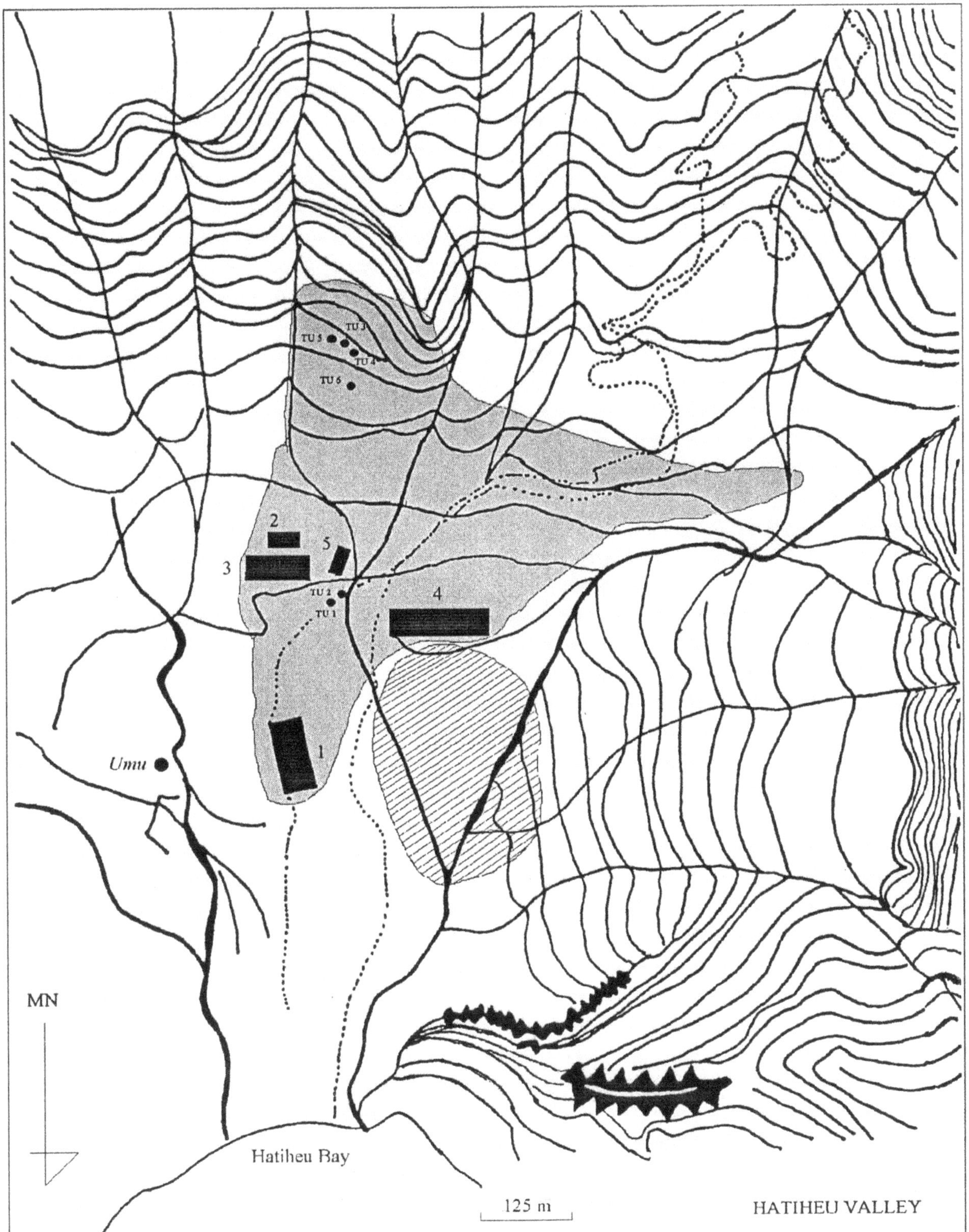

Figure 83. Research area with the location of Tohua Hikokua (1); Kamuihei II (2); Kamuihei I (3), Tahakia (4); Te Iipoka (5), the six test units, and the earth oven. The solid line encloses the chiefly area while the striped section indicates the agricultural section.

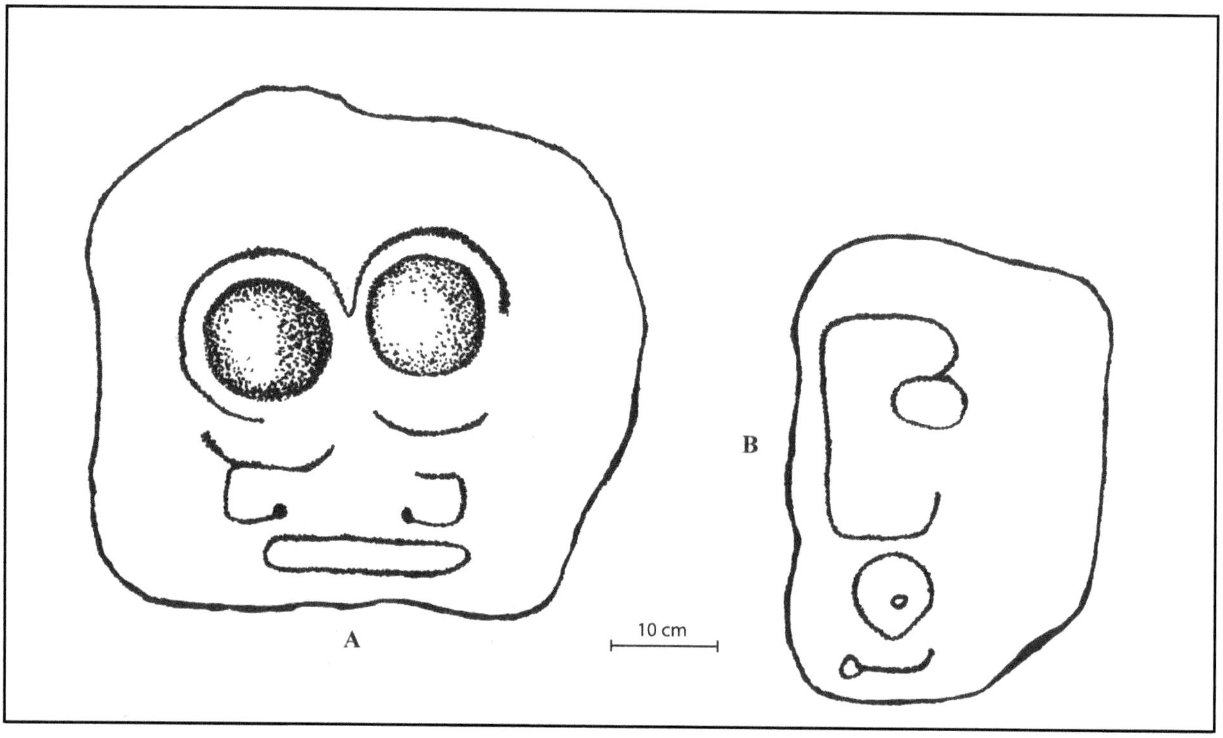

Figure 84 a–b. Tohua Kamuihei I. Portable petroglyph: A) top surface shows a face image with cup-shaped impressions; and B) one side depicts geometric figures (331hth 29).

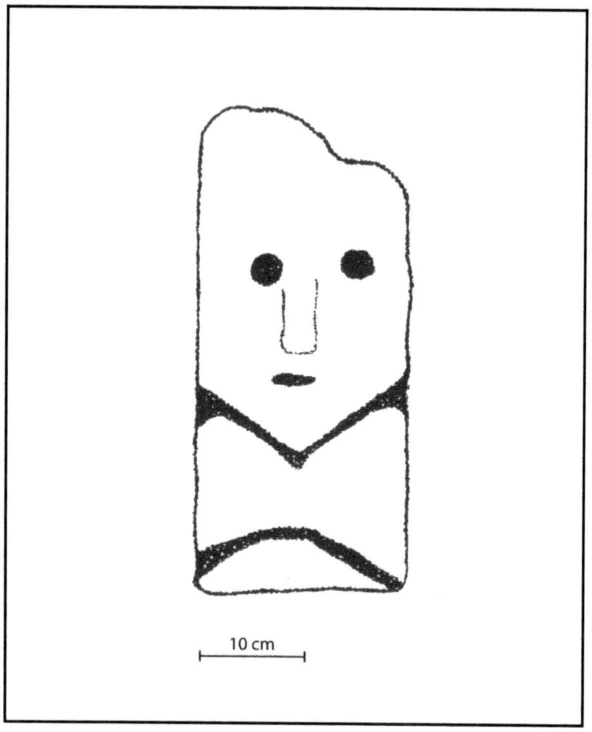

Figure 85. Tohua Kamuihei I. Portable sculpture in red tuff (331hthT1).

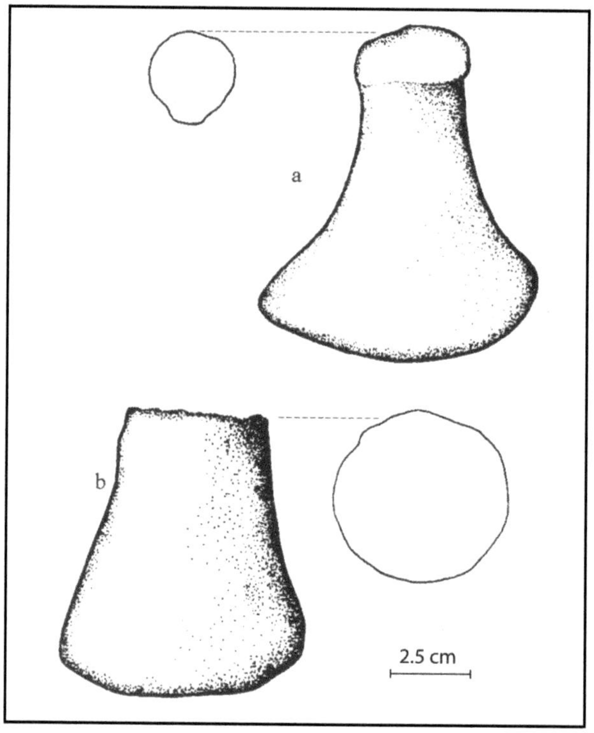

Figure 86 a–b. Basalt pounders from A) Tohua Kamuihei I and B) Tohua Kamuihei II.

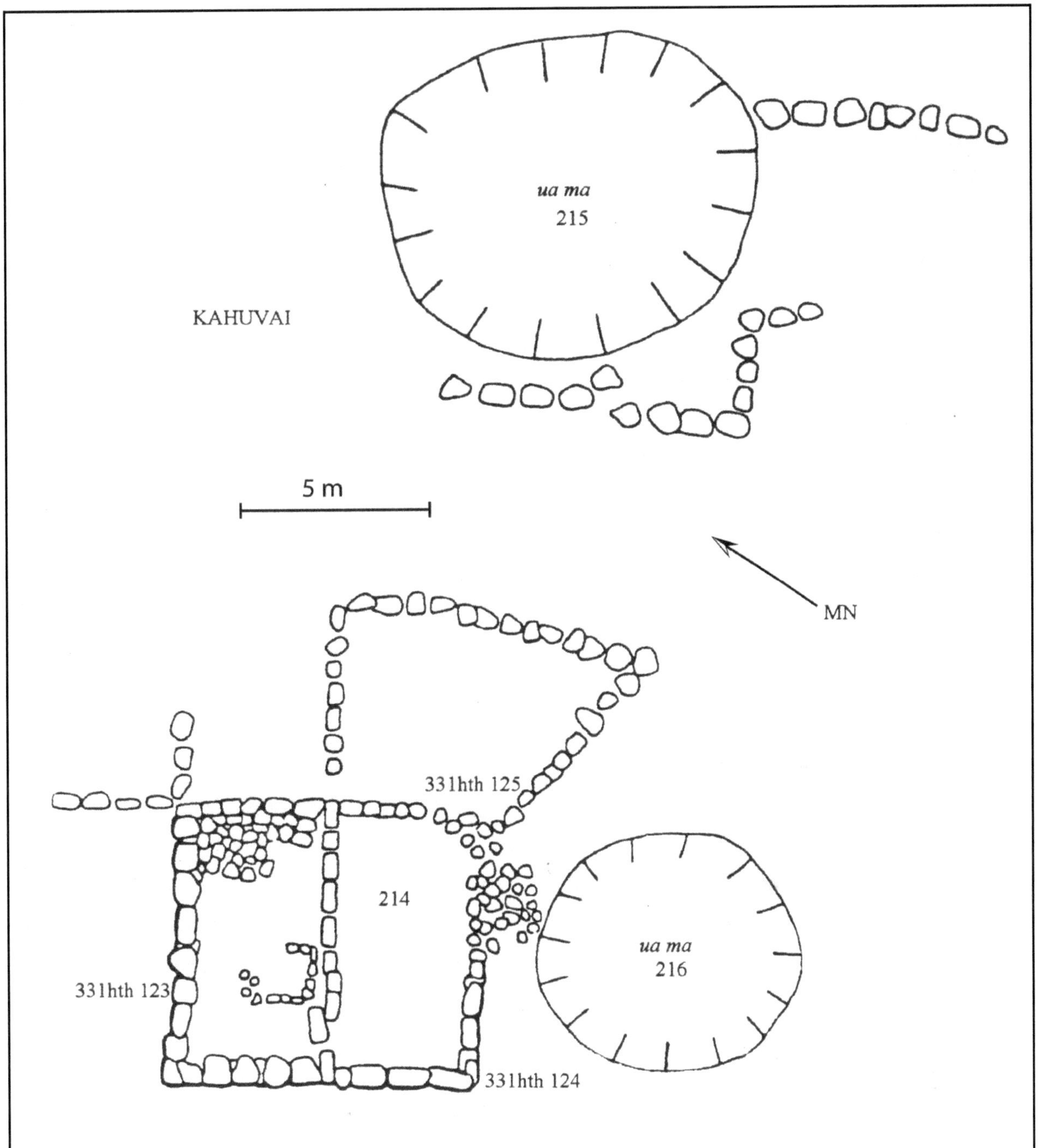

Figure 87. Structure 214 with two breadfruit silos (Structure 215 and 216). Circular geometric and face petroglyphs are placed on the outer walls of the house foundation (331hth 123, 124, and 125).

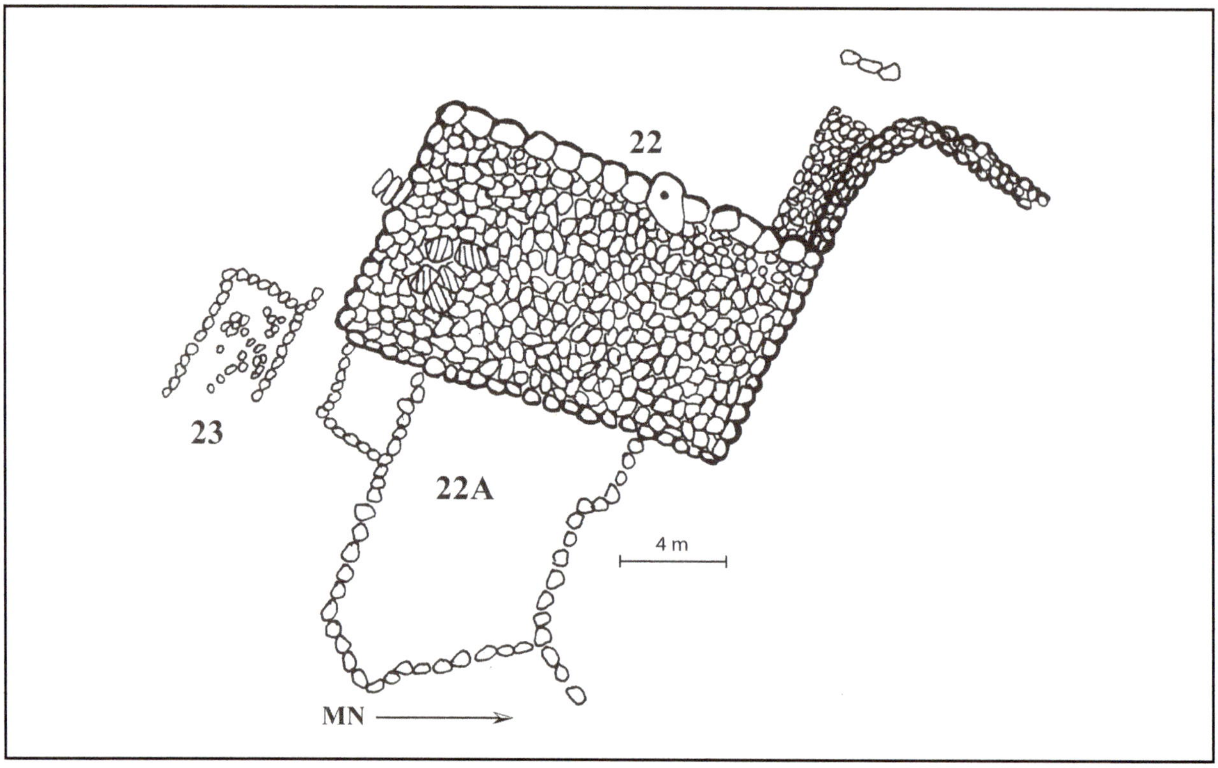

Figure 88. Modified traditional sleeping house foundation (Structure 22).

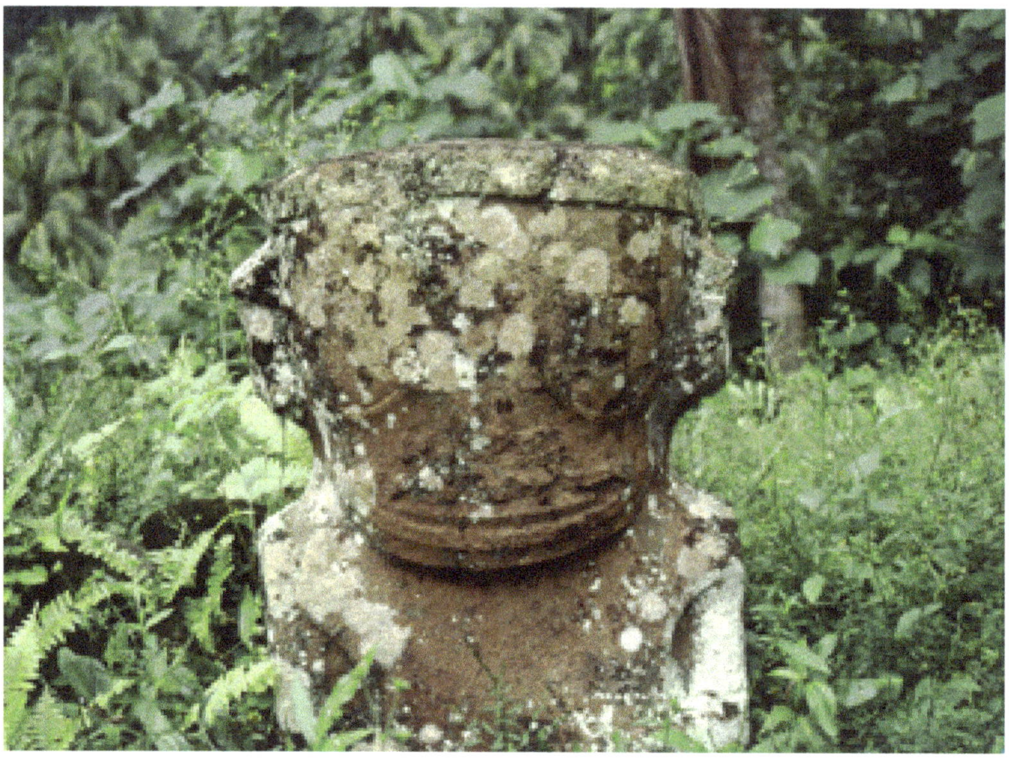

Figure 89. Site Meaiaute, Ua Huka with one of the four tiki.

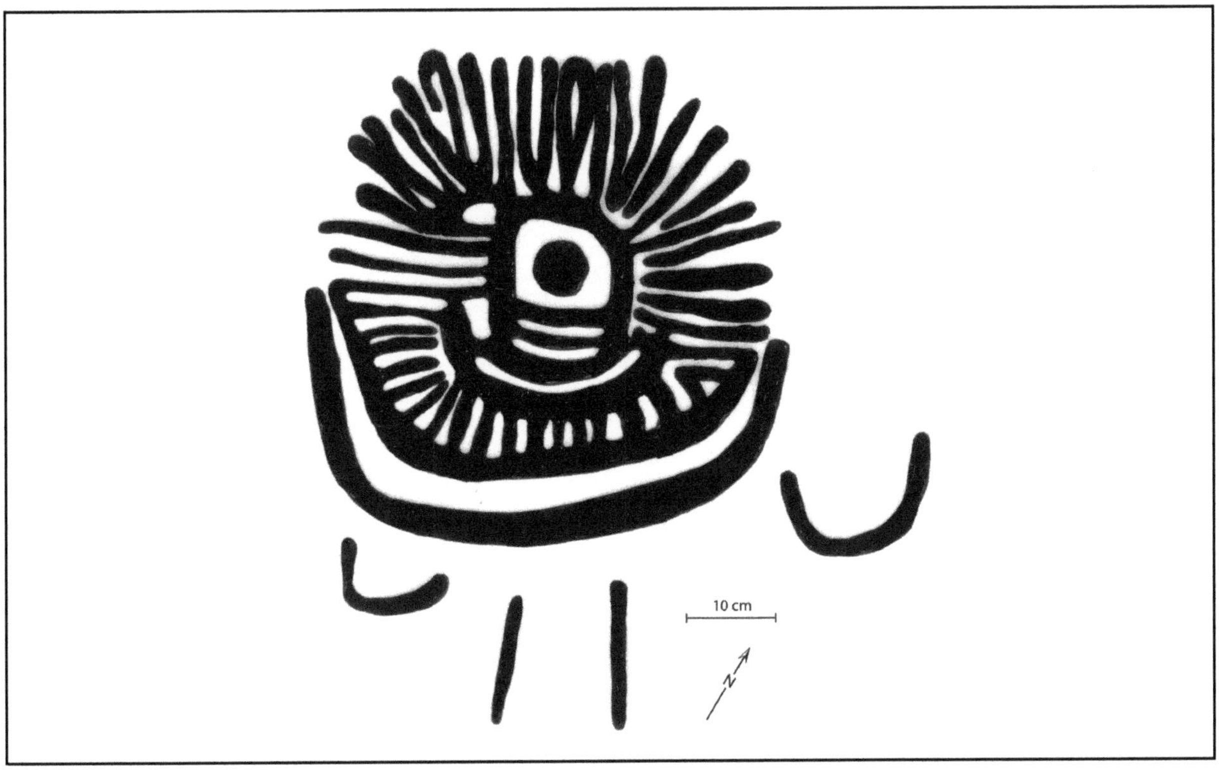

Figure 90. Vaiote Valley, Tautira, Tahiti. Part of a mourning headdress and breast-plate.

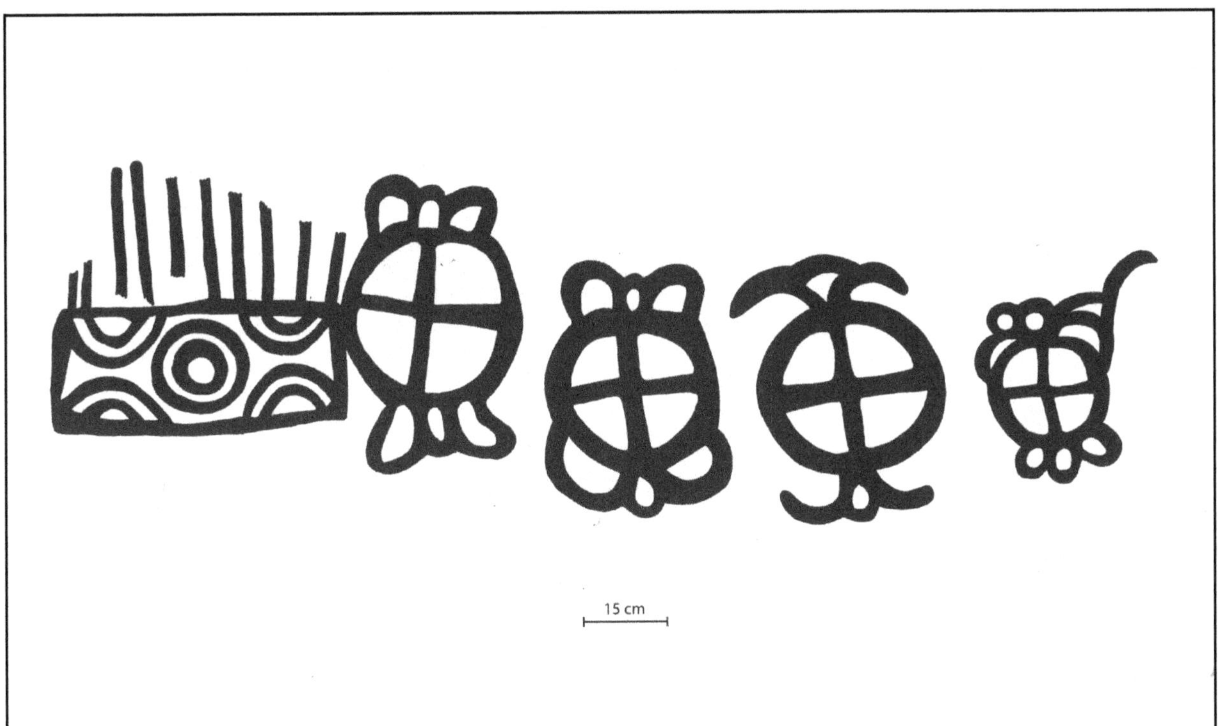

Figure 91. Vaiati Valley, Bora Bora. A petroglyph panel with four turtles and a headdress referred to as Ofai honu or turtle rock (BOB 3a).

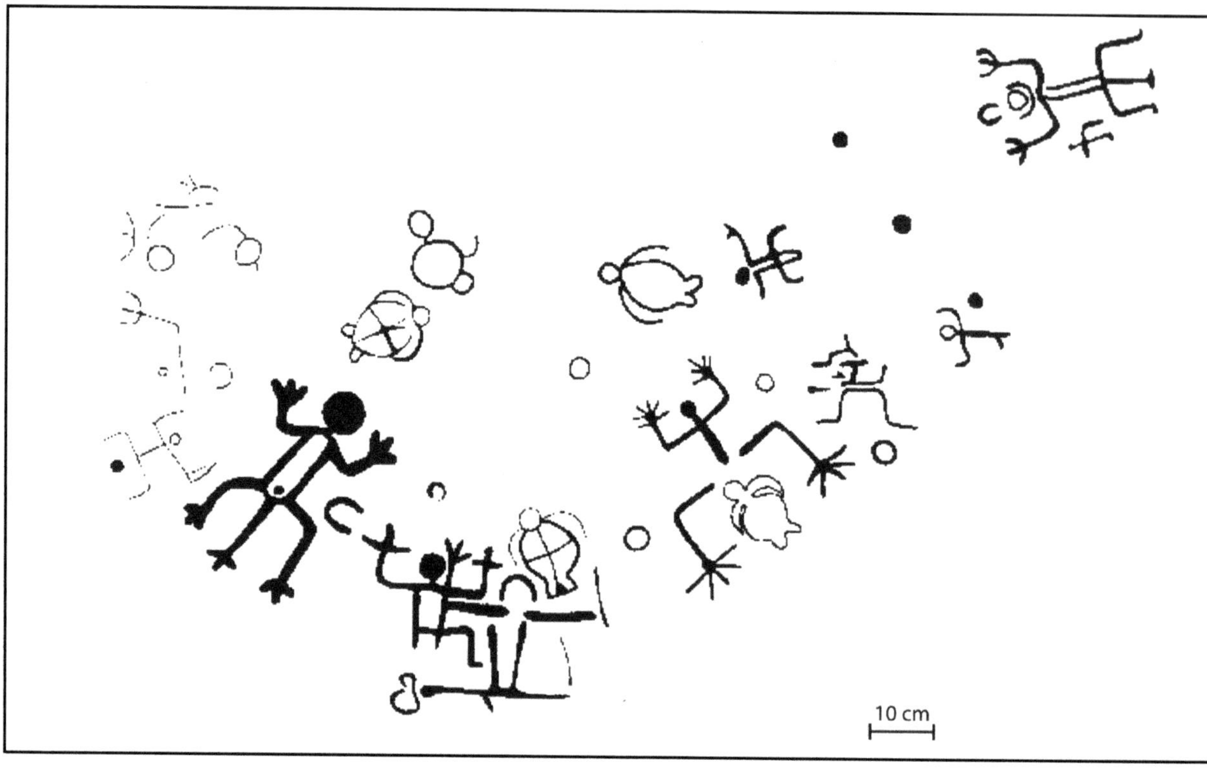

Figure 92. Opoa, Haapapara Valley, Raiatea. Boulder with turtles and anthropomorphs with forked tails (REP 23).

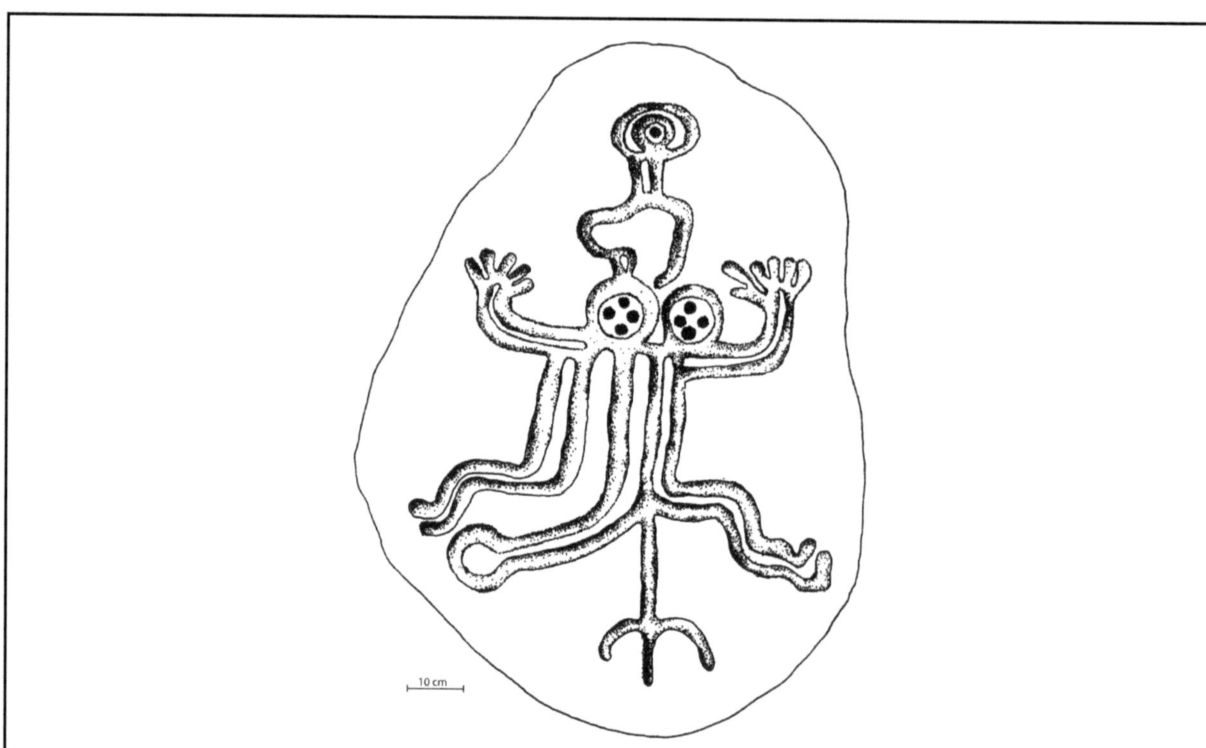

Figure 93. Tipaerui Valley, Tahiti. Exotic figures. The petroglyph boulder is presently housed in the garden of Musée de Tahiti et des îles, Punaauia.

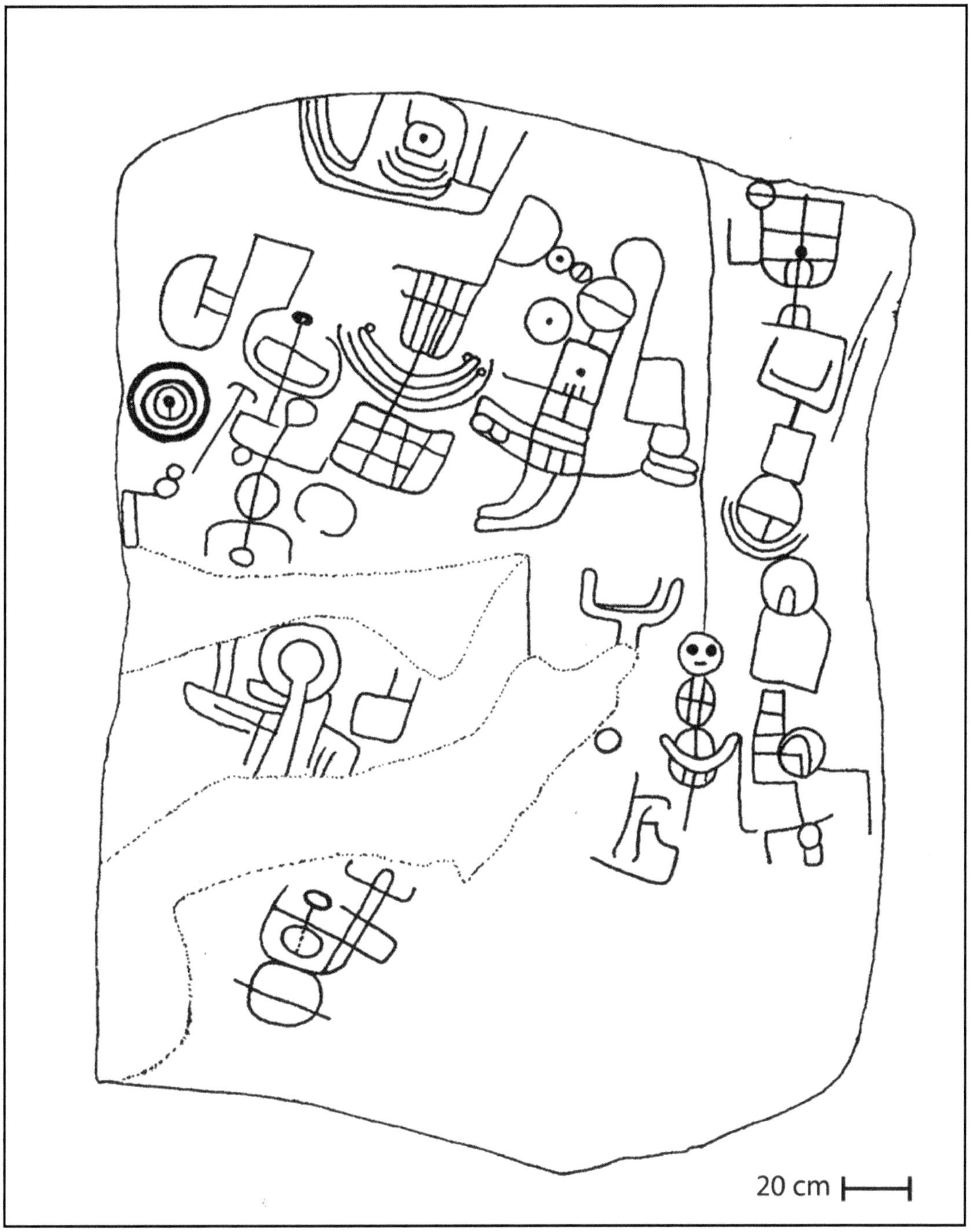

Figure 94. Papenoo Valley. Megalithic petroglyph bolder with numerous geometric figures (scaled drawing by E. Edwards in 1998).

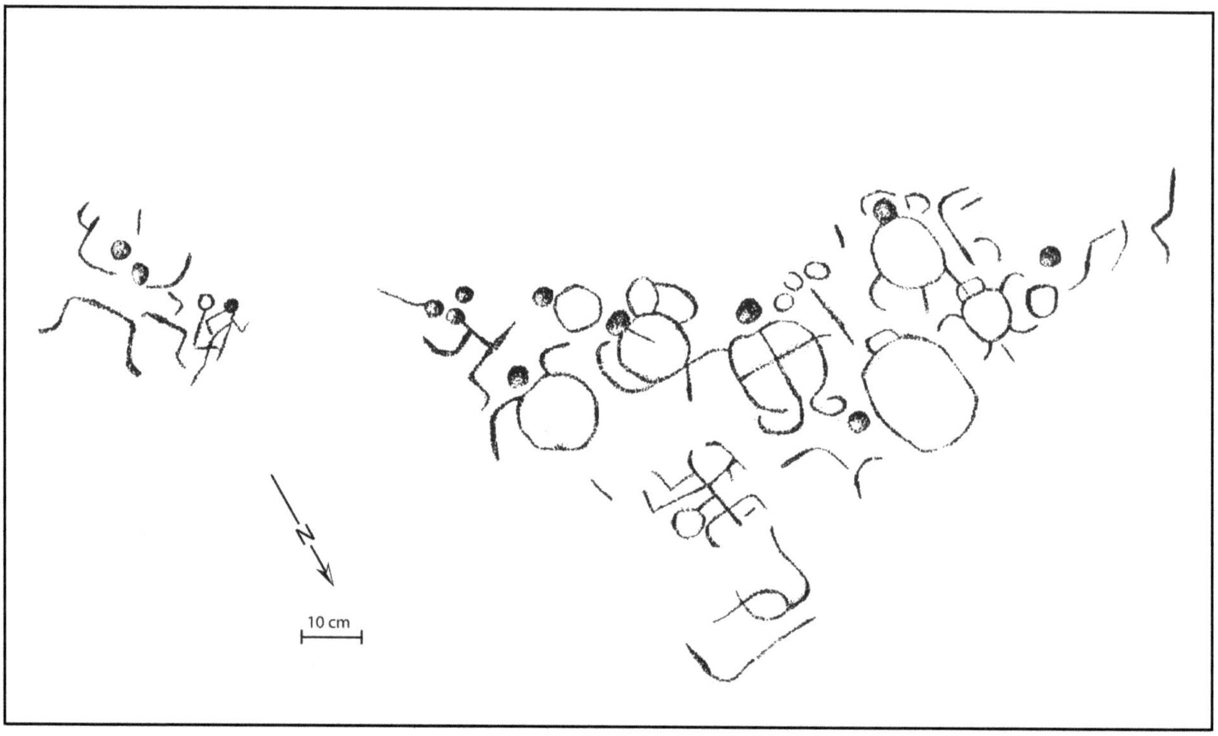

Figure 95. Tefaarahi, Moorea. A petroglyph boulder situated among agricultural terraces (Site 4a).

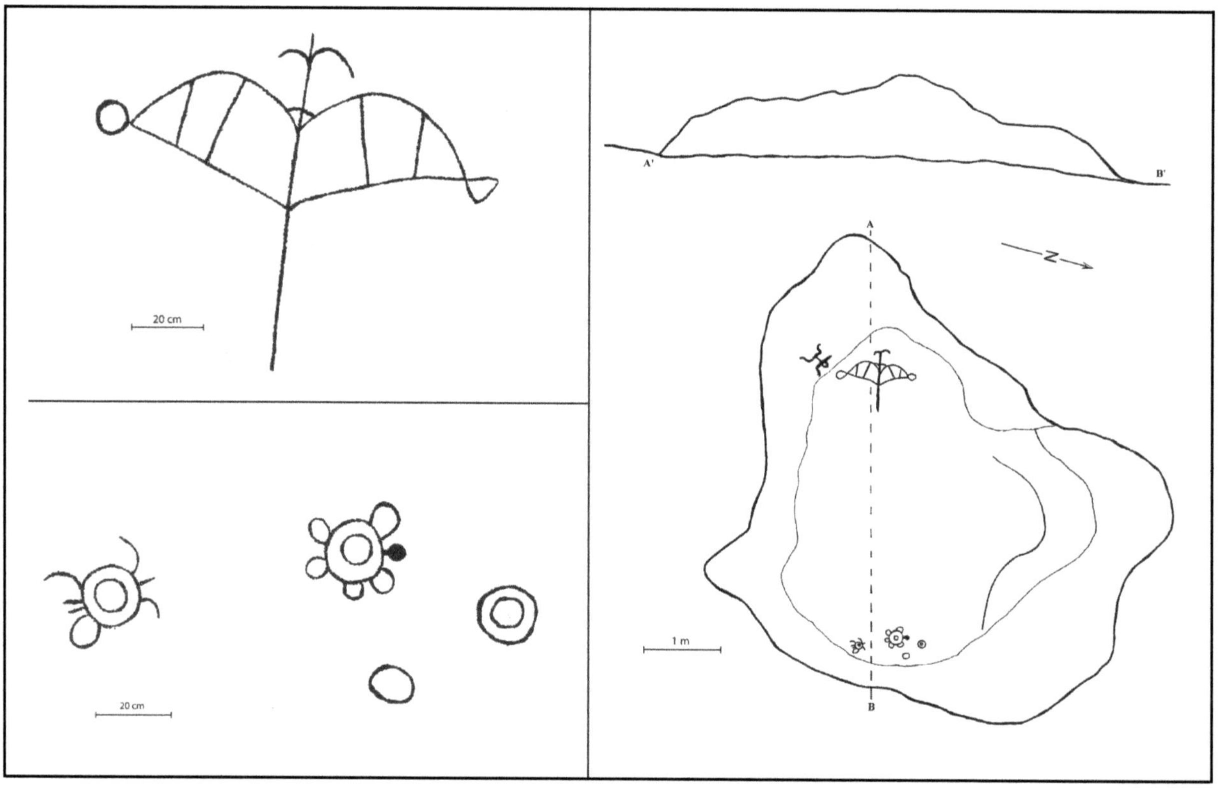

Figure 96. Paopao Valley, Moorea. An anthropomorph, turtles, and an unidentified petroglyph figure are placed on a large flat boulder.

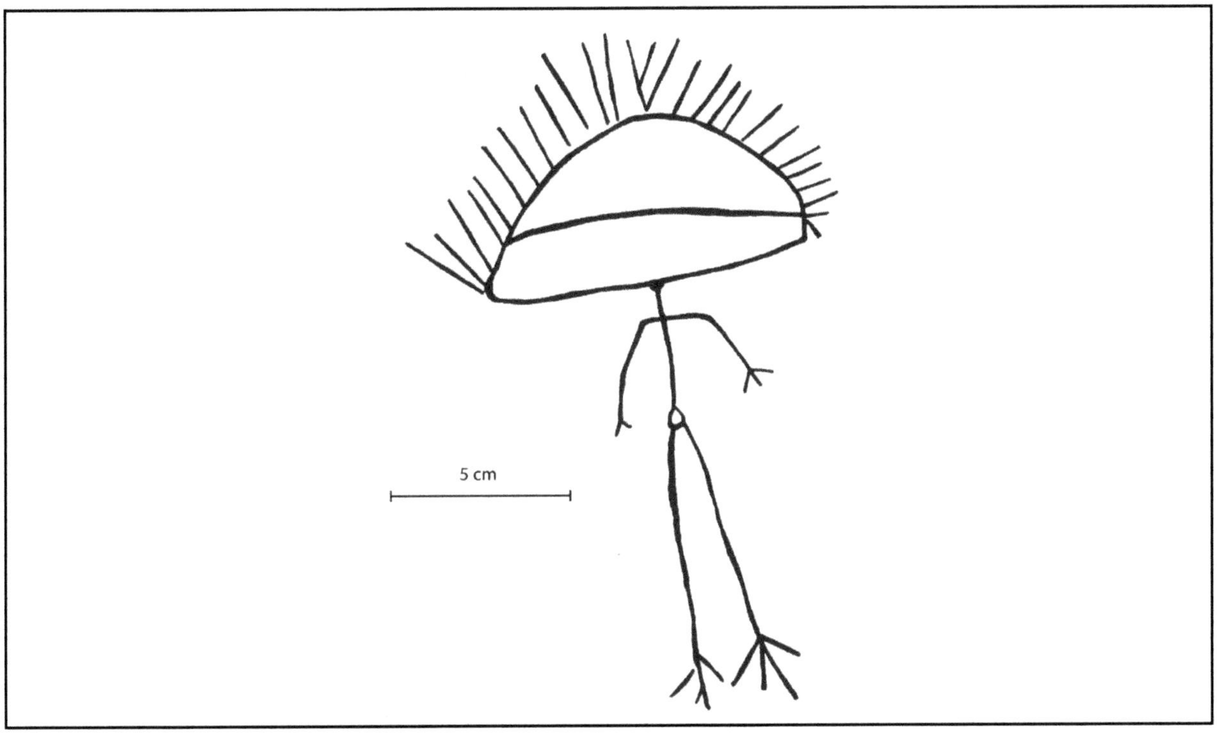

Figure 97. Toroura, Tubuai. Exotic anthropomorphic figure.

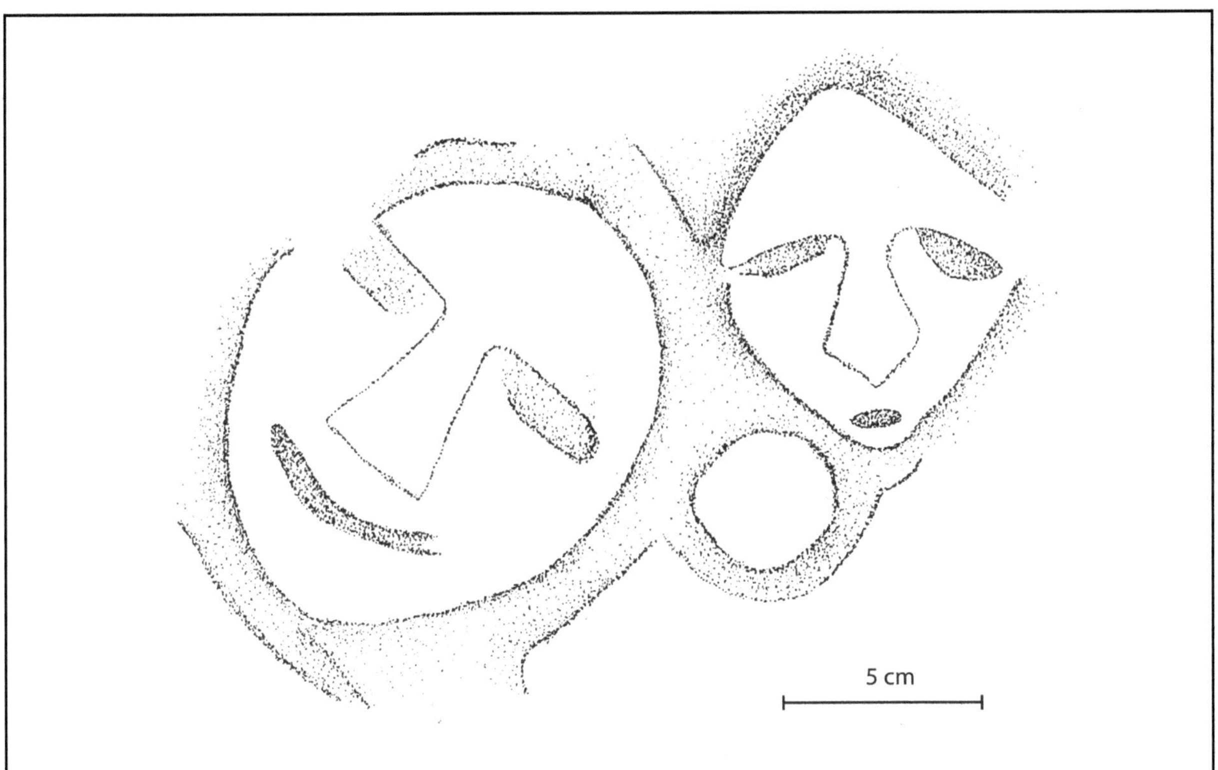

Figure 98. Raivavae. Anthropomorphic faces (RRA 160).

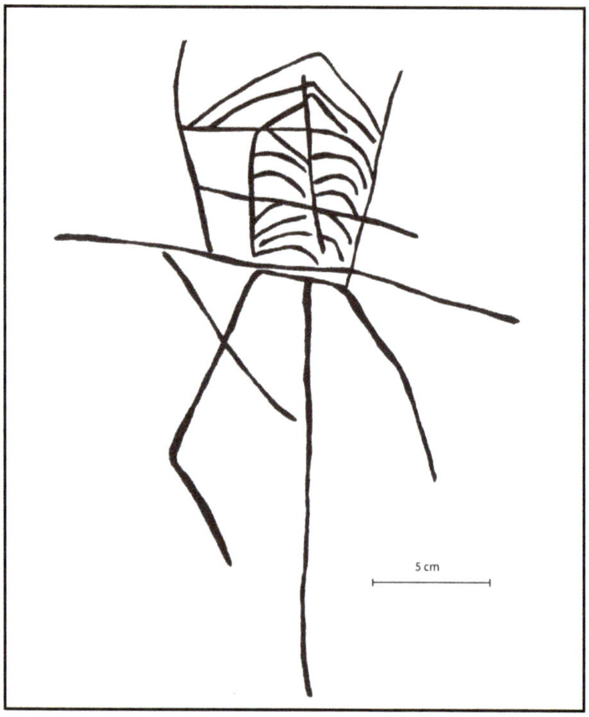

Figure 99. Marae Maunaoto, Raivavae. An unidentified figure placed on an upright stone.

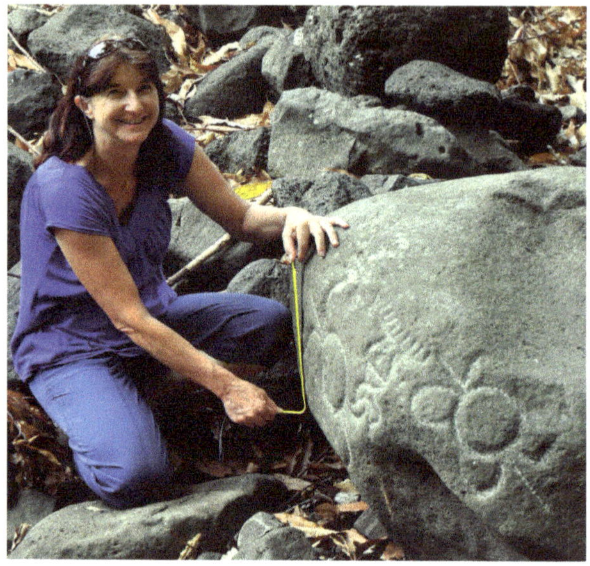

Figure 100. Maupiti. Petroglyph boulder with turtles. Unfortunately, people continuously mark the petroglyphs (photo courtesy Heidy Baumgartner-Lesage, 2009).

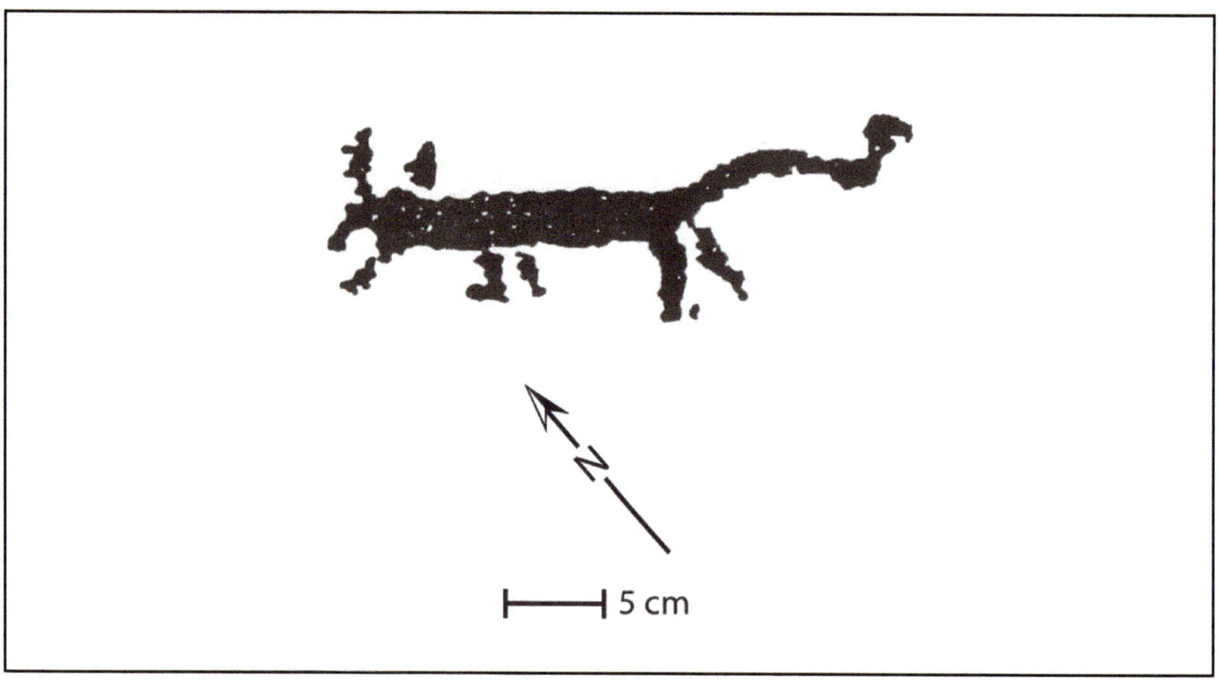

Figure 101. Kahikinui, Maui. Dog petroglyph at site KIP-1197. From Millerstrom and Kirch (2002).

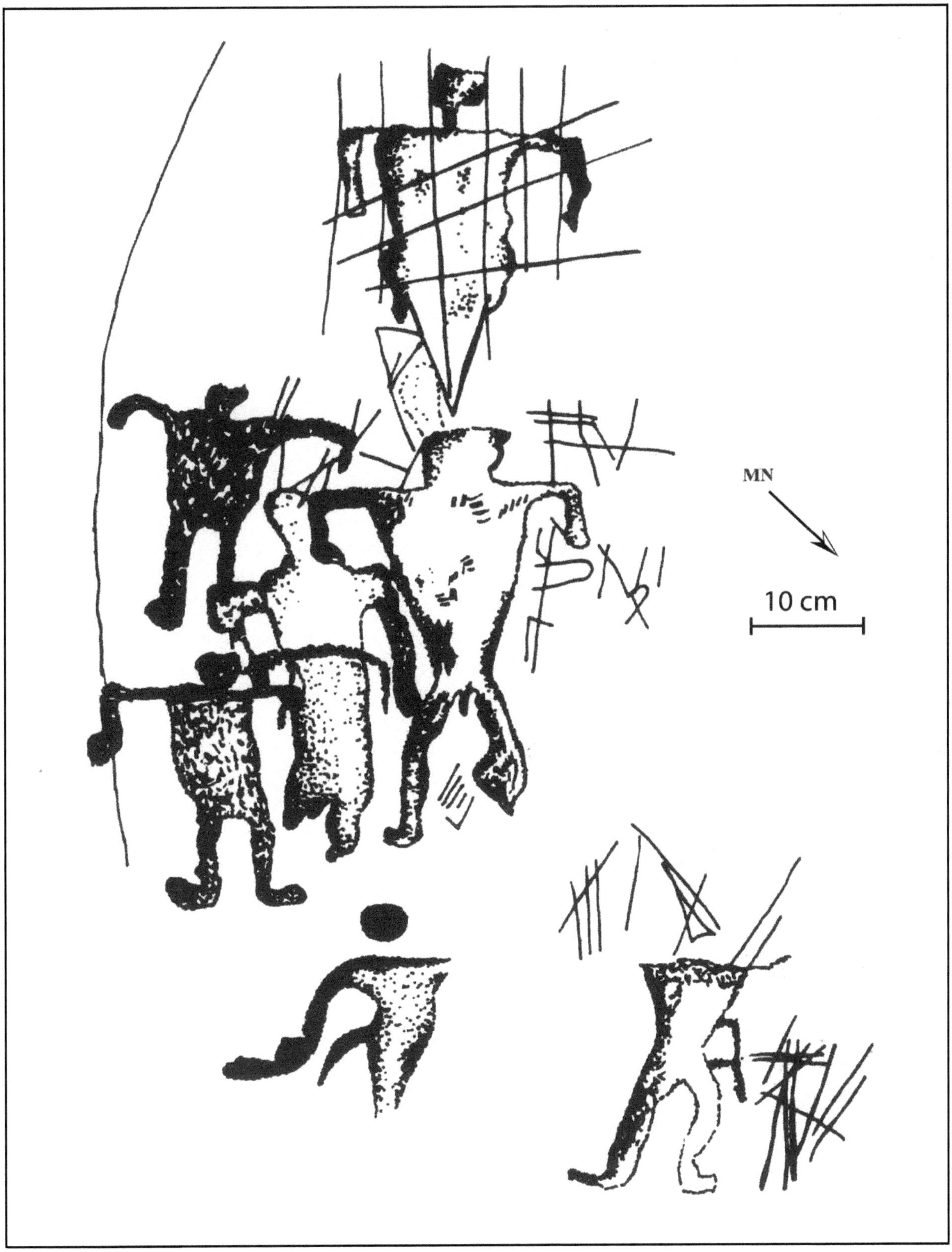

Figure 102. Nuu Mauka Ranch, Maui. Superimposed incised lines on pecked anthropomorphs (NUU-94, panel L).

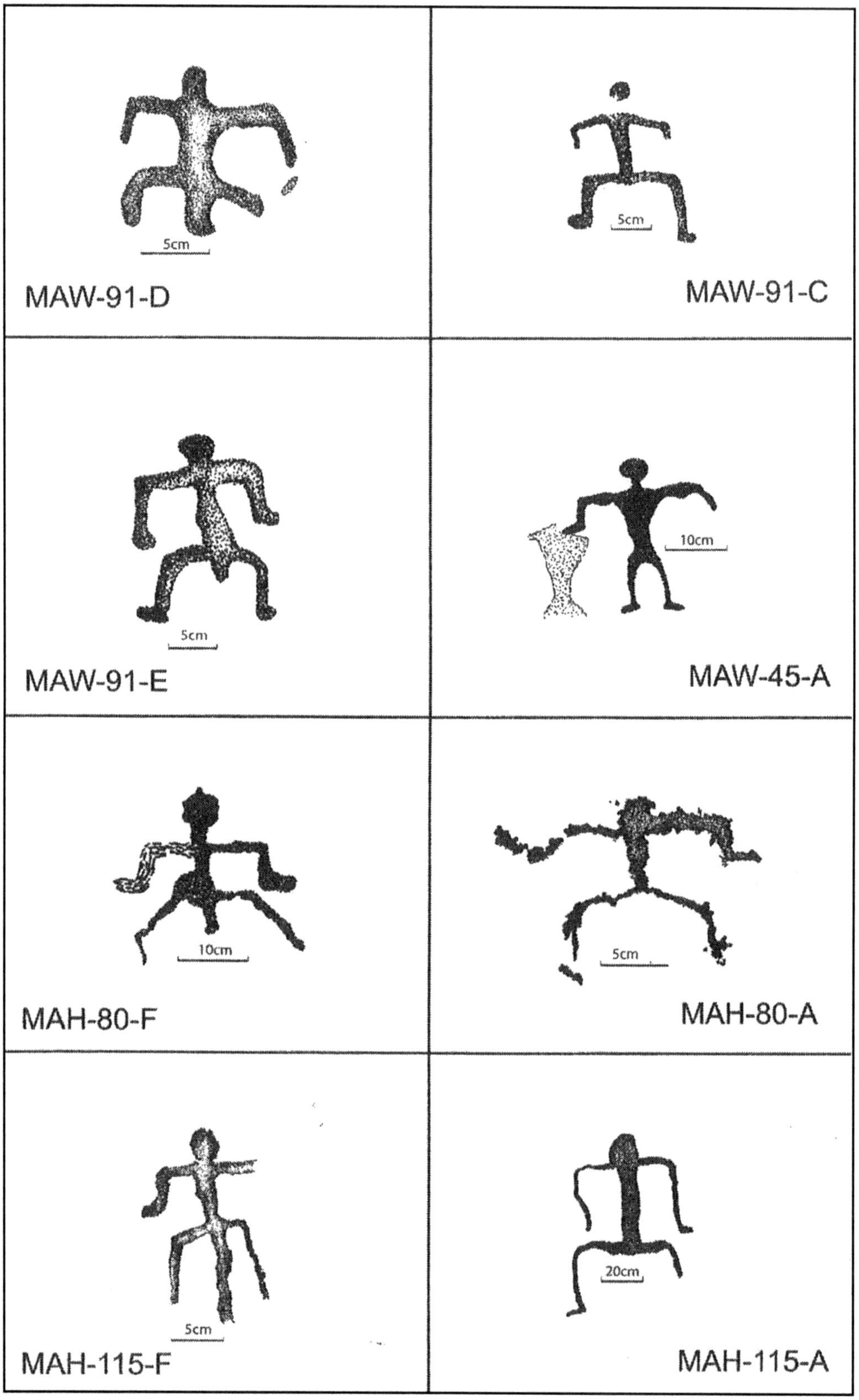

Figure 103. Kahikinui, Maui. Various anthpomorphic figures (from Millerstrom and Kirch 2004).

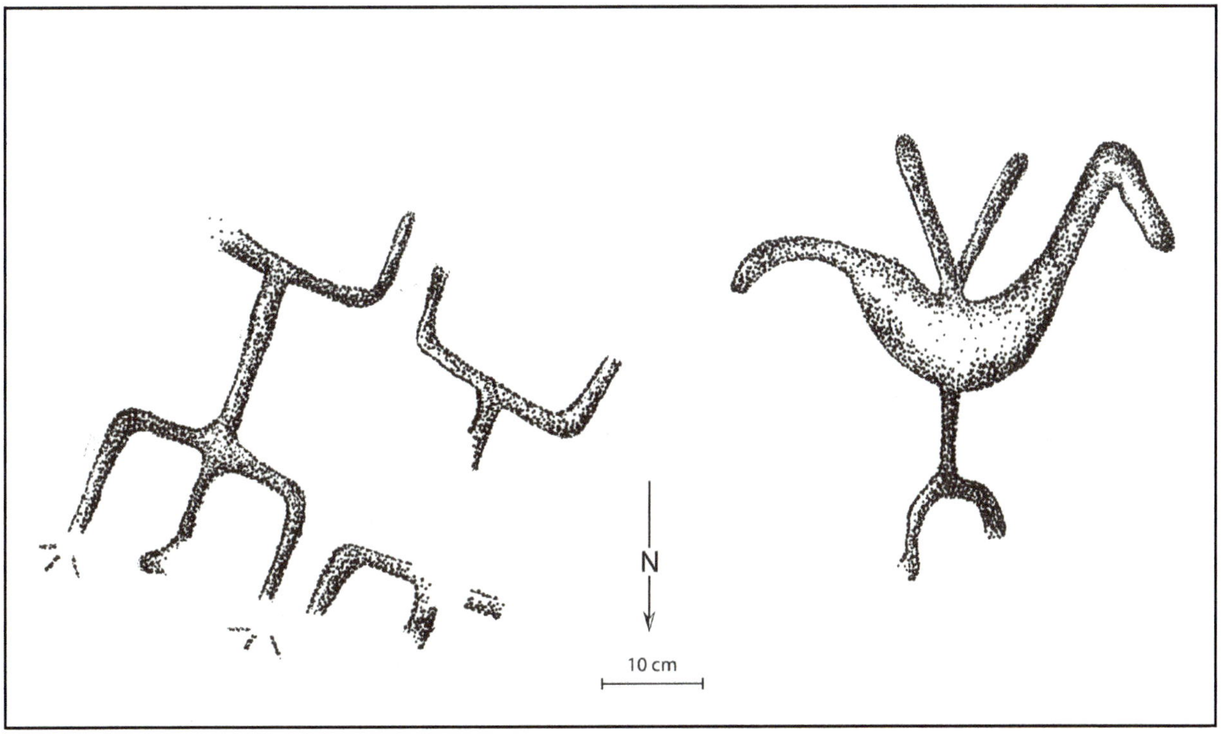

Figure 104. Kahuvai, Hatiheu Valley, Marquesas. Pecked bird and human representation (331hth 138/1).

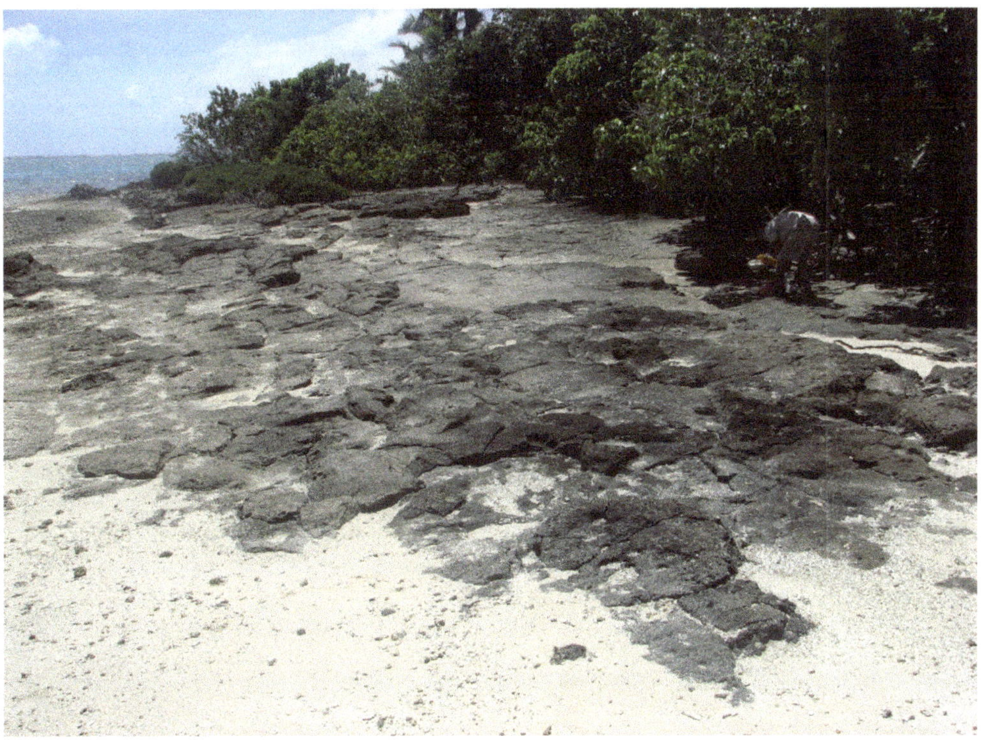

Figure 105. Houmale'eia, Foa Valley Island, Ha'apai Group, Tonga. Petroglyphs panel at low tide (photo courtesy S. Egan, D. Burley, and C. Egan 2011).

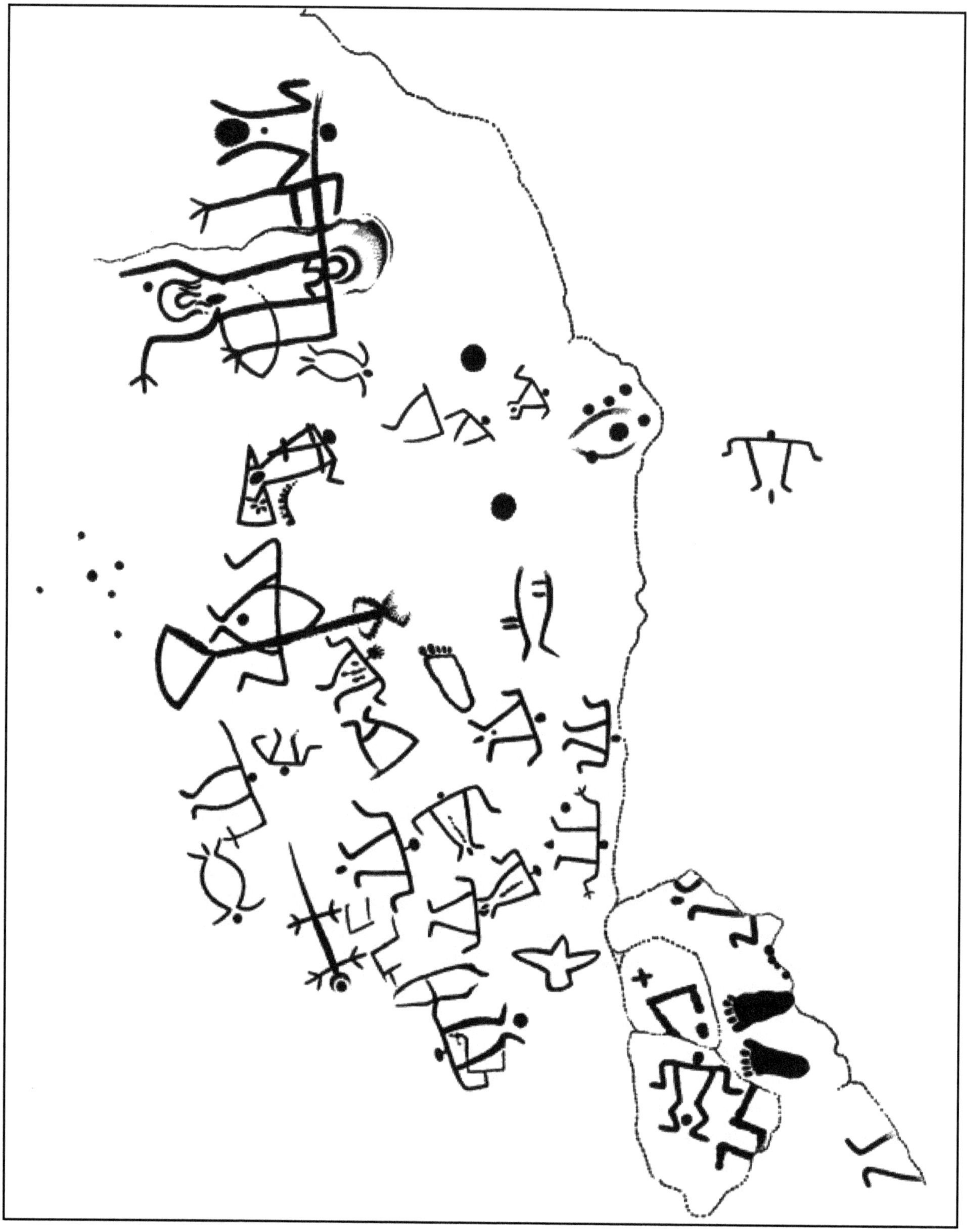

Figure 106. Houmale'eia, Foa Valley Island, Ha'apai Group, Tonga. Upper petroglyphs panel (photo curtsey S. Egan, D. Burley, and C. Egan 2011).

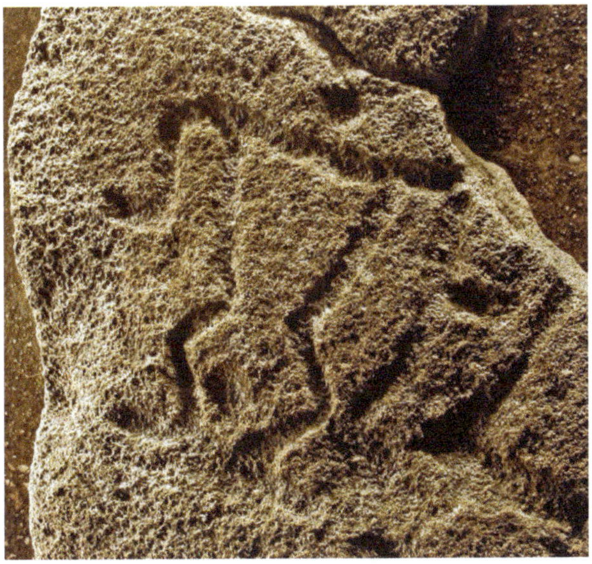

Figure 107. Houmale'eia Foa Valley Island, Ha'apai Group. Open anthropomorphic figure (photo curtsey S. Egan, D. Burley, and C. Egan 2011).

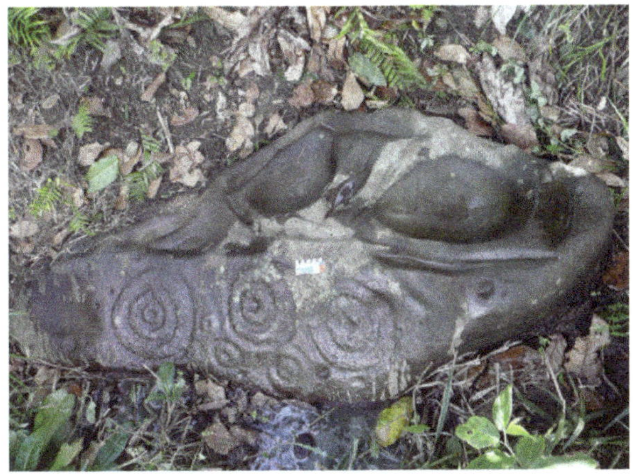

Figure 108. Uluibau, Moturiki Island, Fiji. Petroglyph boulder with concentric circles and polishing grooves.

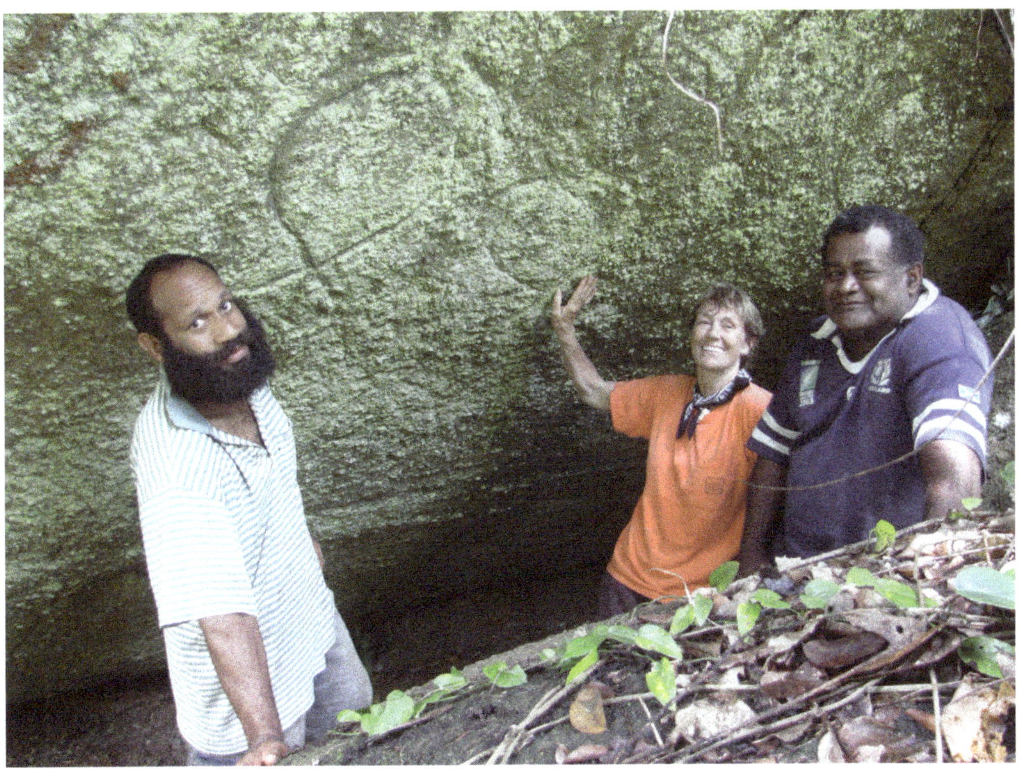

Figure 109. Qaravonu rockshelter, Nailou, Vanua Levu, Fiji, contains pecked and charcoal drawn turtles. From left to right. Tureaga-Ni-Karo, elected head of the village, Maikeli Uludole, SM, and the Tui-ni-Mataqali, chief of Nailou Village, Simione R. Nakasami.

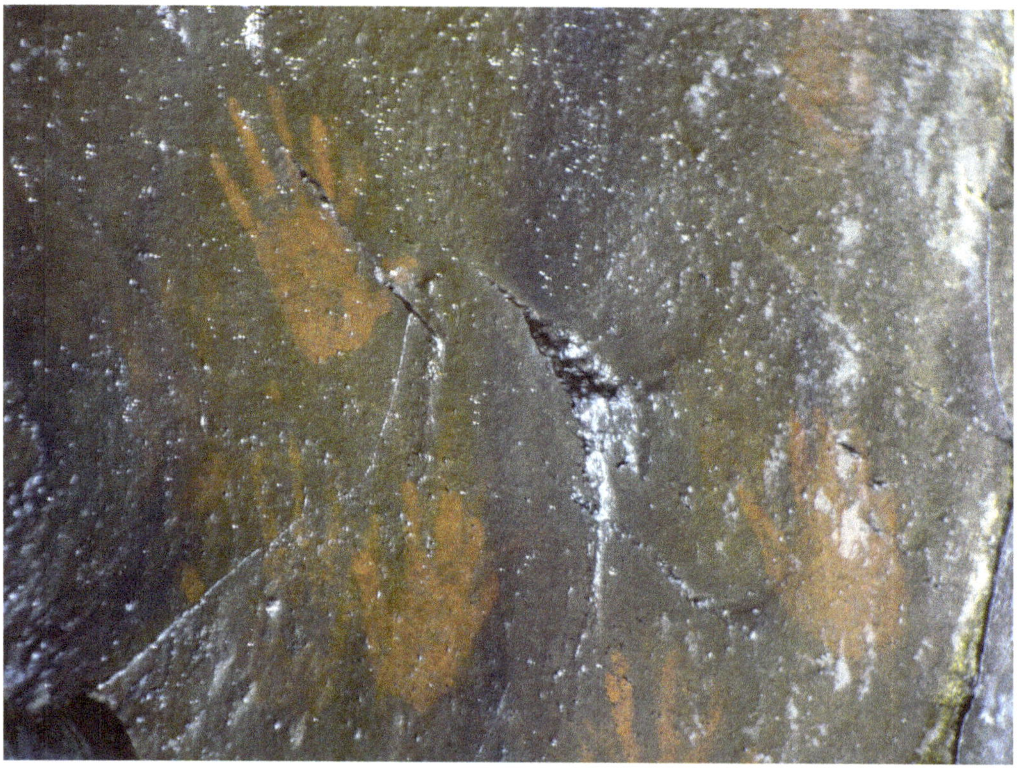

Figure 110. Lalavata site, Taveuni, Fiji. A panel with nine red painted handprints.

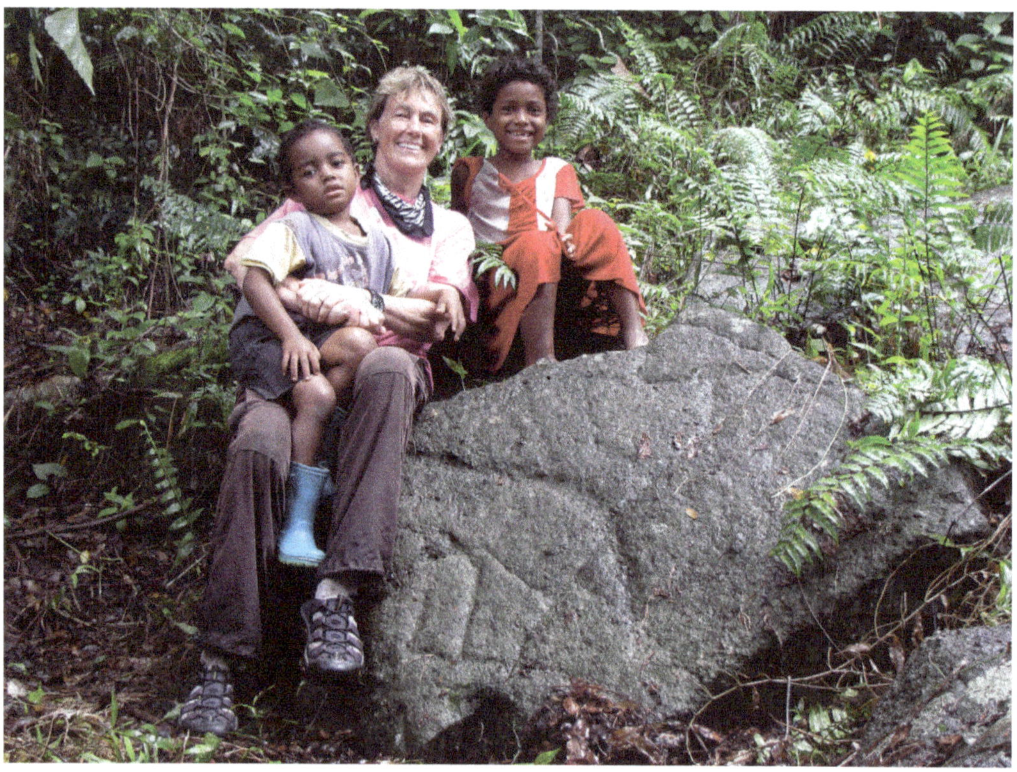

Figure 111. Dakuniba, Vanua Levu, Fiji. Boulder with deeply cut geometrics.

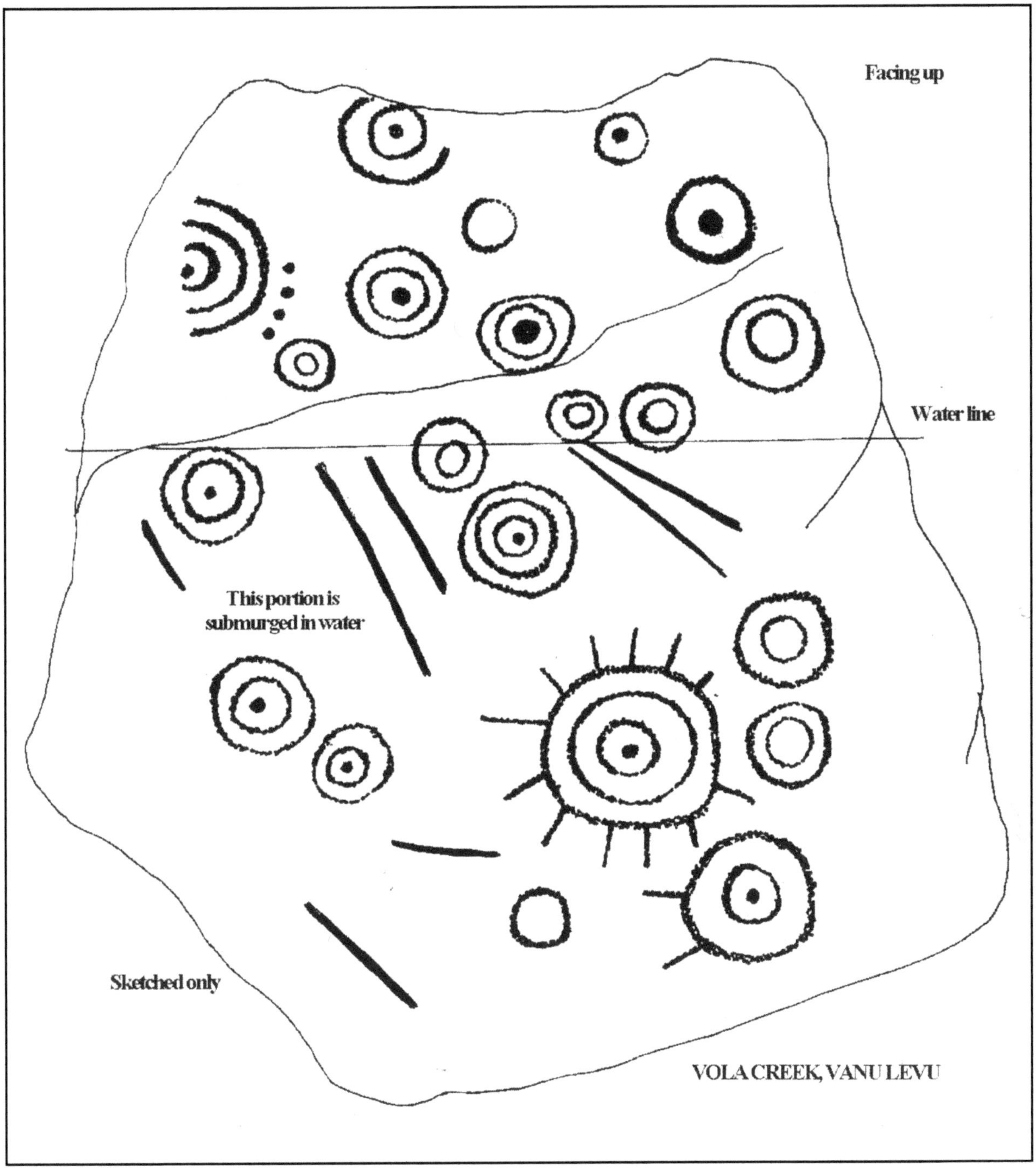

Figure 112. Vola Creek, Vanu Levu, Fiji. Part of the boulder is laying in the river.

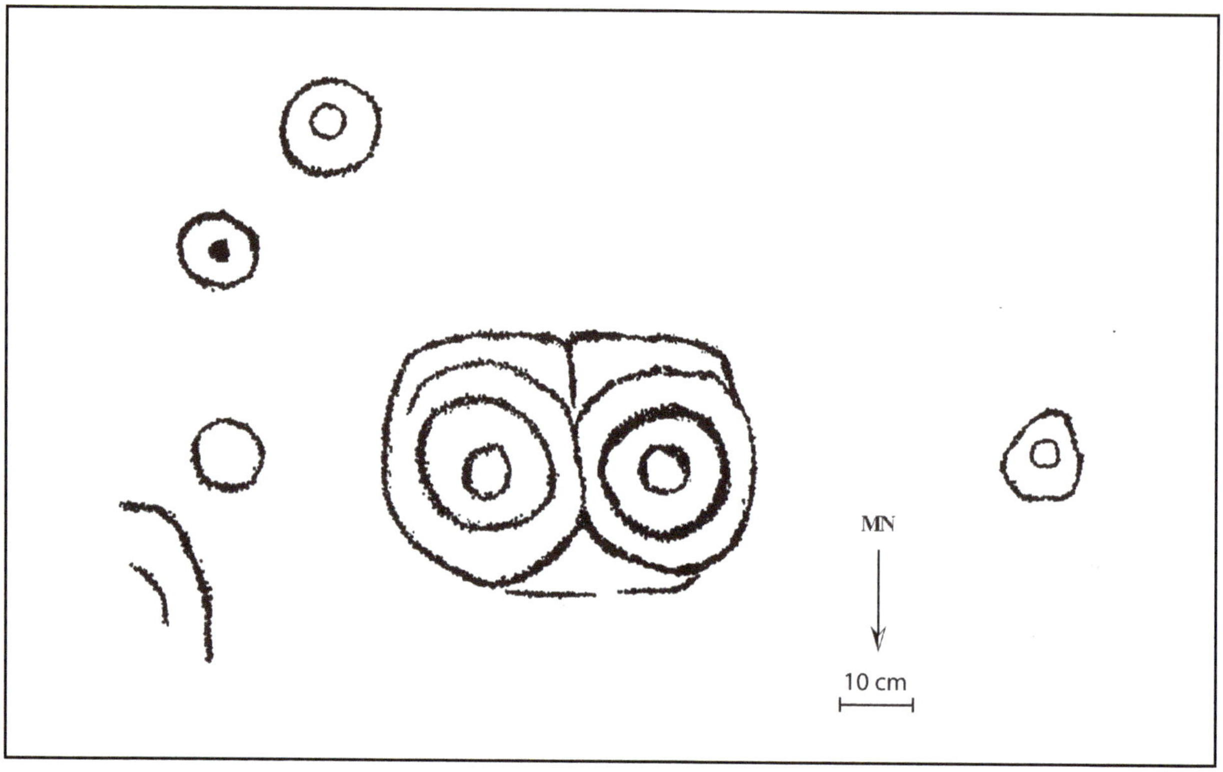

Figure 113. Vatua vola na vu, Vatukuca Village, Vanua Levu, Fiji.

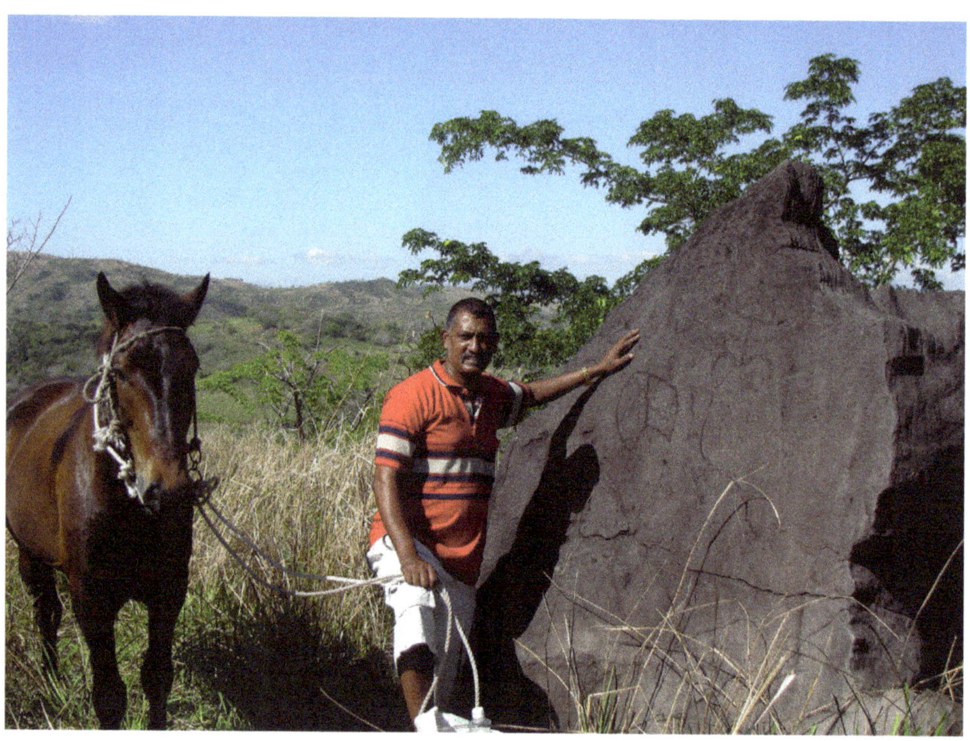

Figure 114. Our guide Vinod Kumar at Maqere, Viti Levu, Fiji, next to a boulder with pecked turtles and an anthropomorphic stick figure.

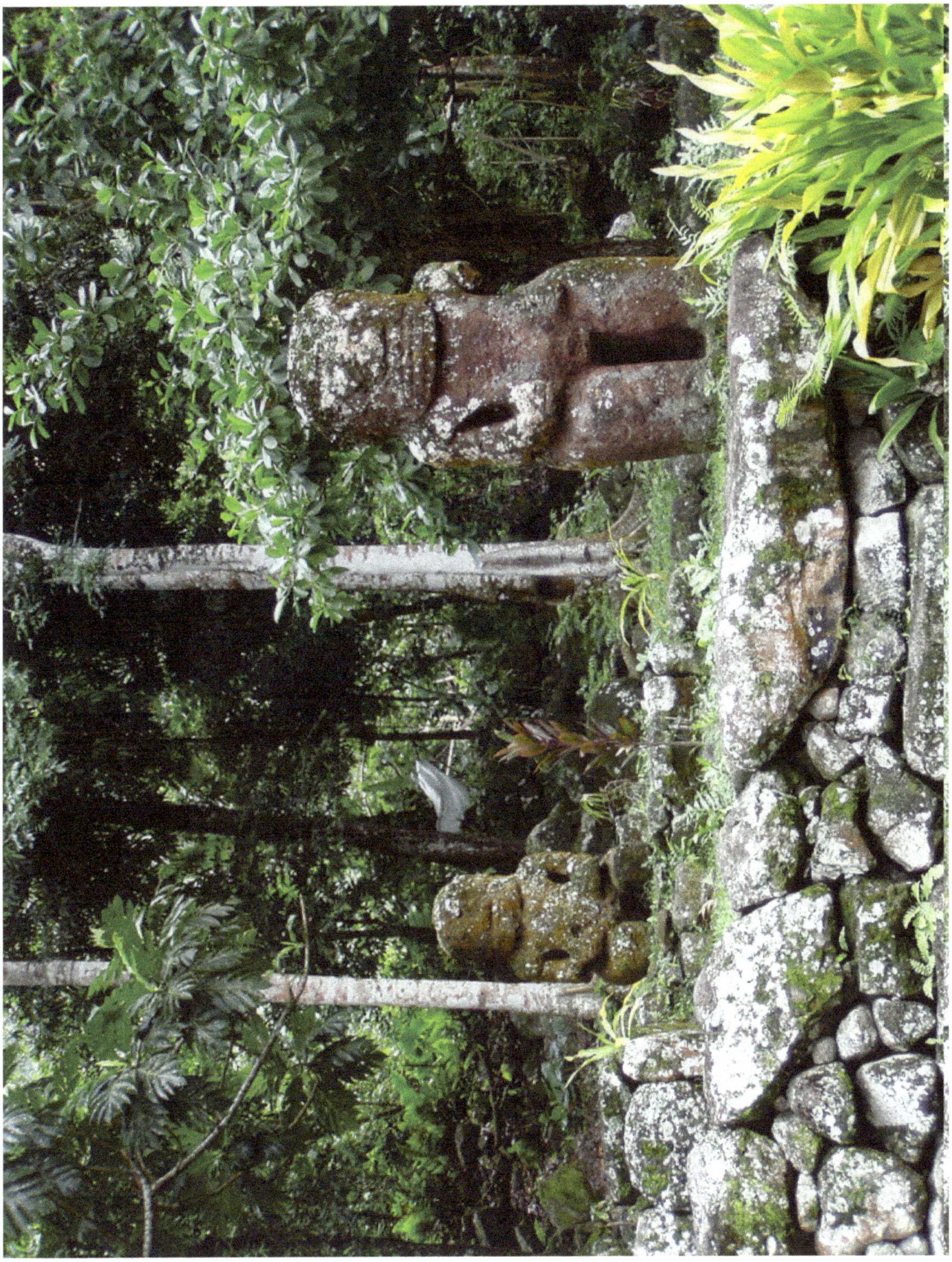

Figure 115. Meae Iipona, Puamau, Hiva Oa has excellent acoustics.

Appendix B

Marquesan Typology

This typology is paying close attention to the typology created by Georgia Lee for Easter Island and Hawaiian Islands (1999:173–179). The same numbering system has has been maintained whenever possible.

Anthropomorphs, Types 1100–2000

Simple stick figures and bi-symmetrical figures with a linear body shown in front view, in various configurations. Arms are depicted in up, down, and opposed positions. Heads are open or closed circles, some with facial features and headdresses. Sometimes, but not often, the figures are connected with some type of tool. When toes and fingers are shown, they almost always depict three fingers. Although not common, when human and animal combinations occur they are always human-lizard, human-dog, human-bird, or human-fish. Anthropomorphs can have square or rectangular torsos, open torsos; naturalistic, or double outline types may or may not have facial features, arm positions are the same as described as above. Only two squatting anthropomorphs in profile has been recorded. Anthropomorphic sculptures (tiki) are carved in the round. A special typology has been constructed for these sculptures.

Disconnected Body Parts, Types 2100–2300

Faces and eyes are the only body parts documented. They are depicted in a great number of variations, e.g., mismatched eyes, and concentric circles as eyes. Similar to some of the *makemake* faces on Easter Island, many of which are placed on the curved part of a boulder giving them a three-dimensional look.

Prehstoric Fauna, Types 3100–3600

Dogs are the most frequent depiction. They are always shown in profile with two or more legs, a long neck, and a long tail that curves over a long body. Birds are shown from above/below and in profile. Few lizards are documented.

Sea Forms, Types 4100–4700

Turtles, although not in great numbers, are the most common sea form. Most fish are unidentifiable, but whale and/or dolphin, and sharks occur. One hammerhead shark has been identified. Throughout the islands three octopuses and three lobsters have been documented.

Material Culture, Types 5100–5700

Canoes are the only type placed in this category. Island wide a total of five canoes have been documented, two of which are found in Eione Valley, Hiva Oa. None occur in Hatiheu Valley.

Historic, Types 6100–6610

Few historic petroglyphs occur in the archipelago. A total of ten historic types are documented island wide, five of which are found on the same boulder located along the old trail from Taipi Valley to Taiohae, Nuku Hiva. A sailing ship is located near the word "poni" with two animals that stylistically look like the island's dog petroglyphs, but with no tail and with hatched lines depicting fur. Modern lettering is found on a large boulder next to a communal breadfruit pit associated with Tohua Kamuihei II, Hatiheu Valley.

Geometric, Types 7100–7700

Geometric figures are by far the most numerous image types. While both linear and curvilinear motif are found, curvilinear motifs dominate. The variations are abundant. Many of the figures are linked to tattoo motifs. Variations of the *ipu* or tattoo motif type are prevalent.

Unidentified, Types 8000

Some of the motifs that defy description are placed in this category.

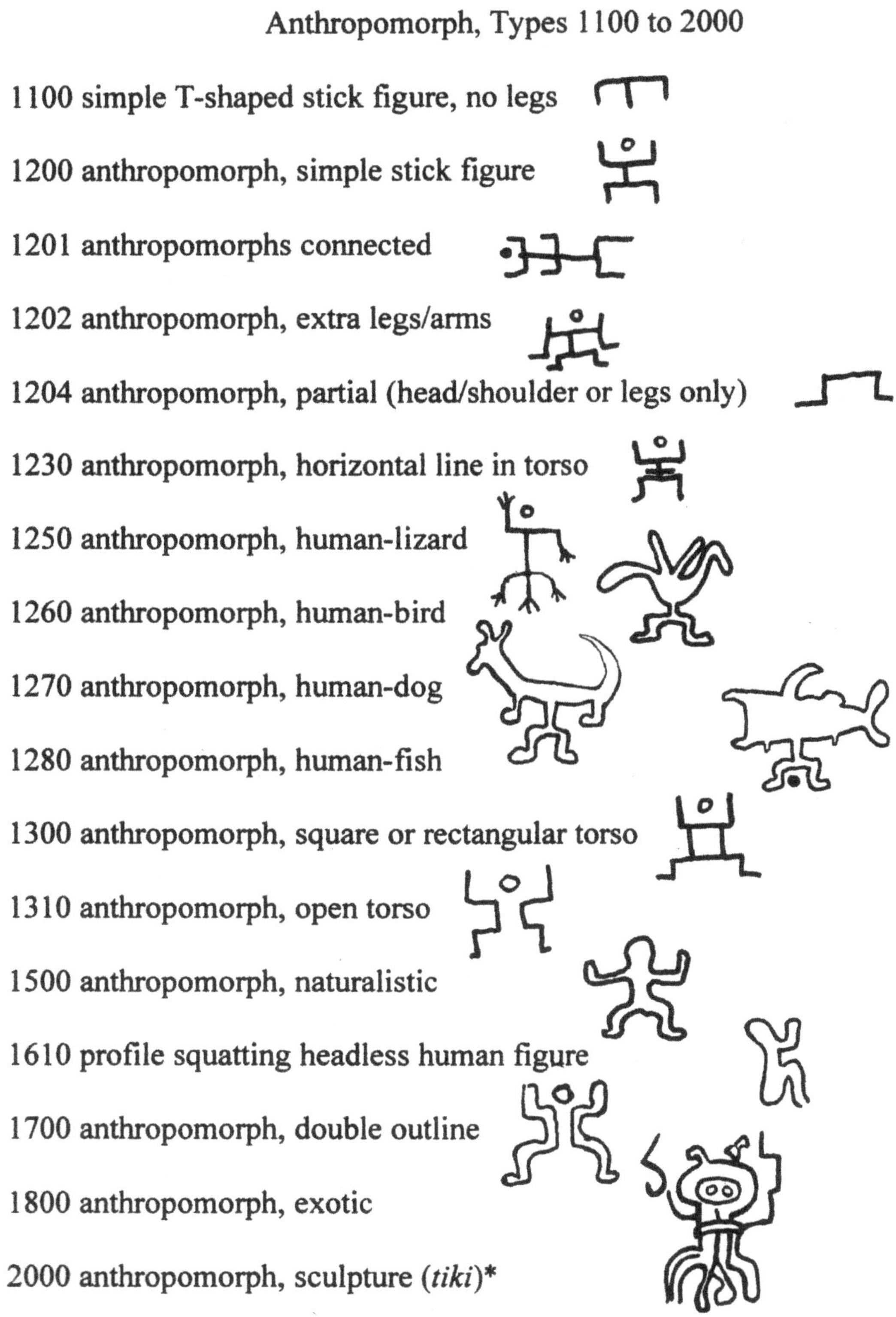

Anthropomorphic features for figure typology

Head

0: not determined

1: absent

2: open circle

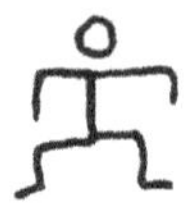

3: closed circle

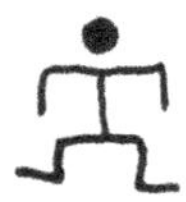

4: face (eyes, mouth indicated)

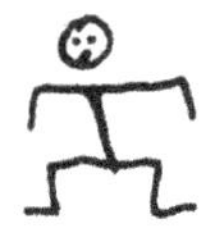

5: headddress

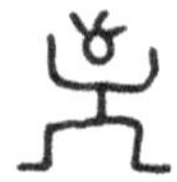

Arms

1: up

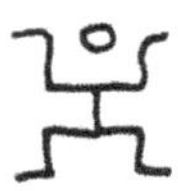

2: down

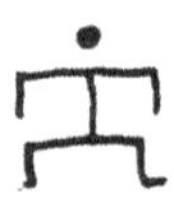

3: opposed

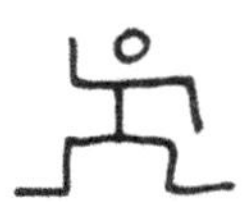

4: straight

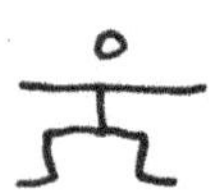

5: object in hand

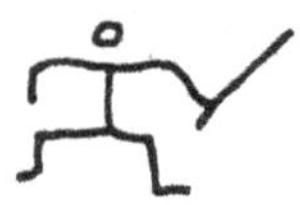

6: digits

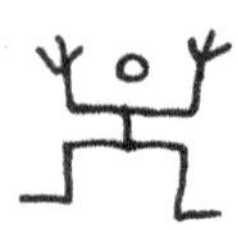

7: other

Legs

1: action

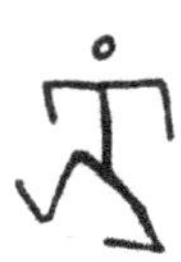

Gender

0: not determined

2: curved 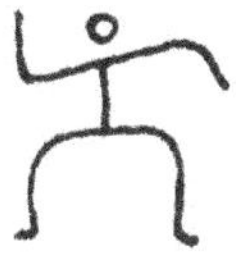

3: regular 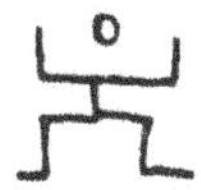

4: digits 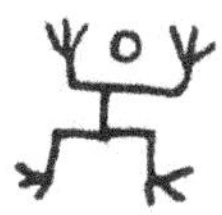

1: male

3: female

4: no indication

Disconnected Body Parts, Types 2100 to 2300

2100 eyes 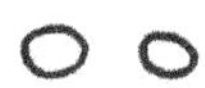

2110 eyes, mismatched 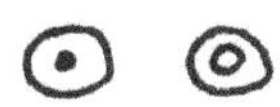

2120 eyes, with brows 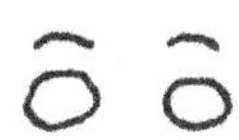

2130 eyes, concentric circles 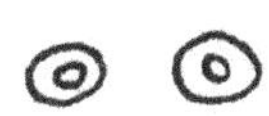

2200 face with eyes and mouth 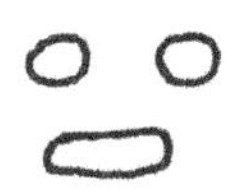

2410 eyes, mouth, nose 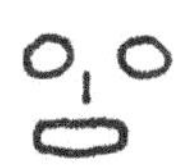

2410 face with mismatched eyes 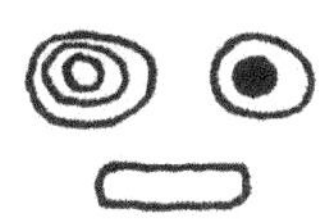

2411 eyes and nose 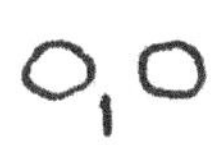

2420 face with concentric-circle eyes 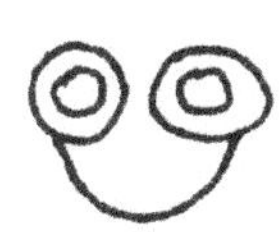

2430 face with cupule eyes 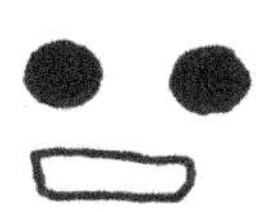

2340 double face 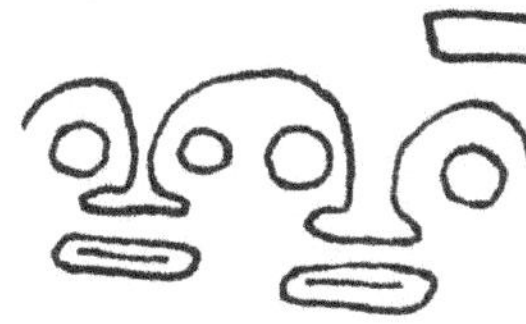

2350 other

Prehistoric Fauna Types 3100 to 3600

3100 dog, general 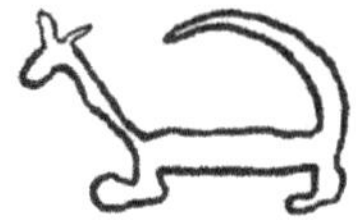

3400 bird 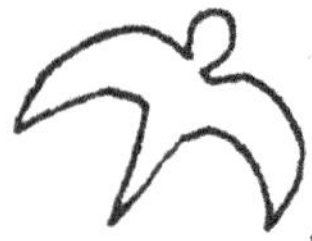

3500 lizard 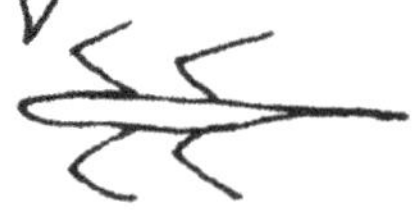

3600 unidentified zoomorph, bug-like figure 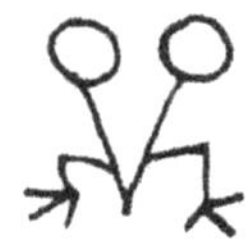

Sea Forms, Types 4100 to 4700

4100 turtle, general shape 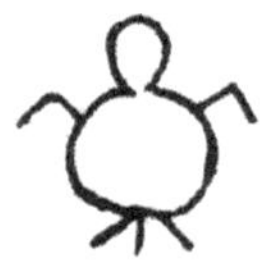

4110 turtle with motif 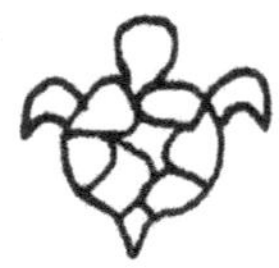

4200 fish, general shape 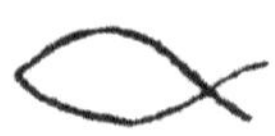

4210 barracuda 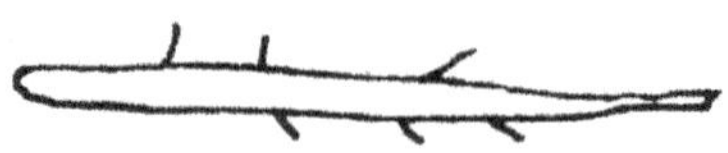

4220 shark 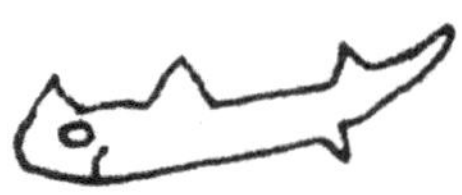

4230 hammerhead 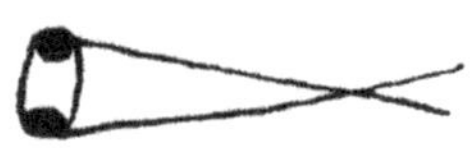

4260 lobster

4270 mahi-mahi 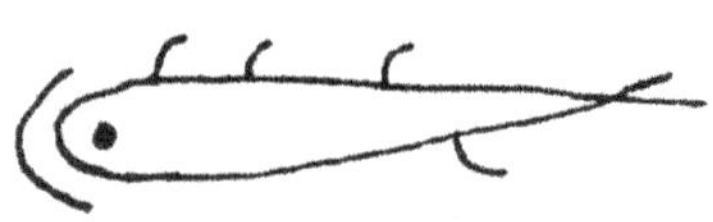

4300 whale/dolphin 

4400 octopus 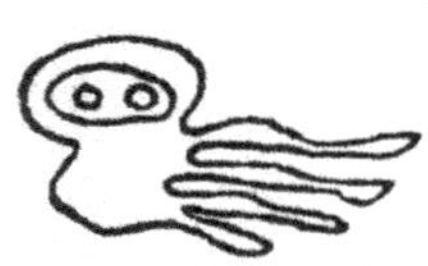

4500 sea-creature, unidentified

Material Culture, Types 5100 to 5700

5100 canoe 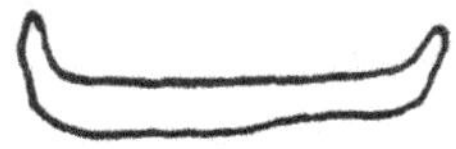

Historic, Types 6100 to 6610

6100 historic ship

6400 gun

6500 horse or "poni"

6600 lettering

6610 script

Geometric, Types 7100 to 7700

7100 parallel straight lines, in series 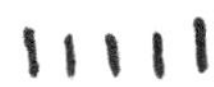

7110 rectangle 

7111 rectangle, segmented 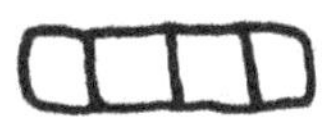

7112 rectangle with line 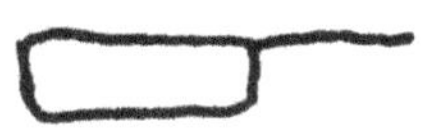

7113 double rectangle with knob 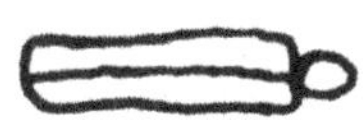

7114 cross 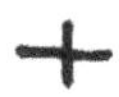

7115 zigzag 

7116 zigzag, nested 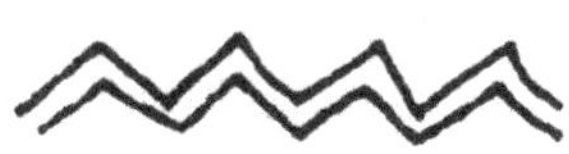

7117 wavy lines 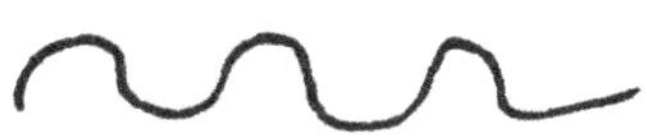

7118 wavy lines nested 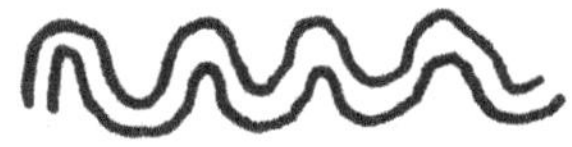

7119 crisscross pattern 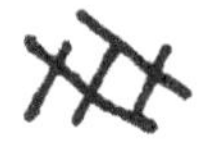

7120 straight line(s), random 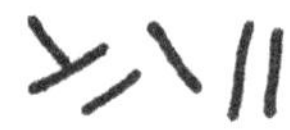

7121 curved line(s) 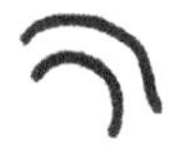

7122 combination straight and curved line

7125 enclosure 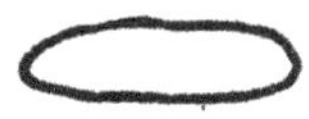

7123 triangle 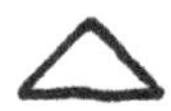

7200 circle 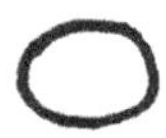

7201 enclosure with circle 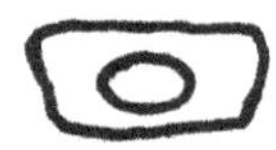

7202 linked circles 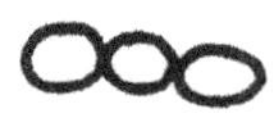

7203 circles in a cluster 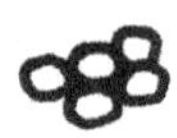

7204 circle with tail (curved or straight) 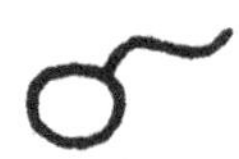

7205 circle with dot(s) inside 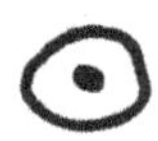

7206 D-shaped figure 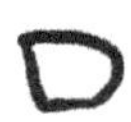

7207 banana shape figure 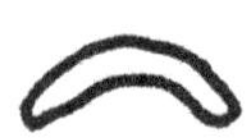

7208 linked concentric circles 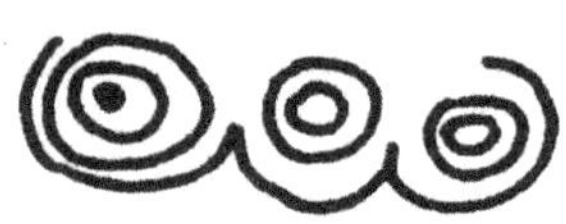

7209 closed circle 

7210 circle with internal divisions 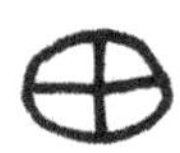

7211 connected double circle 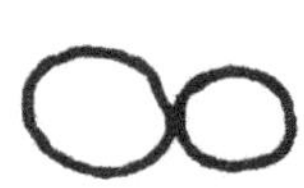

7220 semi-circle 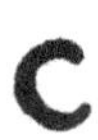

7221 double semi-circle 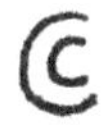

7222 open circle 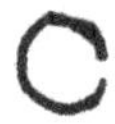

7223 tattoo motif, anthropomorph 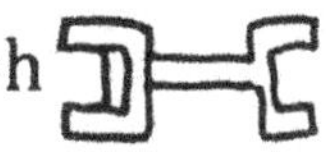

7224 concentric circle with intercepted bar (tattoo motif) 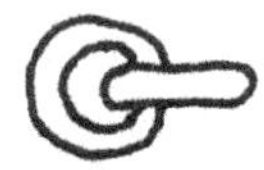

7225 tattoo motif, variation of the *ipu* figure 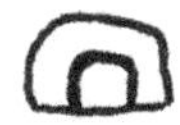

7226 tattoo motif, variation of the *ipu* figure 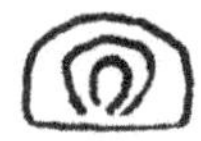

7227 tattoo motif, variation of the *ipu* figure 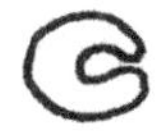

7228 sun shape 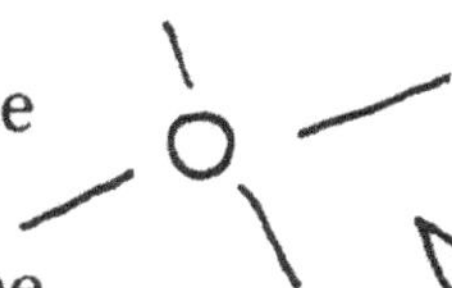

7229 star shape 

7230 concentric circles, 2 rings 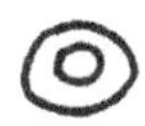

7231 concentric circles, 3 rings 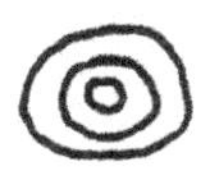

7232 concentric circles, 4 rings

7233 concentric circles, 5 rings

7234 concentric circles, 6 rings

7235 concentric circles, 7 rings

7236 concentric circles, 8 rings

7237 concentric circle with knob(s) on the outside 

7239 cupule with tail 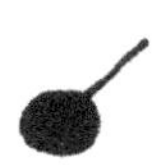

7240 spiral 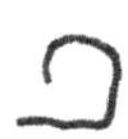

7400 cupules, random

7401 cup and ring 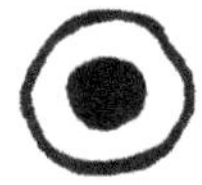

7402 cup with more than one ring 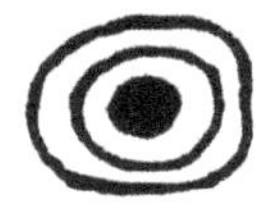

7403 cup and ring with tail 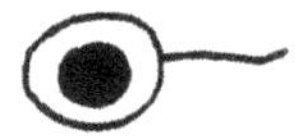

7405 cupules with cross bar 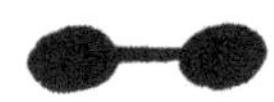

7406 cupule with radiating lines 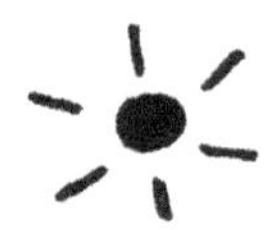

7500 oval open 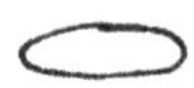

7501 oval with cup inside 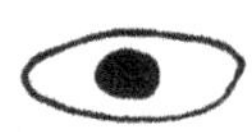

7502 oval filled 

7700 other, geometric

7710 polishing lines, cigar shaped (U-shaped grooves)

7720 polishing lines, straight grove (V-shaped grooves)

7721 superimposed U-shaped and V-shaped grooves

7730 basin

7740 polished area

Unidentified, Type 8000

8000 unidentified motifs

8010 modern fake

*A separate typology has been worked out for the sculptures (*tiki*).

Notes

1. I debated long and hard on the problem of including glottal stops and macrons in the text. Since glottal stops and macrons have not been fully worked out in the Marquesan written language, I decided to omit all glottal stops and macrons in the text. The reference section, as it list titles of articles and books, will include glottal stops but not macrons.
2. A stone monument in Vaitahu, Tahuata, raised in 1995 in commemoration of the discovery of the Marquesas Islands by Europeans in 1595, has the following inscription: "FENUA ENATA TERRES DES HOMMES." En 1595, Elle fut appelée ILES MARQUISES Nom qui la fit connaître Au reste du monde. Qu'aujourd'hui, Le monde connaisse Son nom d'origine. VAITAHU, Le 29 Juillet 1995." (Fenua Enata, the land of men. In 1595 she was called the Marquesas Islands, a name that become known to the rest of the world. Today the world knows the original name. Vaitahu, July 29, 1995.)
3. H. M. Le Cleac'h argued that the name Fatuhiva is correct, as it is written as pronounced (1996:34–36); however, here I use Fatu Hiva, which is the more conventional spelling.
4. I adopt the term *Enana* from the northwest island group rather than Enata from the southeast island group.
5. This was Captain Cook's second voyage to the Pacific between 1772 and 1775.
6. Captain Cook's efforts to obtain provisions ended abruptly when Hitihiti, a native of Bora Bora and Cook's translator, traded red feathers for a pig. Red feathers, something Cook did not personally possess, was what the Enana wanted (Crook 2007:62 [1797-1799]).
7. Nuku Hiva, for instance, received four names between April, 1791, and February, 1793. Marchand named it Baux; Richard Hergest named it Sir Henry Martin's; and Josiah Roberts first called it Adams but then changed it to Washington.
8. A handwritten unpublished manuscript of W. P. Crook's account, deposited in the Mitchell Library, Sydney, Australia, is written in the third person and marked "original MS, writer unknown." However, the handwriting is that of Samuel Greetheed, on the basis of information provided by W. P. Crook and Temoteitei (Thomas 1990:237). From a photocopy, George Maurice Sheanan Jr. transcribed Crook's manuscript in 1952. It is now part of the Marquesan Source Material, Roneo, Quincy, Massachusetts. In 2007 Haere Po Tahiti published Crook's account in English and French.
9. The British captured the ship *Essex* and the annexation was never ratified.
10. According to the museum catalogue published in connection with the exhibit, James Cook and the Exploration of the Pacific in Bonn, Vienna, and Bern (2009–2011), only one club is identified as part of the Marquesan objects, and it was probably collected by the Forsters on Cook's second voyage to the Pacific (2009:218. Item #381).
11. Typically the head of a Marquesan wooden club depicts several sets of anthropological eyes and in some cases faces. Both eyes and faces are found on stone, wood, and a number of petroglyphs. The eyes are most often large and circular.
12. The dates refer to the time period the missionaries spent in the Marquesas.
13. This was Captain Dupetit-Thouars second trip to the Marquesas. In August and September, 1838, on the frigate *Vénus*, he carried out a detailed survey of the Marquesas (Dunmore 1991).

14. Despite W. P. Crook and later Reverend R. Thomson's heroic efforts, the London missionaries largely failed to convert the Enana.
15. The term oono is a word that derived from the proto-Polynesian term *rongorongo. The asterisk denotes a reconstructive term. In the 1870s, rongorongo in Mangarevan referred to "priests who direct ceremonial dancing and chanting" (Fischer 1994:438). In Rapa Nui *rongorongo refers to carvings of figures on a wood tablet. Linguist Steven Roger Fischer argues that some of the *rongorongo were more than a mnemonic device used by singers/priests, but that the inscriptions represented cosmogonic texts or creation chants, which were physically read by the singer/priest, perhaps even "creatively composing it" (Fischer 1994:413–43, 1995:303–310).
16. Crook (2007:119) was told that the Tafati, a tribe living in Hanatekuua Valley on the northern coast of Hiva Oa, did not tattoo themselves. Furthermore, they "are remarkable for not speaking the common language."
17. Since the late 1980s there has been a panoceanic cultural revival. The late Jean-Louis Candelot (J.-L. Teuruarii Tamatoa), an Enana historian, wrote a section on the renaissance of tattooing in Le Tatouage aux îles Marquises (P. and M. N. Ottino-Garanger 1998:177). According to Candelot, Marquesan cultural revival occurred after the first Marquesas Art Festival on May 1, 1987, in Ua Pou. I also believe that the monthly visits of M/V Aranui, a Tahitian freighter that began carrying passengers in the early 1980s, stimulated artists and aided in cultural pride and revival.
18. Ann Stoll photographed a wooden tiki house post in the Musée de Tahiti et des Iles using a DStretch program. Red pigment is visible in the resulting image. The anthropomorphic figure was part of a funeral house and collected on Ua Po in 1947 (Stoll pers. comm., 2013). Mounting evidence strongly suggests that at least some of the sculptures, even those in wood, were painted on special occasions.
19. Every time I have been at Haatuatua beach on the island of Nuku HIva, I have searched for the petroglyphs in the riverbed, but they are no longer visible.
20. Tragically all original rock art data collected was destroyed in the Berkeley/Oakland firestorm of October 20, 1991. While copies of the inventory forms were housed in SCP, maps, field notes, sketches, photos, and slides were lost. Some of the sites have been relocated, photographed, and documented again. However, other sites have since been demolished due to house and road construction.
21. The place name Maatea derives from the proto-Polynesian word **makatea* that means "white stone." A *makatea* is an elevated island formed by tectonic action (Kirch 2000:49–50).
22. The Eiaone paintings have the potential to be dated if the red pigment contains organic material. This could help to place the figures in a relative timeframe. Unfortunately after a tragic accident the landowner has declared the valley tapu and closed the valley indefinitely.
23. A "class of men (which are usually called Kioe [*kaioi*]) and the male attendants of women, may partake of food with the women, and like them are restricted from using any article of food, or utensil into which it has been put, belonging to the tabbu class" (Crook 2007:54[1797–1799]).
24. When Chiefess Titoi Ani died in 1916, she was entombed in Structure no.1. Some years ago when the corrugated iron roof of the tomb collapsed, it was discovered that the chiefess was entombed with her two bicycles (Chester et al. 2004:63–64).
25. Linton (1925:31) writes that the term *meae* was used in the southern Marquesas and in Ua Huka while in Nuku Hiva and Ua Pou ahu was used to designate a sacred place. However, for Dordillon

(1931) and Le Cléac'h (1997) both terms mean a sacred place. The term *meae* is used in this book to designate sacred places.

26. The trade store was abandoned in the early 1970s (Tioka Puhetini, pers. comm. 1995).
27. In 1984, when I first visited Tohua Hikokua, the ceremonial complex was completely hidden in dense vegetation, and barbed wire enclosed part of the *tuu*. In 1989–1990 the Mayor, Yvonne Katupa, restored the *tuu*. The dance plaza of the tribal ceremonial complex has, in the past, been used by the local Hikokua Dance Group to perform for the Aranui passengers that visit once a month, as well as various cruise ship travelers that occasionally call at the Hatiheu Valley. Presently the dance group performs at various places on the Tohua Kamuihei.
28. In the past a sleeping house was located at the present entrance to Tohua Hikokua.
29. The image boulder was at one time removed to the Catholic mission in Taiohae, Nuku Hiva (Suggs 1961:68–76). It presently resides outside the new Catholic Church in Hatiheu Valley.
30. Tohua Hikokua means "wonderful banyan tree" (Suggs 1961:68). Suggs found the remains of a banyan tree, burnt by the former owner, to the north of the ceremonial structure.
31. Three modern tiki made of red tuff carved by the late Uki Haiti of Taiohae are located on various platforms on the *tohua*.
32. The present location of these image stones is in the front garden of Chez Yvonne, Hatiheu Village.
33. The anthropomorphic stick figure (331hth 470) and an anthropomorphic face (331hth 479) were found by Pierre Ottino-Garanger and his crew when they restored the ceremonial complex for the Marquesan Art Festival Year 2000 (Pierre Ottino-Garanger pers. comm. 1999).
34. The landowners regularly collect copra in the vicinity of Tahakia, but the people of Hatiheu were surprised, in the 1990s, when Ottino cleared the brush and identified the place as a tribal community complex (Tioka Puhetini, pers. comm., 1996).
35. The walled structure had partly collapsed and the *keetu* with the anthropomorphic figure was turned over; thus the size and the attributes of the figure could only be estimated.
36. In 1994, Ottino (1998:52–65) recorded a local legend about the well-known historic warrior Chief Keikahanui.
37. Te Heu was, and still is, considered a special place by the inhabitants of Hatiheu. From this peak, the Madonna of Hatiheu watches over the valley. In 1872, French Catholic missionary Frére Michel Blanc built the Madonna out of a breadfruit log. According to the history of the valley, the missionary wanted to build a church, but the inhabitants were too busy engaging in warfare with their neighbors. Thus Brother Blanc climbed the vertical peak and sculpted the Madonna from a mixture of lime and sand using the breadfruit log as a foundation. He crowned her with a large piece of coral (Chester et al. 2004). Once a year someone climbs up to repaint the Madonna in white.
38. In 1988 and 1989, as part of the Marquesas Rock Art Project, we documented the rock art on the western side of the Puhioho stream. In 1990 and 1994, all the architecture associated with the images at Kahuvai were mapped.
39. At a picnic in Hatiheu (1995), I was offered a roasted seeded variety of breadfruit that had been collected in the upper western part of the valley. The seeded variety is rare in Hatiheu Valley today, the reason being, according to Yen (1991:87), that people selected the seed-free breadfruit for cultivation in order to facilitate the manufacture of fermented paste for underground storage.

40. In some literature, the terms "witchcraft" and "sorcery" are frequently interchanged. Generally, sorcery involves the use of magical paraphernalia by an individual to harness supernatural power, while witchcraft is a power inherent in certain individuals without the use of magical charms or other objects (Lehmann and Myers 2001).

41. In an oral tradition from the fifteenth or sixteenth century, collected by Karl von den Steinen, the Naiki apparently once lived in the Puamau Valley (Hiva Oa). After being defeated by a neighboring tribe, they fled to other islands, such as Ua Huka and Nuku Hiva.

42. The reason for alliance shifts in the protohistoric period can be explained by the consequences following Commander Porter's lengthy stay in Nuku Hiva in 1814. After Porter aided Keatanui, the chief of the Teii at Taiohae, to subdue the Taipi people, Keatanui largely gained control of Nuku Hiva (Porter 1822:1–142; Stewart 1832). After Keatanui's death, all the tribes including the Taipi partially acknowledged his son Moana. However, it is not likely that Moana had much authority over the people living on the north coast. Handy (1923:34) believed that Moana was allied with Pua, Akapa, and Haume (Hooumi).

43. Petroglyphs continue to be discovered throughout the island. For example, bas-relief figures were found at Meae Pele, Fatu Hiva, when the site was restored in 2011.

44. When Linton lived in the Marquesas in 1920–1921, stone artifacts "still formed a regular part of the domestic equipment" (Linton 1923:77). Today old stone pounders, as long as they are intact, are still part of kitchen equipment in some households.

45. Both objects were removed from the ceremonial site in 1988. They are now located in front of Chez Yvonne in the center of the village.

46. None of the terms appear in the Marquesan dictionaries by Crook (1790s), Mgr. René I. Dordillon (1931), or Mgr. Hervé Le Cléac'h (1997). It is uncertain if this means that the terms were specific to an island(s) or if the terms pertained only to adzes, and thus were excluded or deemed unimportant in this context.

47. Archaeologists who plan to work in the Marquesas should be mindful of the damage done to some archaeological sites by feral and domestic pigs.

48. A member of the Hikokua dance group used the Cassidae trumpet in Hawaii when the group performed there in 1993.

49. In 1994, the lower section of a *poi* pounder was found on the dance floor while mapping Tohua Maikuku. Another was found near a high-status house in Kahuvai.

50. The petroglyph boulder from Tipaerui, Tahiti, is presently located in the courtyard of the Musee de Tahiti et des Iles, Punaauia.

51. It should be noted that these dates are old and that the range for New Zealand is now forwarded a couple of hundred years (Higham and Jones 2004).

52. Several years ago, after I had presented a talk for tourists with Aranui 3, a cargo/passenger ship that voyages from Papeete to the Marquesas Islands, a French musician/conductor approached me and asked me if I was aware of the unusual acoustic at Meae Iiopona. He said that even though he was standing quite a distance from where I was talking he heard me perfectly.

Glossary

Ahu	Temple or sacred place
Ahupuaa	Traditional Hawaiian land boundary
Alii	Chief (Hawaii)
Atua/etua	Deity, god, spirit
Eka/ena	Dye from the turmeric plant (*Curcuma longa*)
Enana/enata	Traditional name of a Marquesan
Fataa	Sacred store house, men's house
Hakaiki	A general term for chief
Hakatepeiu/hattepeiu	Chiefess
Hae (fae)	Habitation house
Hae konini	House of pleasure
Hae tukau	Inspirational priest's house
Hae tumau	Cooking house
Hami/maro	Men's loin cloth
Hau/fau	*Hibiscus tiliaceus*
Hei	Necklace, crown, garland
Heiau	Ritual place (Hawaii)
Henua /fenua	The name of the Marquesas, land
Hiapo/heapo	Cloth made from the sacred banyan tree
Hoka	Courageous
Ii	A general name for shells, a cowry shell scraper
Ii/mape	*Inocarpus fagifer*
Ipu	A small container, a bowl or gourd. Also the name of a tattoo motif
Ipu mata	Eye orbit
Kaioi	Youthful libertines (Handy 1923); sexually oriented adolescent society (Thomas 1990). Also a warrior (Karl von den Steinen 1925a 1925b, 1928)
Kava	Polynesian ritual beverage made from the *Piper methysticum* root
Kea	Rock, stone, boulder
Kea tuki popoi	Stone to beat a vegetable paste
Keetu	Cut blocks of red tuff
Kehika	Mountain apple (*Eugenia malaccensis*)
Kikino	Landless people
Kiva	Smooth water worn pebbles and boulders
Koika/koina	Feast or festival
Konane	Game (Hawaii)
Langi	Tomb (Tonga)
Liku	Women's dress or girdle (Fiji)
Ma	Fermented breadfruit paste

Makatea	White stone, reef limestone. It refers to elevated islands formed by tectonic uplift at plate margins. It is also the proper name of an island in the Tuamotu
Makemake	Face or eye images linked to the Miru clan (Easter Island). The ancient creator god of Easter Island
Mana	Supernatural power, skill
Marae	Temple or sacred place (Society Islands)
Mata	Eye, face, lineage, genealogy. The cup-shaped impressions in stones are also referred to as *mata*
Mataeinaa/mataeinana	Tribe, clan, people, lineage
Mata hoka	Courageous warriors
Mata komoe	Face or eye motifs. Possibly the proper name of a chief
Mau	Funeral feast
Meae	Temple, ritual place, tomb. Marae in Tahitian, ahu in Rapa Nui
Mei nui	The first breadfruit harvest
Mei	Breadfruit (*Artocarpus altilis*)
Moai	Monolithic anthropomorphic stone carvings (Easter Island)
Moemoe	A gesture of respect reserved for the high chief (Tonga)
Moku	Traditional Hawaiian district
Mounu/momo	Bait used by fishermen, also used as the medium for sorcery (Handy 1923:272)
Naonao	Checkerboard pattern in tattoos (Society Islands)
Nani kaha/nati kaha	Sorcery (Handy 1923:272), Spell, charm
Opu	Stomach, a type of pounder
Paepae/upe	Raised stone platform
Paepae hiamoe	Sleeping house
Paheeka	Land boundary, limit
Pandanus (haa/faa)	Screw pine (*Pandanus tectorius*)
Papa	Flat rock
Papamu	Game board used for playing *konane* (Hawaii)
Papatea	White beach rock (coquina)
Pakeho	Stone-lined rectangular pit found in house platforms
Patu tiki	Tattoo, to strike a tattoo
Pekio	Secondary husband
Piko	Umbilical cord, navel, navel string (Hawaii)
Popoi	A paste of breadfruit, taro and other vegetables
Puu	Natural lava formation or volcanic dome (Hawaii)
Tao	Taro (*Colocasia esculenta*)
Tama	Shout of respect and reverence uttered by people when approaching a chief (Fiji)

Taua	A shamanistic priest/priestess, inspirational priest/priestess
Tapa	Bark cloth
Tapu/taboo	Chiefly,, divine, sacred, restricted or prohibited
Ti/auti	(*Cordyline fruticosa*)
Tiki	Idol, image, anthropomorphic figure
Toa	Warrior, brave, ironwood, male, strong
Tohua/Tohua koina	Tribal ceremonial complex, tribal feasting place
Toki	Adz
Tuhuka taai tiki	Image maker
Tuhuka /tuhuna	Expert, professional
Tuhuka/tuhuna ooko, Tuhuna oono	Ceremonial priest
Tuhuna nani kaha	Someone who practice witchcraft
Tuu	Sacrificial display platform (Suggs 1961:68)
Ua ma	Underground breadfruit pit or silo
Uhu board	Decorated wooden board placed on *marae* (Society Islands)
Umu	Earth oven
Ute/outi	Paper mulberry (*Broussonetia papyrifera*)
Uu	War club, usually carved with anthropomorphic eyes and faces
Vaka	Canoe, coffin

References

Adamson, A. M.

1936 *Marquesan Insects: Environment*. Bernice P. Bishop Museum Bulletin 139. Honolulu: Bishop Museum Press.

Addison, D. J.

2006 Traditional Marquesan Agriculture and Subsistence: The Historic Evidence, Part I of IV. General Descriptions, Garden Locations, the Agricultural Calendar, Hydrology and Soils, Cultigens, and Agricultural Techniques. *Rapa Nui Journal* 20(2):111–124.

Addison, D. J.

2007 Traditional Marquesan Agriculture and Subsistence: Cultivation and Processing of Specific Agricultural Products, Part II of VI. *Rapa Nui Journal* 21(1):33–57.

Allen, M. S.

2004 Revisiting and Revising Marquesan Culture History: New Archaeological Investigation at Anaho Bay, Nuku Hiva Island. *Journal of Polynesian Society* 113(2):143–196.

Allen, M. S.

2010 Morphological Variability and Temporal Patterning in Marquesan Domestic Architecture: Anaho Valley in Regional Context. *Asian Perspectives* 48(2):342–382.

Allen, M., W. R. Dickinson, and J. M. Huebert

2012 The Anomaly of Marquesan Ceramics: A Fifty Year Retrospective. *Journal of Pacific Archaeology* 3(1):90–104.

Allen, M. S., and A. McAlister

2013 Early Marquesan Settlement and Patterns of Interaction: New Insights from Hatiheu Valley, Nuku Hiva Island. *Journal of Pacific Archaeology* 4:90–109.

Allingham, B.

1990 South Island Maori Rock Art Project. South Island Maori Rock Art Project (SIMRAP). www.ngaitahu.iwi.nz (accessed July, 2016).

Allen, T.

1992 A Re-evaluation of Handy's Analysis of Marquesan Tattooing. Unpublished M.A. thesis, Department of Anthropology, University of Hawai'i, Manoa.

Ambrose, W.

1970 Archaeology and Rock Drawings from the Waitaki Gorge, Central South Island. *Records of the Cantebury Museum* 8(5):383–437.

Anderson, A. J.

1988 The Art of Concealment: Maori Rock Art in the South Island. *Ka Tuhituhi o Nehera,* 4–8 Wellington: National Museum Publication.

Anderson, A. J., and Y. H. Sinoto

2002 New Radiocarbon Ages of Colonization Sites in East Polynesia. *Asian Perspectives* 41:242–257.

Anderson, R. L.

1990 *Calliope's Sisters: A Comparative Study of Philosophies of Art*. New Jersey: Prentice Hall.

Antoni, J.

1997 Un oiseau enigmatique aux Marquises. *Bulletin de la Société des Etudes Océaniennes* 273(4):88–90.

Ashmore, W., and R. J. Sharer
1996 *Discovering Our Past: A Brief Introduction to Archaeology*. Mountain View: Mayfield Publishing Company.

Ashmore, W., and A. B. Knapp (editors)
1999 *Archaeologies of Landscape: Contemporary Perspectives*. Malden, MA: Blackwell

Bahn, P. G.
1998 *The Cambridge Illustrated History of Prehistoric Art*. Cambridge: Cambridge University Press.

Bahn, P. G., and J. Vertut
1988 *Images of the Ice Age*. London: Windward.

Bain, P.
1982 Maori Rock Drawings: A Stylistic Analysis of Drawings in North Otago and South Canterbury. Abstract. Australasian Digital Theses Program. University of Otago. Electronic document, adt.otago.ac.nz/public/adt-NZDU20070619.125027/index.html (accessed July, 2016).

Bain, P.
1985 Geographic and Temporal Variation in Maori Rock Drawings in Two Regions of Southern New Zealand. *New Zealand Journal of Archaeology* 7:39–59.

Barrau, J.
1961 *Subsistence Agriculture in Polynesia and Micronesia*. Bernice P. Bishop Museum Bulletin 223. Honolulu: Bishop Museum Press.

Barrow, T.
1956 Maori Decorative Carving-an Outline. *Journal of the Polynesian Society* 65(4):305–331.

Barrow, T.
1979 *The Art of Tahiti*. London: Blacker Calmann Cooper Ltd.

Beaglehole, J. C.
1974 *The Life of Captain James Cook*. Stanford: Stanford University Press.

Bellwood, P.
1972 A Settlement Pattern Survey, Hanatekua Valley, Hiva Oa, Marquesas Islands. *Pacific Anthropological Records* 17. Honolulu: Bishop Museum Press.

Bellwood, P.
1978 The Polynesians, Prehistory of an Island People. London: Thames and Hudson Ltd.

Bellwood, P.
1979 *Man's Conquest of the Pacific*. New York: Oxford University Press.

Binford, L. R.
1982 Meaning, Inference, and the Material Record. In *Ranking, Resource and Exchange*, edited by C. Renfrew and S. Shennan, pp. 160–163. Cambridge: Cambridge University Press.

Blanshard, A.
2005 A Stylistic Chronology of Maori Rock Art in the South Island of New Zealand. In *Making marks: Graduate Studies in Rock Art Research at the New Millennium*, edited by J. K. K. Huang and E. V. Culley, pp. 41–50. Phoenix: American Rock Art Association.

Bourdieu, P.
1973 The Berber House. In *Rules and Meanings*, edited by M. Douglas, pp. 98–110. Harmondsworth: Penguin.

Brousse, R., J.-P. Chevalier, M. Denizot, and B. Salvat.
1978 Etude géomorphologique des Iles Marquises 21:9–74. *Les Iles Marquises; Géomorphologie. Climatologie. Faune et Flore. Cahier du Pacifique*. Paris: Publication de la Fondation Singer-Polignac.

Burley, D. V.
1994 As a Prescription to Rule: The Royal Tomb of Mala'e Lahi and the 19th-century Tongan Kingship. *Antiquity* 68:504–517.

Burrows, E. G.
1938 Western Polynesia: A Study in Cultural Differentiation. *Etnologiska Studier.* Goteborg, Sweden: Walter Kaudern.

Butikofer, M.
2005 Unpublished Site Conservation Report on the Takiroa Rock Art Shelter. Manuscript on file, Ngai Tahu Maori Rock Art Trust. Dunedin.

Calder, A.
1996 The Temptations of William Pascoe Crook: An Experience of Cultural Difference in the Marquesas, 1797–1798. *The Journal of Pacific History* 31(2):144–161.

Candelot, J.-L. (Teuruarii Tamatoa)
1980 Note préliminaire sur les pierres gravées de Tubuai. *Journal de la Société des Océanistes* 66–67:xxxvi.

Candelot, J.-L. (Teuruarii Tamatoa)
1998 La renaissance l'art du tatouage. In P. and M. N. Ottino-Garanger *Le Tatouage aux Iles Marquises.* Tahiti.

Cartmail, K. S.
1997 *The Art of Tonga, Ko e ngaahi'aati' o Tonga*. Honolulu: University of Hawai'i Press.

Cauchard, G., and J. Inchauspe
1978 Climatologie de l'archipel des Marquises. In *Les îles Marquises; Géomorphologie. climatologie. faune et flore. Cahier du Pacifique*, Vol. 21:75–105. Paris: Publication de la Fondation Singer-Polignac.

Chang, K. C.
1967 *Settlement, Society, and Archaeology*. Palo Alto: National Press.

Charleux, M.
2011 Rapport préliminaire des trois missions archéologiques sur Eiao. Missions "EIAO. 2010", EIAO, 2010.2". et "EIAO. 2011". SCP (Service de la Culture et du Patrimoine) and CIRAP (Collogue pour Règlement Administratif des Prestations), Tahiti, French Polynesia.

Chaulet, P.
1853–1888 Unpublished manuscripts and letters. Archives générales, Congrégation des Sacrés Cæurs, Rom.

Chaulet, P.
1858–1897 Unpublished manuscripts and letters. Archives générales, Congrégation des Sacrés Cæurs, Rom.

Chavaillon, K., and E. Olivier
2007 Le patrimoine archéologique de l'île de Hiva Oa (archipel des Marquises). No 5 (Dossier d'Archéologie Polynésienne. Ministére de la culture de Polynésie Française, Service de la culture et du patrimoine.

Chester, S., H. Baumgartner, D. Frechoso, and J. Oetzel.
2004 *Maeva Mai; The Marquesas Islands*. San Mateo: Wandering Albatross.

Chevalier, J-P.
1978 Les coraux des iles Marquises. In *Les îles Marquises; Géomorphologie. climatologie. Faune et Flore*. Cahier du Pacifique, Vol. 21:243–283. Paris: Publication de la Fondation Singer-Polignac.

Christian, F. W.
1910 *Eastern Pacific Lands: Tahiti and the Marquesas*. London: Robert Scott.

Clarke, D. (editor)
1977 *Spatial Archaeology*. London: Academic Press.

Clavel, M.
1884 Le dépopulation aux îles Marquises. *Bulletin de la Société d'Anthropologie de Paris*, pp. 490–500.

Cleghorn, P.
1986 Organizational Structure at the Mauna Kea Adze Quarry Complex, Hawai'i. *Journal of Archaeological Science* 13:375–387.

Cleghorn, P. L.
1980 *The Hilina Pali Petroglyph Cave, Hawai'i Island: A Report on Preliminary Archaeological Investigation*. Department of Anthropology. Report Series 80-1. Honolulu: Bernice P. Bishop Museum.

Clottes, J.
1993 *Theory and Record*. Cambridge Archaeological Journal 3(1):121–23.

Clottes, J.
1996 Paleolithic Images at Cosquer. www.bradshawfoundation.com (accessed August, 2016).

Clottes, J.
2003 *Return to Chauvet Cave, Excavation the Birthplace of Art: The First Full Report*. London: Thames & Hudson.

Clottes, J., and D. Lewis-Williams
1998 *The Shamans of Prehistory: Trance and Magic in the Painted Caves*. New York. Harry N. Abrams.

Conkey, M. W.
1978 Style and Information in Cultural Evolution: Towards a Predictive Model for the Paleolithic. In *Social Archaeology: Beyond Subsistence and Dating*, edited by C. L. Redman, M. J. Berman, E. V. Curtin, W. T. Langhorne, M. N. Versaggi, J. C. Wanser, pp. 61–85. New York: Academic Press.

Conkey, W. M.
1980 The Identification of Prehistoric Hunter-Gatherer Aggregation Sites: The Case of Altamira. *Current Anthropology* 21:609–630.

Conkey, M. W.
1981 A Century of Paleolithic Cave Art. *Archaeology* 34(4):20–28.

Conkey, M. W.
1983 On the Origins of Paleolithic Art: A Review and Some Critical Thoughts. In *The Mousterian Legacy: Human Biocultural Change in the Upper Pleistocene*, edited by E. Trinkhaus, pp. 201–207. International Series 164. Oxford: British Archaeological Reports.

Conkey, M. W.
1984 To Find Ourselves: Art and Social Geography of Prehistoric Hunter-Gatherers. In *Past and Present in Hunter-Gatherer Studies*, edited by C. Schrire, pp. 253–276. New York: Academic Press.

Conkey, M. W.
1997 Beyond Art and Between the Caves: Thinking About Context in the Interpretive Process. In *Beyond Art: Pleistocene Image and Symbol*, edited by M. W. Conkey, O. Soffer, D. Stratmann, and N. G. Jablonski, pp. 343–367. San Francisco, California: Wattis Symposium Series in Anthropology, Memoirs of the California Academy of Sciences, No. 23

Conkey, M. W.
2001 Hunting for Images, Gathering up Meanings: Art for Life in Hunting-Gathering Societies. In *Hunter-Gatheres, An Interdisciplinary Perspective*, edited by C. Panter-Brick, R. H. Layton, and P. Rowley-Conwy, pp. 267–290. Biosocial Society Symposium Series. Cambridge: Cambridge University Press.

Conte, E.
1991 Identification des structures et des sites archéologiques. Département d'Archéologie. Centre Polynésien des Sciences Humaines. Te Anavaharau. Papeete: Tahiti.

Conte, E.
2002 Current Research on the Island of Ua Huka, Marquesas Archipelago. *French Polynesia. Asian Perspectives* 2:258–268.

Conte. E., and G. Molle
2014 Reinvestigating a Key-Site for Polynesian Prehistory. New Results from the Hane Dune Site, Ua Huka (Marquesas). *Archaeology in Oceania* 49(3):121–136.

Cox, H. J., and E. Stasack
1970 *Hawaiian Petroglyphs*. Special Publication 60, Bernice. P. Bishop Museum, Honolulu.

Craig, R. D. (editor)
1980 *The Marquesas Islands, Their Description and Early History by Reverend Robert Thomson (1816–1851)*. Hawaii: The Institute for Polynesian Studies, Brigham Young University.

Crocombe, R. G.
1964 *Land Tenure in the Cook Islands*. New York: Oxford University Press.

Crook, W. P.
2007 [1797–1799] *An account of the Marquesas Islands 1797–1799*. Edited by Haere Po. Tahiti.

Danielsson, B.
1954 Contributions of Marquesan Archaeology. *Journal of the Polynesian Society* 63:75–76.

Danielsson, B.
1967 Kia Ora Keneti. In *Polynesian culture history: Essays in honor of Kenneth P. Emory*, edited by G. A. Highland, R. W. Force, A. Howard, M. Kelley, and Y. H. Sinoto, pp. 1–41. Special Publication 56. Bernice P. Bishop Museum. Honolulu.

David, R.
2012 The landscape of Klamath Basin Rock Art. Unpublished dissertation. Department of Anthropology, University of California, Berkeley.

David, R.
2010 The archaeology of myth: Rock art, ritual objects, and mythical landscapes of the Klamath Basin. *Archaeologies*, 6(2):372-400.

David, B., and M. Wilson
2002 Space of Resistance: Graffiti and Indigenous Place markings in the Early European Contact Period of Northern Australia. In *Inscribed Landscapes, Marking and Making Place*, edited by B. David and M. Wilson, pp. 42–60. Honolulu: University of Hawai'i Press.

Davidson, J. M.

1984 *The Prehistory of New Zealand*. Auckland: Longman Paul.

DeBoer, W. R.

1991 The decorative Burden: Design, Medium, and Change in Ceramic. In *Ethnoarchaeology*, edited by W. A. Longacre, pp. 144–161. Tucson: The University of Arizona Press.

Decantes, C.

1993 Simple *marae* of the Opunohu Valley, Mo'orea, Society Islands, French Polynesia. *Journal of the Polynesian Society* 102(2):187–216.

Delmas, S.

1889–1902. Letters. Archives générales, Congrégation des Sacrés Cæurs. Rom.

Delmas, S.

1927 La Religion ou la Paganisme des Iles Marquisiens. Paris, France.

Dening, G.

1966 Ethnohistory in Polynesia. *Pacific History Series* 1:23–42.

Dening, G. (editor)

1974 *The Marquesan Journal of Edward Robarts 1797–1824*. Honolulu: University Press of Hawai'i.

Dening, G.

1980 *Islands and Beaches, Discourse on a Silent Land: Marquesas 1774–1880*. Honolulu: University Press of Hawai'i.

Dickinson, W. R., and R. Shutler Jr.

1974 Probably Fijian Origin of Quartzose Temper Sands in Prehistoric Pottery from Tonga and the Marquesas. *Science* 185:454–57.

Dickinson, W. R., B. V. Rolett, Y. H. Sinoto, M. E. Rosenthal, and R. Shutler

1998 Temper Sands in Exotic Marquesan Pottery and the Significance of their Fijian Origin. *Journal de la Société des Océanistes* 107:119–33.

Dodge, E. S.

1940 An Account of the Marquesas Islands in 1825. *Journal of the Polynesian Society* 49:382–392.

Donaldson, E.

2004 Vanishing Artifacts of the South Seas. *Journal of the Polynesian Society* 113(4):349–367.

Donley, L. W.

1982 House Power: Swahili Space and Symbolic Markers. In *Symbolic and Structural Archaeology*, edited by I. Hodder, pp. 63–73. Cambridge: Cambridge University Press.

Dordillon, R. I.

1931 *Grammaire et dictionnaire de la langue des îles Marquises*. Marquisien-Français. Paris.

Dunmore, J.

1991 *Who's Who in Pacific Navigation*. Honolulu: University of Hawaii Press.

Dupetit-Thouars

1840–1841 *Voyages autour du monde sur le frégate la Vénus pendant les annés 1836–1839*. Four volumes and atlas. Paris: Gide.

Dupetit-Thouars

1995 [1842] *Bulletin de la Société des Etudes Océaniennes* (268):59.

Dye, T.
1990 The Causes and Consequences of a Decline in the Marquesan Fishing Industry. In *Pacific Production Systems: Approaches to Economic Prehistory*, edited by D. E. Yen, pp. 70–84. Canberra: Department of Prehistory, Australian National University.

Earle, T. K.
1990 Style and Iconography as Legitimation in Complex Chiefdoms. In *The uses of Style in Archaeology*, edited by M. Conkey and C. Hastorf, pp. 73–81. New Directions in Archaeology. Cambridge: Cambridge University Press.

Earle, T. K.
1994 Preface, In *New Light on Old Art; Recent Advances in Hunter-Gatherer Rock Art Research*, edited by D. S. Whitley and L. L. Loendorf. Monograph 36. Los Angeles: The Cotsen Institute of Archaeology, University of California.

Eddowes, M.
1991 *Marae Nuurua; Travaux partieles de restauration, partial restoration of the Marae Complex. Département Archéologie*. Centre Polynesien des Sciences Humaines, "Te Anavaharau". Tahiti.

Eddowes, M.
1998 La vallée de la Papeno'o livre ses secrets. *Dixit d'or*. Imprimaie Officielle du territoire. Tahiti.

Edwards, E.
1985a *Field Notes from the Marquesas Islands.* Manuscript in the possession of Service de la Culture et du Patrimoine, Tahiti.

Edward, E.
1985b *Archaeological survey of Hiva Oa, Ua Pou and Ua Huka*. Tahiti: Service de la Culture et du Patrimoine, Tahiti.

Edwards, E.
1985c *Elements de description du tohua Pehe Kua, vallee de Puamau, Hiva Oa*. Tahiti: Service de la Culture et du Patrimoine.

Edwards, E.
1988 *Archeological Survey of Tahinu River, Papeno'o Valley*. Département d'Archéologie. Centre Polynésien des Sciences Humaines, Te Anavaharau. Tahiti.

Edwards, E.
1989 Peintures rupestres et petroglyphs des Iles Marquises. Tahiti: *Te Ana* 1:65–71.

Edwards, E.
1991 Field Report from Tubuai and Ra'ivavae. Département d'Archéologie. Centre Polynésien des Sciences Humaines, Te Anavaharau. Tahiti.

Edwards, E.
1998 The Archaeological Survey of Ra'ivavae, Austral Islands, French Polynesia. In *Easter Island and East Polynesian Prehistory*, edited by P. C. Vargas, pp. 131–141 Santiago: Universidad de Chile. Facultad de Arquitectura y Urbanismo. Instituto de Estudios Isla de Pascua.

Edwards, E.
1999 The Archaeological Survey of Raivavae, Austral Islands, French Polynesia. Preliminary results. In *Easter Islands and East Polynesian Prehistory. Proceedings of the Second international congress on Easter Islands and East Polynesia*, pp 31–58, edited by P. Vargas. Santiago: Universidad de Chile, Instituto de Estudios Isla de Pascua,

Edwards, E.

2005 [2003] *Ra'ivavae. Archaeological Survey of Ra'ivavae, French Polynesia*. Easter Island Foundation. Los Osos.

Edwards, E. and S. Millerstrom

1986 *Archaeological survey of Nuku Hiva*. Tahiti: Service de la Culture et du Patrimoine.

Edwards, E. and S. Millerstrom

1987 *Archaeological survey of Hatiheu Valley, Nuku Hiva*. Tahiti: Service de la Culture et du Patrimoine, Tahiti.

Edwards, E., and S. Millerstrom

1995 Peintures Rupestres de la Vallée de Eiaone a Hiva Oa. *Bulletin de la Société des Etudes Océaniennes* 267:5–17.

Egan, S., and D. V. Burley

2009 Triangular Men on One Very Long Voyage: The Context and Implication of a Hawai'ian-style Petroglyph Site in the Polynesian Kingdom of Tonga. *Journal of Polynesian Society* 118(3):209–232.

Ellis, W.

1969 [1831] *Polynesian Researches*. Rutland, Vermont: Charles E. Tuttle Co.

Emory, K. P.

1922 Rock Carvings and Paintings of Kaupo, Maui. Manuscript on file at the Bernice P. Bishop Museum. Honolulu.

Emory, K. P.

1924 *The Island of Lanai, a Survey of Native Culture*. Bernice P. Honolulu: Bishop Museum Bulletin 12.

Emory, K. P.

1927 L'art Tahitien. Bulletin de la Société des Etudes Océaniennes 19:236–239.

Emory, K. P.

1928 *Archaeology of Nihoa and Necker Islands*. Bernice P. Honolulu: Bishop Museum Bulletin 53.

Emory, K. P.

1933 *Stone Remains in the Society Islands*. Honolulu: Bernice P. Bishop Museum Bulletin 116.

Emory, K. P.

1934 *Tuamotuan Stone Structures*. Honolulu: Bernice P. Bishop Museum Bulletin 118.

Emory, K. P.

1943 *South Sea Lore*. Honolulu: Bernice P. Bishop Museum Special Publication 36.

Emory, K. P.

1947 *Tuamotuan Religious Structures and Ceremonies*. Honolulu: Bernice P. Bishop Museum Bulletin 191.

Emory, K. P.

1975 *Material Culture of the Tuamotu Archipelago*. Pacific Anthropological Records. Honolulu: P. B. Bishop Museum.

Emory, K. P.

1979 The Societies. In *The Prehistory of Polynesia*, edited by J. D. Jennings, pp. 200–221. Cambridge: Harvard University Press.

Emory, K. P., and Y. H. Sinoto

1964 Eastern Polynesian Burials at Maupiti. *Journal of the Polynesian Society* 73:143–167.

Eyriaud des Vergnes
1877 *L'Archipel des Iles Marquises.* Paris, Berger-Levrault et Cie.

Ewins, R.
1995 Proto-Polynesian art? The Cliff Paintings of Vatulele, Fiji. *Journal of the Polynesian Society* 104(1):23–74.

Ferdon, E. N. J.
1965 Surface Architecture of the Site of Paeke, Taipi Valley, Nukuhiva. In *Reports of the Norwegian Archaeological Expedition to Easter Island and the East Pacific*, edited by T. Heyerdahl and E. N. Ferdon, pp. 117–122. Monographs of the School of American Research and the Kon-Tiki Museum, Vol. 2. Stockholm: Forum Publishing House.

Firth, R.
1936 *We, the Tikopia. A Sociological Study of Kinship in Primitive Polynesia*. Stanford: Stanford University Press.

Firth, R.
1967 *Tikopia Ritual and Belief.* Boston: Beacon Press.

Fischer, S. R.
1994 Rapanui's "Great Old Words": E timo te akoako. *Journal of the Polynesian Society* 103(4):413–443.

Fischer, S. R.
1995 Preliminary Evidence for Cosmogonic Texts in *Rapanui's Rongorongo Inscriptions. Journal of the Polynesian Society* 104:303–321.

Fomison, A., and R. Fyfe (editors)
2014 Maori Rock Art in North Otago and South Canterbury: A Guide to Interpretation of its Styles and Subject Matter. *Records of the Canterbury Museum* 27:47–95.

Forge, A.
1970 Learning to See in New Guinea. In *Socialization: The Approach from Social Anthropology*, edited by P. Mayer, pp. 269–291. London: Cass.

Forge, A. (editor)
1973 *Primitive Art and Society*. New York: Oxford University Press.

Forge, A.
1991 Hand Stencils: Rock Art or Not Art. In *Rock Art and Prehistory*, edited P. G. Bahn and A. Rosenfeld, Papers presented to symposium G of the AURA Congress, Darwin 1988. Oxford: Oxbow Monograph 10:39–44.

Forster, J. R.
1778 *Observations Made During a Voyage Round the World on Physical Geography, Natural History, and Ethic Philosophy*. London: G. Robison.

Forster, G.
1777 *A Voyage Round the World*. Resolution, Commanded by Captain James Cook. Vol. 2. London: Whik.

Fowke, G.
1922 Archaeological Investigation. *Archaeological Work in Hawaii Bulletin* 76:174–195.

Freeman Moulin, J.
1997 Gods and Mortals: Understanding Traditional Function and Usage in Marquesan Musical Instruments. *Journal of the Polynesian Society* 106:250–283.

Fritz, J. M.

1978 Paleopsychology Today: Ideational Systems and Adaptation in Prehistory. In *Social Archaeology: Beyond Subsistance and Dating*, edited by C. Redman, M. J. Berman, E. V Curtin, W. T. Langhorne Jr., N. M. Versaggi, J. C. Wanser, pp. 37–59. New York: Academic Press.

Garanger, J.

1967 *Pilons Polynésiens. Catalogues du Musée de l'Homme*. Paris: Muséum National d'Histoire Naturelle.

Garanger, J.

1980 Prospection archéologiques de l'îlot Fenuaino et des vallées Aiuru et Vaiote a Tahiti. *Journal de la Société des Océanistes* 35:77–104.

Gathercole, P.

1979 New Zealand Maori. In *The Art of the Pacific Islands*, edited by P. Gathercole, A. L. Kaeppler, and D. Newton, pp. 97–100. Washington, D.C.: National Gallery of Art.

Gell, A.

1993 *Wrapping in Images, Tattooing in Polynesia*. Oxford: Clarendon Press.

Gifford, E. W.

1929 *Tongan Society*. Bernice P. Bishop Museum Bulletin 61. Honolulu: Bernice P. Bishop Museum.

Gillette, D. L., M. Greer, M. H. Hayward, and W. B. Murray (editors)

2015 *Rock Art and Sacred Landscapes*. New York: Springer.

Goldman, I.

1970 *Ancient Polynesian Society*. Chicago: The University of Chicago Press.

Gombrich, E. H.

1961 *Art and Illusion*. New York: Princeton University Press.

Gosden, C., and L. Head

1994 Landscape–A Usefully Ambiguous Concept. *Archaeology in Oceania* 29(3):113–116.

Gover, E.

2010 *Twelve Days at Nuku Hiva, Russian Encounters and Mutiny in the South Pacific*. Honolulu: University of Hawai'i Press.

Gracia, P. M.

1843 Lettres sur les îles Marquises ou mémoires pour servir a l'étude religieus, morale, politique, et statistique des îles Marquises et de l'Océanie orientale. Paris: Chez Gaume Fréres, Libraires.

Green, R. C.

1961 Mo'orea Archaeology: A Report. *Man* 61:169–73.

Green, R. C.

1979 Early Lapita Art from Polynesia and Island Melanesia: Continuities in Ceramic, Bark-Cloth, and Tattoo Decorations. In *Exploring the Visual Art of Oceania, Australia, Melanesia, Micronesia, and Polynesia*, edited by S. M. Mead, pp. 13–31. Honolulu: The University Press of Hawaii.

Green, R. C.

1980 Makaha Before 1880 A.D. *Pacific Anthropological Records* 31.

Green, R. C.

1998 Rapanui origins prior to European Contact. The View from Eastern Polynesia. In *Easter Island and*

East Polynesian Prehistory, edited by P. V. Casanova, pp.87–110. Santiago: Universidad de Chile. Facultad de Arquitectura y Urbanismo. Instituto de Estudio Isla de Pascua.

Green, R. C.
2000a Origins for the Rapanui of Easter Island Before European Contact: Solutions from Holistic Anthropology to an Issue no Longer much of a Mystery. *Rapa Nui Journal* 14(3):71–76.

Green, R. C.
2000b A Range of Disciplines Support a Dual Origin for the Bottle Gourd in the Pacific. *Journal of the Polynesian Society* 110(2):191–97.

Green, R. C., K. Green, R. Rappaport, A. Rappaport, and J. Davidson
1967 Archaeology of the Island of Mo'orea, French Polynesia. *Anthropological Papers of the American Museum of Natural History* 51(2). New York: American Museum of Natural History.

Green, R. C., and J. Davidson (editors)
1969 *Archaeology in Western Samoa*. Vol. 1. Bulletin of the Auckland Institute and Museum 6. Auckland, New Zealand: Auckland Institute and Museum.

Green, R. C., and J. Davidson
1974 *Archaeology in Western Samoa*. Vol. 2. Bulletin of the Auckland Institute and Museum 7. Auckland, New Zealand: Auckland Institute and Museum.

Green, R. C., and M. Weisler
2000 Mangarevan Archaeology: Interpretations Using New Data and 40 Year Old Excavations to Establish a Sequence from 1200 to 1900 AD. *University of Otago Studies in Prehistoric Anthropology* No. 19. Otago: Department of Anthropology, University of Otago, Dunedin, New Zealand.

Greiner, R. H.
1923 *Polynesian Decorative Designs*. Bernice P. Bishop Museum Bulletin 7. Honolulu: Bernice P. Bishop Museum.

Haddon, A. C., and J. Hornell
1975 [1936, 1937, 1938] *Canoes of Oceania*. Bernice P. Bishop Museum Special Publications 27, 28, and 29. Honolulu: Bernice P. Bishop Museum Press.

Handy, E. S. C.
1923 *The Native Culture in the Marquesas*. Bernice P. Bishop Museum Bulletin 9. Honolulu: B. P. Bishop Museum.

Handy, E. S. C.
1927 *Polynesian Religion*. Bernice P. Bishop Museum Bulletin 34. Honolulu: B. P. Bishop Museum.

Handy, E. S. C.
1930a *Marquesan Legends*. Bernice P. Bishop Museum Bulletin 69. Honolulu: Bernice P. Bishop Museum.

Handy, E. S. C.
1930b *History and Culture in the Society Islands*. Bulleting 79. Honolulu: Bernice P. Bishop Museum.

Handy, E. S. C.
1943 Two Unique Petroglyphs in the Marquesas which point to Easter Island and Malaysia. Studies in the Anthropology of Oceania and Asia. *Papers of the Peabody Museum of American Anthropology and Ethnology* 20:22–31.

Handy, E. S. C., and J. L. Winne
1925 *Music in the Marquesas Islands*. Bernice P. Bishop Museum. Bulletin 17. Honolulu: Bernice P. Bishop Museum.

Handy, W. C.
1922 *Tattooing in the Marquesas*. Bernice P. Bishop Museum Bulletin 1. Honolulu: B. P. Bishop Museum.

Handy, W. C.
1925 *String Figures from the Marquesas and Society Islands*. Bernice P. Bishop Museum Bulletin 18. Honolulu: Bernice P. Bishop Museum Press.

Handy, W. C.
1938 *L'Art des Iles Marquises*. Paris: Les Editions d'Art et d'Histoire.

Handy, W. C.
1965 *Forever the Land of Men: An Account of a Visit to the Marquesas Islands*. New York: Dodd, Mead and Company.

Handy, W. C.
1973 *Thunder from the Sea*. Honolulu: The University Press of Hawaii.

Harman, J.
2011 Using DStretch to Investigate Paintings in Cueva San Borjitas, BCS, Mexico. Paper presented at the ARARA session at the 76th Annual Meeting of the Society for American Archaeology, Sacramento, California.

Hastorf, C. A.
1991 Gender, Space, and Food in Prehistory. In *Engendering Archaeology; Women and Prehistory*, edited by J. M. Gero and M. W. Conkey, pp. 132–159. Oxford: Basil Blackwell Publishers Ltd.

Hawkes, C.
1954 Archaeological Theory and Method: Some Suggestions from the Old World. *American Anthropologist* 56:155–168.

Henry, T.
1928 *Ancient Tahiti*. Based on material recorded by J. M. Orsmond. Bernice P. Bishop Museum Bulletin 48. Honolulu.

Heyerdahl, T., and E. N. Ferdon (editors)
1965 *Reports of the Norwegian Archaeological Expedition to Easter Island and the East Pacific*. Monographs of the School of American Research and the Kon-Tiki Museum (Oslo, Norway). Vol. 2. Stockholm: Forum Publishing House.

Heyerdahl, T.
1965 The statues of the Oipona *me'ae*, with comparative analysis of possibly related stone monuments. In *Reports of the Norwegian Archaeological Expedition to Easter Island and the East Pacific*, edited by T. Heyerdahl and J. E. N. Ferdon. Monographs of the School of American Research and the Kon-Tiki Museum (Oslo, Norway) 24(2):123–51. Stockholm: Forum Publishing House.

Heyerdahl, T., and A. Skjølsvold
1965 Notes on the Archaeology of Pitcarin. In *Reports of the Norwegian Archaeological Expedition to Easter Island and the East Pacific*, edited by T. Heyerdahl and A. Skjolsvold. Monographs of the School of American Research and the Kon-Tiki Museum (Oslo, Norway) 24(2):3–7. Stockholm, Sweden: Forum Publishing.

Higham, T., and D. Johns
2004 Chronology and Settlement. In *Change Through Time: 50 Years of New Zealand Archaeology*, edited by L. Furey and S. Holdaway, pp 215–234. Auckland: New Zealand Archaeology Association.

Hill, Rev. W. R.

1956 Some Rock Carvings of Natewa Bay and Savusavu. *Transactions and Proceedings of the Fiji Society* 6(2):74–84.

Hiroa, Te Rangi (P. H. Buck)

1930 *Samoan Material Culture*. Bernice P. Bishop Museum Bulletin 75. Honolulu: Bernice P. Bishop Museum.

Hodder, I.

1984 Archaeology in 1984. *Antiquity* 58:25–32.

Hodder, I.

1986 *Reading the Past: Current Approaches to Interpretation in Archaeology*. Cambridge: Cambridge University Press.

Hoerman, R.

2016 Pacific Rock Art from 2010–2014: Research Trends, Conservation and Losses. In *Rock Art Studies: News of the World* V, edited by P. Bahn, N. Franklin, M. Strecker and K. Devlet. Oxford: Oxbow Books.

Holcomb, R. T.

1987 Eruptive History and Long-term Behavior of Kilauea Volcano. In *Volcanism in Hawaii*, edited by R. W. Decker, T. L. Wright, and P. H. Stauffer, pp. 261–350. Washington D. C.: U.S. Geological Survey.

Hughes, H. G. A., and S. R. Fischer (editors)

1998 *An Essay Towards a Dictionary and Grammar of the Lesser-Australian Language, According to the Dialect Used at the Marquesas (1799)*. Auckland: The Institute of Polynesian Languages and Literatures.

Huebert, J. M.

2012 Evaluating Forest Transformation and the Rise of Arboreal Subsistence Strategies in Highly Variable Environments: The Anthropogenic Forests of the Maruqesas Islands, French Polynesia. Paper presented at the 8th International Conference on Easter Island and the Pacific, Santa Rosa, California (USA) July 8–13, 2012.

Hubert, J. M. and M. S. Allen

2016 Six Centuries of Anthropogenic Forest Change on a Polynesian High Island: Archaeological Charcoal Records from the Marquesas Islands. *Quaternary Science Reviews* 137:79–96.

Huebert, J. M., M. S. Allen, and R. T. Wallac

2010 Polynesian Earth Ovens and their Fuels. *Journal of Polynesian Society*. 119(11):61–97.

Hurst Thomas, D.

1979 *Archaeology*. New York: Holt, Rinehart and Wilston.

Ingraham, J.

1790–1792 Journal of the Voyage of the Brigantine 'Hope' from Boston the North-West Coast of America 1790 to 1792. Vol. 3. Honolulu: Hawai'ian Historical Society.

Inventaire Archeologique de Polynesie Française.

1989 *Petroglyphs. Ra'iatea, Bora-Bora, Huahine*. Tahiti: Département Archéologie. Centre Polynésien des Sciences Humaines Te Anavaharau.

Irwin, G.

1992 *The Prehistoric Exploration and Colonisation of the Pacific*. Cambridge: Cambridge University Press.

Ivory, C. S.

1990 Marquesan Art in the Early Contact Period 1774–1821. Unpublished Ph.D. Dissertation, Division of Art History, University of Washington.

Ivory, C. S.
1993 Re-viewing Marquesan art. In *Artistic Heritage in a Changing Pacific*, edited by P. J. C. Dark and R. G. Rose, pp. 63–73. Honolulu: University of Hawaii Press.

Ivory, C. S.
1994 Marquesan *'u'u*: Stylistic and Historical Review. *Pacific Arts* July 9–10:53–63.

Ivory, C. S.
1995 Late Nineteenth Century Marquesan Clubs: A Preliminary Analysis. *Pacific Arts* July 11–12:20–28.

Ivory, C. S.
1998 Painted Tapa from the Marquesas Islands. Easter Islands in Pacific Context, South Seas Symposium. *Proceedings of the Fourth International Conference on Easter Island and East Polynesia*, pp. 51–54. University of New Mexico, Albuquerque.

Jardin, M. E.
1862 Essai sur l'histoire naturelle de l'archipel des Marquises. Paris: J.-B. Bailliére.

Jennings, J., R. Holmer, and G. Jackmond
1982 Samoan Village Patterns: Four Examples. *Journal of the Polynesian Society* 91:81–102.

Jones, T. J., A. A. Storey, E. A. Matisoo-Smith, andJ. M. Ramirez-Aliaga (editors)
2011 *Polynesians in America; Pre-Columbian Contacts with the New World.* AltaMira Press. New York.

Joppien, R., and B. Smith.
1985 *The Art of Captain Cook's Voyages, the Voyage of the Resolution and Adventure 1772–1775*. Vol. 2. New Haven and London: Yale University Press.

Jouan, H.
1890 La dépopulation aux îles Marquises (Océanie). Cherbourg, impr. Le Maout, 1890:24.

Kaeppler, A. L.
1979 The Survey of Polynesian art. In *Exploring the Visual Art of Oceania- Australia, Melanesia, Micronesia, and Polynesia*, edited by S. M. Mead, pp.180–191. Honolulu: The University Press of Hawaii.

Kahn, J. G.
1996 Prehistoric Stone Tool Use and Manufacture at the Ha'atuatua Dune Site, Marquesas Islands, French Polynesia. Unpublished M.A. thesis, Department of Archaeology, University of Calgary.

Kahn, J. G.
2005 Household and Community Organization in the late Prehistoric Society Islands (French Polynesia). Unpublished Ph.D. dissertation, Department of Anthropology, University of California at Berkeley.

Kahn, J. G.
2006 Society Islands (Central Eastern Polynesia) Chronology: 11 Radiocarbon Dates for the Late Prehistoric Expansion and Proto-Historic Periods in the late 'Opunohu Valley, Mo'orea. *Radiocarbon* 48(3):409–419.

Kahn, J. G.
2010 A Spatiol-Temporal Analysis of 'Oro Cult *Marae* in the 'Opunohu Valley, Mo'orea, Society Islands. *Archaeology in Oceania* 45:103–110

Kahn, J. G., and P. V. Kirch
2014 *Monumentality and Ritual Materialization in the Society Islands: The Archaeology of a Major Ceremonial Complex in the 'Opunohu Valley, Mo'orea*. Bishop Museum Bulletin in Anthropology, 13. Honolulu: Bernice. P. Bishop Museum Press.

Linton, R.
1939 *The Individual Land His Society; The Psychodynamics of Primitive Social Organization*. New York: Columbia University Press.

Kellum, M.
1966 *Depopulation in the Marquesas*. University of Hawai'i at Manoa. Honolulu: Pacific Studies.

Kellum-Ottino, M.
1968 Sites and Settlement in Hane Valley, Marquesas. Unpublished M.A. thesis, Pacific Island Studies, University of Hawaii. Honolulu.

Kellum-Ottino, M.
1971 Archéologie d'une vallée des iles Marquises, évolution des structures de l'habitat a Hane, Ua Huka. Société des Océanistes, No 26. Paris: Musée de l'Homme.

Kikuchi, W. K.
1964 Petroglyphs in American Samoa. *Journal of the Polynesian Society* 73(2):163–66.

Kikuchi, W. K.
1967 Additional Petroglyphs from American Samoa. *Journal of the Polynesian Society* 76(3):372–373.

Kirch, P. V.
1973a Archaeological Reconnaissance Survey of Kalahuipua'a and Portions of Wailoloa, Lalamilo, and Aneaeho'omalu, South Kohala, Hawai'i Island. Report on file at the Department of Anthropology, Honolulu: Bernice P. Bishop Museum.

Kirch, P. V.
1973b Prehistoric Subsistence Pattern in Northern Marquesas Islands, French Polynesia. *Archaeology and Physical Anthropology in Oceania* 8:24–40.

Kirch, P. V.
1982 Advances in Polynesian Prehistory: Three Decades in Review. *Advances in World Archaeology* 1:51–97.

Kirch, P. V.
1984 *The Evolution of the Polynesian Chiefdoms*. New Studies in Archaeology. Cambridge: Cambridge University Press.

Kirch, P. V.
1985 *Feathered Gods and Fishhooks: An Introduction to Hawaiian Archaeology and Prehistory*. Honolulu: University of Hawaii Press.

Kirch, P. V.
1986 Rethinking East Polynesian Prehistory. *Journal of the Polynesian Society* 95:9–40.

Kirch, P. V.
1989a. Prehistory. In *Developments in Polynesian Ethnology*, edited by A. Howard and R. Borofsky, pp. 13–46. Honolulu: University of Hawai'i Press.

Kirch, P. V.
1989b *Prehistoric Hawaiian Occupation in the Anahulu Valley, O'ahu Island: Excavation in Three Island Rockshelters*. Berkeley: Archaeological Research Facility. Department of Anthropology, University of California.

Kirch, P. V.
1990 Regional Variation and Local Style: A Neglected Dimension in Hawaiiian Prehistory. *Pacific Studies* 13:41–54.

Kirch, P. V.
1991a Chiefship and Competitive Involution: The Marquesas Islands of Eastern Polynesia. In *Chiefdoms, Power, Economy and Ideology*, edited by T. Earle, pp. 119–145. Cambridge: Cambridge University Press.

Kirch, P. V.
1991b Polynesian Agricultural Systems. In *Islands, Plants, and Polynesian: An Introduction to Polynesian Ethnobotany*, edited by P. A. Cox and S. A. Banack, pp. 113–133. Portland: Dioscorides Press.

Kirch, P. V.
1994a The Pre-Christian Ritual Cycle of Futuna, Western Polynesia. *Journal of the Polynesian Society* 103(3):255–298.

Kirch, P. V.
1994b *The Wet and the Dry; Irrigation and Agricultural Intensification in Polynesia*. Chicago: The University of Chicago Press.

Kirch, P. V.
1996 Tikopia Social Space Revisited. In *Oceanic Culture History, Essays in Honour of Roger Green*, edited by J. Davidson, G. Irwin, F. Leach, A. Pawley, and D. Brown, pp. 257–274. New Zealand: New Zealand Journal of Archaeology Special Publication.

Kirch, P. V.
1997a Changing Landscapes and Sociopolitical Evolution in Mangaia, Central Polynesia. In *Historical Ecology in the Pacific Islands: Prehistoric Environmental and Landscape Change*, edited by P. V. Kirch and T. L. Hunt, pp. 147–65. New Haven, Connecticut: Yale University Press.

Kirch, P. V. (editor)
1997b *Na Mea Kahiko o Kahikinui, Studies in the Archaeology of Kahikinui, Maui*. Special Publication No 1. Oceanic Archaeology Laboratory, Archaeological Research Facility, University of California, Berkeley.

Kirch, P. V.
2000 *On the Road of the Winds: An Archaeological History of the Pacific Islands efore European Contact*. Berkeley: University of California Press.

Kirch, P. V.
2014 *Kuaaina Kahiko: Life and Land in Ancient Kahikinui, Maui*. Honolulu: University of Hawai'i Press.

Kirch, P. V., and M. Kelly (editors)
1975 Prehistory and Ecology in a Windward Hawai'ian Valley: Halawa Valley, Molokai. *Pacific Anthropological Records* 24. Honolulu: B. P. Bishop Museum.

Kirch, P. V., and D. E. Yen
1982 *Tikopia. The Prehistory and Ecology of a Polynesian Outlier*. Bernice P. Bishop Museum Bulletin 238. Honolulu:.

Kirch, P. V., and R. C. Green
1987. History, Phylogeny, and Evolution in Polynesia. *Current Anthropology* 28(4):431–56.

Kirch, P. V., W. R. Dickinson, and T. L. Hunt
1988 Polynesian Plainware Sherds from Hivaoa and Their Implications for Early Marquesan Prehistory. *New Zealand Journal of Archaeology* 10:101–107.

Kirch. P. V., and D. Lepofsky
1993 Polynesian Irrigation: Archaeological and Linguistic Evidence for Origins and Development. *Asian Perspectives* 32:183–204.

Kirch, P. V., and M. I. Weisler
1994 Archaeology in the Pacific Islands: An Appraisal of Recent Research. *Journal of Archaeological Research* 2:285–328.

Kirch, P. V., and J. Ellison
1994 Paleoenvironmental Evidence for Human Colonization of Remote Oceanic Islands. *Antiquity* 68:310–21.

Kirch, P. V., and T. L. Hunt (editors)
1997 *Historical Ecology in the Pacific Islands: Prehistoric Environmental and Landscape Change*. New Heaven and London: Yale University Press.

Kirch, P. V., and E. Conte (editor)
2004 *Archaeological Investigations in the Mangareva Islands (Gambier Archipelago) French Polynesia*. No. 62. Archaeological Research Facility, University of California, Berkeley.

Kirch, P. V., and J-L. Rallu. (editors)
2007 *The Growth and Collapse of Pacific islands Societies: Archaeological and Demographic Perspectives*. Honolulu: University of Hawai'i Press.

Kooijman, S.
1972 *Tapa in Polynesia*. Honolulu: Bernice P. Bishop Museum Bulletin 234.

Krauss, B.
1988 *Keneti: South Seas Adventures of Kenneth Emory*. Honolulu: University of Hawai'i Press.

Krusenstern, A. J. V.
1968 [1813] *Voyage Round the World in the Years 1803, 1804, 1805, and 1806*. Vol. 1. Bibliotheca Australiana No. 38. New York: Da Capo Press.

Kunst- und Ausstellungshalle der Bundesrepublik Deutschland, Bonn
2009 *James Cook and the Exploration of the Pacific*. Exhibition book, August 27, 2009 to February 28, 2010. London: Thames & Hudson.

Kwiatkowski, P. F.
1991 Na Ki'i Pohaku: A Hawaiian Petroglyph Primer Honolulu: Ku Pa'a, Inc.

Lavachery, H.
1936 Contribution a l'étude de l'Archéologie de l'île de Pitcarin. *Bulletin de la Société des Américanistes de Belgique* 19:3–42.

Lavachery, H.
1939 *Les Pétroglyphes de l'île du Paques*. 2 vols. Antwerpen: De Sikkel.

Lawson, T.
1861–1867 Collection of manuscript letters and notes pertaining to the customs, language, genealogies, chants, etc. of the Marquesas. On file at the Bernice P. Bishop Museum. Honolulu.

Layton, R.
1991 *The Anthropology of Art*. Cambridge, UK: Cambridge University Press.

Le Cleac'h, H.
1996 Fatuiva ou Fatu Hiva? In *Marquises*. L'Association des Historiens et Géographes de Polynésie Française et le Centre Territorial de Recherche et de Documentation Pédagogiques de Polynésie Française, pp. 34-36, Papeete.

Le Cleac'h, H.
1997 Pona Te'ao Tapapatina. Lexique Marquisien-Français. STP Multipress. Papeete.

Lee, G.
1986 The birdman Motif of Easter Island. *Journal of New World Archaeology* 7(1):39–49. Los Angeles: UCLA Institute of Archaeology.

Lee, G.
1988 *The Petroglyphs of Lanai, Hawaii*. University of California Research Expeditions Program.

Lee, G.
1989 The Petroglyphs of Puako, Hawai'i. *Rapa Nui Journal* 3:4–6.

Lee, G.
1990/91 Hawai'i, the Big Island. *Rapa Nui Journal* 4:63–64.

Lee, G.
1992 *The Rock Art of Easter Island, Symbols of Power, Prayers to the Gods. Monumenta Archaeologica 17*. The Cotsen Institute of Archaeology, University of California, Los Angeles.

Lee, G.
1993a *The Petroglyphs of Pu'uloa (HV-225) Hawai'i Volcanoes National Park*. National Park Service, Hawaii Volcanoes National Park and Historic Sites Section, Department of Land and Natural Resources, State of Hawai'i.

Lee, G.
1993b The Petroglyphs of Tongariki. *Rapa Nui Journal* 7:12–16.

Lee, G.
1995 Wahi Pana o Hawai'i Nei. *Rapa Nui Journal* 9:47–54.

Lee, G.
1996 Rock Art of Polynesia. In *Rock Art Studies, News of the World I*, edited by P. G. Bahn and A. Fossati, pp. 163–172. Oxford: Oxbow Monograph 72.

Lee, G.
2002 Wahi Pana, Legendary Places on Hawai'i Island. In *Inscribed Landscapes, Marking and Making Place*, edited by B. David and M. Wilson, pp 79–92. Honolulu: University of Hawai'i Press.

Lee, G., and E. Stasack
1999 *Spirit of Place: Petroglyphs of Hawai'i*. Los Osos, California: Easter Island Foundation, Bearsville and Cloud Mountain Presses.

Lehmann, A. C., and J. E. Meyers (editors)
2001 *Magic, Witchcraft, and Religion: An Anthropological Study of the Supernatural*. California: Mayfield Publishing Company.

Lepofsky. D. S.
1994 Prehistoric Agricultural Intensification in the Society Islands, French Polynesia. Unpublished Ph.D. dissertation, Department of Anthropology, University of California, Berkeley.

Leroi-Gourhan, A.
1965 *Treasures of Paleolithic Art*. New York: Harry Abrams.

Lidin, G.
1932 Excursion dans la vallée Vaiote au Pari. *Bulletin de la Société des Etudes Océaniennes* 5(2)43:47–52.

Linton, R.
1923 *The Material Culture of the Marquesas Islands*, Vol. 8, No. 5. Memoirs of the Bernice P. Bishop Museum. Honolulu.

Linton, R.
1925 *Archaeology of the Marquesas Islands*. Honolulu: Bernice P. Bishop Museum Bulletin 23.

Linton, R., and P. S. Wingert
1946 *Arts of the South Seas*. New York: The Museum of Modern Art.

Lisiansky, U.
1968 [1814] *A Voyage Round the World in the Years 1803, 4, 5, and 6*. Israel: Da Capo.

Lund University.
2014 Digital Heritage Forum, the Faculties of Humanities and Technology http://projekt.ht.lu.se/digital-heritage/projects/petroglyphics-virtual-rock-arts-experiences/ (accessed August 9, 2016).

Magne, M. P. R.
1985 Lithics and Livelihood: Stone Tools Technologies of Central and Southern Interior British Columbia. *Archaeological Survey of Canada Mercury Series* Paper 133. National Museum of Man, Ottowa.

Malmer, M. P.
1993 A Poet's Ambiguity. *Cambridge Archaeological Journal* 3(1):113–18.

Mantell, W.
1852 Sketchbook No. 1. Wellington: Turnbull Library. www.teara.govt.nz (accessed August 2, 2016).

Markham, S. C.
1904 *The Voyages of Pedro Fernandez de Quiros, 1595 to 1606*. 1(4):15–29. London: The Hakluyt Society.

Marshack, A.
1972 *The Roots of Civilization*. New York: McGraw Hill.

Marshack, A.
1997 Paleolithic Image Making and Symboling in Europe and the Middle East: A Comparative Review. In *Beyond Art: Pleistocene Image and Symbol*, edited by M. W. Conkey, O. Soffer, D. Stratmann, and N. G. Jablonski, pp. 53–91. San Francisco: Memoirs of the California Academy of Sciences.

Marshall, D.
1961 *Ra'ivavae: An Expedition to the Most Fascinating and Mysterious Island in Polynesia*. New York: Doubleday and Company, Inc.

Matisoo-Smith, E., R. M. Roberts, G. J. Irwin, J. S. Allen, D. Penny, and D. M. Lambert
1998 Patterns of Prehistoric Human Mobility in Polynesia Indicated by mtDNA from the Pacific rat. *PNAS Online* 95(25):15145–15150. Electronic document, available at http:/www.pnas.org/cgi/content/full/95/25/15145.

McArthur, N.
1968 *Island Populations of the Pacific*. Honolulu: University of Hawaii Press.

McCoy, P. C.

1976 *Easter Island Settlement Patterns in the Late Prehistoric and Proto-Historic Periods. International Fund for Monuments.* Easter Island Committee, Bulletin 5. New York: International Fund for Monuments.

McBride, L. R.

2004 *Petroglyphs of Hawaii.* Hilo: Petroglyph Press, Ltd.

McEwen, J. M.

1966 Maori Art. In *An Encyclopedia of New Zealand*, edited by A. H. McLintock, pp. 408–429. Wellington: Government Printer.

McKern, W. C.

1929 *Archaeology of Tonga*. Honolulu: Bernice P. Bishop Museum Bulletin 60.

Mead, S. M., L. Birks, H. Birks, and E. Shaw.

1975 *The Lapita Pottery Style of Fiji and its Associations*. Polynesian Society Memoir No. 38. Wellington, New Zealand: Polynesian Society.

Melville, H.

1964 [1846] *Typee*. New York: Signet Classic.

Métraux, A.

1937 Relief Carvings in Stone in Polynesia. Stockholm: Ethos 2:340–344.

Métraux, A.

1940 *Ethnology of Easter Island*. Bernice P. Bishop Museum Bulletin 160. Honolulu: Bishop Museum Press.

Meyer, A. J. P.

1995 *Oceanic Art*. New Jersey: Knickerbocker Press.

Miller, B. D.

2002 *Cultural Anthropology*. Allyn and Bacon. Boston.

Millerstrom, S.

1985a Rock Art in the Marquesas Islands: A Preliminary Report. *La Pintura*, American Rock Art Research Association, Newsletter 12:7.

Millerstrom, S.

1985b Up-date on Marquesan Rock Art. *La Pintura*, American Rock Art Research Association Newsletter 12:5–6.

Millerstrom, S

1985c Marquesan Rock Art Project on Hiva Oa, in the Valleys of Taa Oa, Atuona, Tahauku, Punaei, Hanaiapa, Muto Ua, Nahoe, Hanapaaoa, and Puamau. On file in the Département d'Archéologie. Centre Polynésien des Sciences Humaines, Te Anavaharau. Tahiti.

Millerstrom, S.

1985b Marquesan Rock Art Projects on Hiva Oa (Eiaone, Puamau, Hanaiapa, Tahauku, Taaovea Ridge), Ua Huka (in the valleys of Vaipaee, Hane, Hokatu, Haave, Hatuana, Vaikivi) and Tahuata (Hapatoni Valley). On file in the Département d'Archéologie. Centre Polynésien des Sciences Humaines, Te Anavaharau. Tahiti.

Millerstrom, S.

1988 Experimental Archaeology in Rock Art. *Rapa Nui Journal* 3.

Millerstrom, S.
1990 Rock art of Marquesas Islands, French Polynesia: A Case Study of Hatiheu Valley, Nuku Hiva. Unpublished M.A. thesis, Department of Anthropology, San Francisco State University.

Millerstrom, S.
1991 Field Report from Tubuai and Ra'ivavae. Département d'Archéologie. Centre Polynésien des Sciences Humaines, Te Anavaharau. Tahiti.

Millerstrom, S.
1992 Report on the Marquesas Islands Rock Art Project. *The Journal of the Pacific Arts Association, Pacific Arts*,19–25.

Millerstrom, S.
1993 Archaeological Field Reports from the Hatiheu Valley, Marquesas Islands. On file at the Département d'Archéologie. Centre Polynésien des Sciences Humaines, Te Anavaharau. Tahiti.

Millerstrom, S.
1995 Archaeological Field Reports from the Western Side of Hatiheu Valley, Marquesas Islands. On file at the Département d'Archéologie. Centre Polynésien des Sciences Humaines, Te Anavaharau. Tahiti

Millerstrom, S.
1996a Ceremonial Complexes and Taro Production on the North Coast of Nuku Hiva, Marquesas Islands, Eastern French Polynesia. Paper presented at the 61st Annual Meeting of the Society for American Archaeology, New Orleans.

Millerstrom, S.
1996b *The Tefa'arahi Rock Art Site, Mo'orea. Département d'Archéologie.* Centre Polynésien des Sciences Humaines, Te Anavaharau. Tahiti.

Millerstrom, S.
1997a Carved and Painted Rock Images in the Marquesas Islands, French Polynesia. *Archaeology in Oceania* 32(3):181–196.

Millerstrom, S.
1997b Archaeological Art; Research Project on Mo'orea (French Polynesia). Report to Dean of the Graduate School and the Richard Gump Research Station (No. 54), University of California, Berkeley.

Millerstrom, S.
2002 Facts and Fantasies: The Archaeology of the Marquesan Dog. In *Behavior Behind Bones*, edited by S. Jones O'Day, Wim Van Neer and A. Ervynck, pp. 144–152. Oxbow Books. UK.

Millerstrom, S.
2003a Rock Art Research in Polynesia 1995–1999. In *Rock Art Studies, News of the World 2*, edited by P. G. Bahn and A. Fossati, pp. 145–164. Oxford, UK: Oxbow Books.

Millerstrom, S.
2003b Ritual Architecture in the Northern Marquesas Archipelago. *Bilan de la recherche archéologique en Polynésie francaise 2001–2002. No 2 dossier d'Archéologie polynésienne*, pp. 103–110. Ministére de la Culture de Polynésie francaise Service de la Culture et Patrimoine. Tahiti.

Millerstrom, S.
2003c Gravures rupestres et archéologie de l'habitat de Hatiheu à Nuku Hiva (îles Marquises, Polynésie française). Cahiers du Patrimoine 6 [Archéologie]. Tahiti: Service de la Culture et du Patrimoine. Ministère de la Culture de Polynésie française.

Millerstrom, S.

2005a Archaeological Research on Ritual Architecture in the Northern Marquesas Islands 2003. *Bilan de la recherche archéologique en Polynésie francaise 2003–2004. No 4 dossier d'Archéologie polynésienne*, pp. 117–126. Ministére de la Culture de Polynésie francaise Service de la Culture et Patrimoine. Tahiti.

Millerstrom, S.

2005b Avholdte Forfedre in Stein (In Norwegian). www.sydhav.no/artikler/sidsel_tiki.htm

Millerstrom, S.

2006a Ritual and domestic Architecture, Sacred Places and Images; Archaeology in the Marquesan Archipelago, French Polynesia. In *Archaeology of Oceania; Australia and the Pacific*, edited by I. Lilley, pp. 284–301. Oxford: Blackwell Publishing.

Millerstrom, S.

2006b What is New in Polynesian Rock Art? In *Rock Art Studies, News of the World 3*, edited by P. G. Bahn, N. Franklin, and M. Strecker, pp. 213–225. Oxford, UK: Oxbow Books.

Millerstrom, S.

2006c The Nu'u petroglyphs Analyzed Within a Polynesian Settlement Landscape. Paper presented at the Society for Hawai'ian Archaeology, Maui, Hawaii. University of Hawai'i-Maui Community College.

Millerstrom, S.

2012 Polynesian rock art research, 2005–2009. In *Rock Art Studies, News of the World IV*, edited by P. G. Bahn and N. Franklin, M. Strecker, pp. 235–243. Oxford, UK: Oxbow Books.

Millerstrom, S., and E. Edwards

1998 Stone Sculptures of the Marquesas Islands (French Polynesia). In *Easter Island in Pacific Context, South Seas Symposium*, edited by C. M. Stevenson, G. Lee, and F. J. Morin, pp. 55–62. Proceedings of the Fourth International Conference on Easter Island and East Polynesia. University of New Mexico, Easter Island Foundation.

Millerstrom, S., and H. Baumgartner

1998 Archaeological Art on Mo'orea, French Polynesia: An Overview. *Rapa Nui Journal* 12(2)35–39.

Millerstrom, S., and R. W. Rogers

2005 Sailing Ship and Dog-Poni: Historic Petroglyphs on Nuku Hiva, The Marquesas Islands. *Rapa Nui Journal* 19(2):120–126.

Millerstrom, S., and T. Allen

2006 Carved Images and Punctured Skins; Rock Carvings and Tattoos in the Marquesas Islands, Eastern Polynesia. In 1994 IRAC Proceedings, Rock Art-World Heritage, edited by G. Lee. *American Indian Rock Art*, 21:131–138. American Rock Art Association.

Millerstrom, S., and J. H. Coil

2008 Pre-contact Arboriculture and Vegetation in the Marquesas Islands, French Polynesia; Charcoal Identification and Radiocarbon Dates from Hatiheu Valley, Nuku Hiva. *Asian Perspectives* 47(2):330–351.

Millerstrom, S., and M. Cruz Berrocal

2009 Fijian Rock Art Sites Revisited. *Rapa Nui Journal* 23(1):40–47.

Millerstrom, S. and M. Cruz Berrocal

2010 Fiji Revisited: A Field report of the Rock Art Survey Project, 2009. *Rapa Nui Journal*, 24(2):27–35.

Millerstrom, S., and H. Baumgartner-Lesage.
Forthcoming. Petroglyphs of the Society Islands and How They are Situated Within the Polynesian Rock Art Repertoire. Accepted by The e-book project Encyclopaedia of Historiography Africa, America, Asia (EHA), Paris.

Millerstrom, S., and P. V. Kirch
2002 History on Stones: A Newly-discovered Petroglyph Site at Kahikinui, Maui. *Hawaiian Archaeology, Society for Hawaiian Archaeology* 8:3–12.

Millerstrom, S., and P. V. Kirch
2004 Petroglyphs of Kahikinui, Maui, Hawai'ian Islands: Rock Images Within a Polynesian Settlement Landscape. Cambridge: *Proceedings of the Prehistoric Society* 70:107–127.

Mills, G.
1971 Art: An Introduction to Qualitative Anthropology. In Anthropology and Art, edited by C. M. Otten, pp. 66-92. Austin: University of Texas Press.

Molle, G.
2011 Ua Huka, une ile dans l'Histoire. Histoire pré- et post-européenne d'une société marquisienne. Thèse de Doctorat de l'Université de la Polynésie française, 2 vols.

Molle, G.
2013 Archéologie et function rituelle de la maison tapu: Les poteaux anthropomorhes sculptés des îles Marquises. *Bulletin de la Sociéte des Etudes Océaniennes* 329:72–103.

Molle, G., and E. Conte
2015 Nuancing the Marquesan Post-contact Demographic Decline: An Archaeological and Historical Case Study on Ua Huka Island. *The Journal of Pacific History* 50(3):253–274.

Monnin, J., and C. Sand
2004 Kibo, le serment grave. Essai de synthése sur les pétroglyphes calédoniens. Noumea: Les Cahiers de l'Archélogie en Nouvelle-Calédonie 16. Département Archéologie, Service des Musées et du Patrimoine.

Naroll, R.
1962 Floor Area and Settlement Population. *American Antiquity* 27:587–589.

Neich, R., and M. Pendergrast
1997 *Traditional Tapa Textiles of the Pacific*. New York: Thames and Hudson.

Oliver, D. L.
1975 *Ancient Tahitian Society*. Vol I. Hololulu: The University Press of Hawaii.

O'Regan, G.
1997 Te Rūnanga o Ngāi Tahu. Ngai Tahu Development Corporation website, New Zealand. www.ngaitahu.iwi.nz (accessed May, 2009).

O'Regan, G.
2003 The History and Future of New Zealand Maori Rock Art – A Tribal Perspective. In *Before Farming: The Archaeology and Anthropology of Hunter-Gatherers* pp. 1–9. University of Bristol, Center for Human Evolutionary Research. Bristol: England.

O'Regan, G.
2006 Regaining Authority: Setting the Agenda in Maori Heritage through the control of shaping of data. *Public History Review* 13:95–107.

O'Regan, G.
2007 Wahi Tapu and Rock Art: An Intra-site Analysis of Two Southern Maori Rock Shelters. Unpublished M.A. thesis. Department of Anthropology, University of Auckland.

O'Regan, G.
2008 The Shifting Place of Ngai Tahu Rock Art. In *Islands of Inquiry; Colonisation, Seafaring and the Archaeology of Maritime Landscapes*, edited by G. Clark, F. Leach, and S. O'Connor, pp. 411–422. Terra Australis 29. Canberra: Australian National University Press.

O'Reilly, P.
1954 Abri sous roche a gravures, Nailou, district de Cakaudrove, Vanua Levu. *Journal de la Société des Océanistes* X(10):178–79.

ORSTOM
1993 Atlas de la Polynésie Française. Editions de L'Orstom. Paris.

Ottino, P.
1985 Un site ancien aux îles Marquises: l'abrisous-roche d'Anapua, à Ua Pou. *Journal de la Société des Océanistes* 81:33–37.

Ottino, P.
1990 L'habitat des anciens Marquisiens: architecture des maisons, évolution et symbolisme des formes. *Journal de la Sociéte des Océanistes* 90:3–15. Paris: Musée de l'Homme.

Ottino, P.
1992 Anapua: Abri-sous-roche de pécheurs; Etude des hameçons. *Journal de la Sociététe des Océanistes* 94:57–79; 95:201–226.

Ottino-Garanger, P.
1996 Archéologie et restauration a Hiva Oa: le "me'ae" Iipona de Puamau, aux îles Marquises. In *Mémoire de pierre, mémoire d'homme; Tradition et archéologie en Océanie*, edited by M. Julien, M. Orliac, C. Orliac, B. Gérard, A. Lavondés, J. Lavondés, and C. Robineau. Paris: Publications de la Sorbonne.

Ottino, P.
1998 Carnet de terrain; le *Tohua* Tahakia à Hatiheu, Nuku Hiva, îles Marquises. *Bulletin de la Societe des Etudes Oceaniennes* 276:52–65.

Ottino P.
2005 Resherches sur Kamuihei-Teiipoka, vallée de Hatiheu, Nuku Hiva, Iles Marquisess. *Bilan de la recherche archéologique en Polynésie Française 2003–2004*, pp. 127–136. Service de la Culture et du Patrimoine, Punaauia, Tahiti.

Ottino-Garanger, P.
2006. Archéologie chez les Taipi. Hatiheu, un projet partagé aux îles Marquises. Au vent des iles editions Tahiti.

Ottino-Garanger, P., F. Valentin, F. Guiot
2003 Researches à Hatiheu, ile de Nuku Hva, archipel des Marquises. Archéologie identité et développment. Bilan de la recherche archéologique en Polynésie française, 2001–2002. *Dossier d'Archéologie polynésienne*, No 2, pp. 79–100. Ministére de la culture de Polynésie française, Service de la culture et du patrimoine, Tahiti.

Ottino-Garanger, P., et M. N. Ottino-Garanger
1998 *Le Tatouage aux Iles Marquises. Te Patu Tiki*. Edited by Ch. Gleizal. Tahiti.

Ouzman, S.

1998 Towards a Mindscape of Landscape: Rock-Art as Expression of World-Understanding. In *The Archaeology of Rock-Art*, edited by C. Chippindale and P. C. Taçon, pp. 30–41. Cambridge: Cambridge University Press.

Pages, J.

1995 Voyage aux iles Marquises conference de Karl von den Steinen du 3 décembre 1898 a Berlin (premiere partie). *Bulletin de la Société des Etudes Océaniennes* 268:55–62.

Pages, J.

1996 Voyage aux iles Marquises conference de Karl von den Steinen du 3 décembre 1898 a Berlin (2éme partie). *Bulletin de la Société des Etudes Océaniennes* 269–270:2–17.

Paine, R. W.

1929 Some Rock Painting in Fiji. *Man* 29(109):148–151.

Palmer, B., and F. Clunie

1970 Rock Paintings and Engravings in Fiji. *Fiji Museum Educational Series* 1.

Panoff, M.

1995 *Trésors des îles Marquises*. Paris: Musée de l'Homme.

Parson, J. R.

1972 Archaeological Settlement Patterns. *Annual Review and Anthropology* (1):127–151.

Parry, J. T., and D. Watling

1988 Petroglyphs at Maqere, Tavua, Vitilevu. *Domodomo* 6(1–4):106–111.

Pechberty, D.

1996 *La vie aux îles Marquises; Récits de missionnaires (1797–1842)*. Vol. 6. Uvea, Wallis: Te Fenua Foou.

Petard, P.

1986 [1958] *Plantes Utiles de Polynésie. Raau Tahiti*. Papeete: Haere Po No Tahiti.

Phillipps, W. J.

1951 Rock Spirals of Fiji. *Journal of the Polynesian Society* 60(1):51–52.

Porter, C. D.

1970 [1822] *Journal of a Cruise made in the Pacific Ocean by David Porter, in the United States Frigate Essex, in the years 1812, 1813, and 1814*. Vol. II. New York: Wiley and Halsted.

Radcliff-Brown A. R.

1952 *Structure and Function in Primitive Society*. New York: The Free Press.

Radiguet, M.

1860 *Les derniers sauvages: la vie et les moeurs aux îles Marquises (1842–1859)*. Paris: Hachette.

Ragone, D.

1991 *Ethnobotany of Breadfruit in Polynesia. Islands, Plants, and Polynesians*. Portland: Dioscorides Press.

Rallu, J.-L.

1990 *Les Populations Océaniennes aux XIXe et XXe Siécles. Travaux et Documentes Cahier No 128*. Paris: Institut National d'Études Démographiques.

Ramirez-Aliaga, J. M.

2011 The Mapuche Connection, In *Polynesians in America; Pre-Columbian Contacts with the New World*, edited by T. L. Jones, A. A. Storey, E. A. Matisoo-Smith, and J. M. Ramirez-Aliaga, pp. 95–138. AltaMira Press. New York.

Renfrew, C., and P. Bahn
1991 *Archaeology: Theories, Methods, and Practice*. New York: Thames and Hudson.

Roberts, S.
1929 *Population Problems of the Pacific*. London: George Routledge and Sons. Ltd.

Rolett, B. V.
1986 Turtles, Priests, and the Afterworld: A Study in the Iconographic Interpretation of Polynesian Petroglyphs. In *Island Societies: Archaeological Approaches to Evolution and Transformation*, edited by P. V. Kirch, pp. 78–87. Cambridge, England: Cambridge University Press.

Rolett, B. V.
1992 Faunal extinctions and depletions linked with prehistory and environmental change in the Marquesas Islands (French Polynesia). *Journal of the Polynesian Society*, 101:86–94.

Rolett, B. V.
1993 Marquesan Prehistory and the Origins of East Polynesian Culture. *Journal de la Société des Océanistes* 96:29–47.

Rolett, B. V.
1996 Colonisation and Cultural Change in the Marquesas. In *Oceanic Culture History: Essays in Honour of Roger Green*, edited by J. Davidson, G. Irwin, F. Leach, A. Pawley, and D. Brown, pp. 531–540. Dunedin: New Zealand Journal of Archaeology Special Publication.

Rolett, B. V.
1998 *Hanamiai: Changing Subsistence and Ecology in the Prehistory of Tahuata (Marquesas Islands, French Polynesia)*. Yale University Publications in Anthropology No. 84. New Haven, Connecticut: Department of Anthropology, Yale University.

Rolett, B. V.
1997 *Marquesan Prehistory and the East Polynesian Homeland*. Fourth Interim Report: Marquesan Chiefdoms and the Emergence of Monumental Architecture.

Rolett, B. V.
2010 Marquesan Monumental Architecture: Blurred Boundaries in the Distinction Between Religious and Residential Sites. *Archaeology in Oceania* 45:94–102.

Rolett, B. V., and E. Conte
1995 Renewed Investigation of the Ha'atuatua Dune (Nukuhiva, Marquesas Islands): A Key Site in Polynesian Prehistory. *Journal of the Polynesian Society* 104(2):195–228.

Rollin, L.
1974 *Moeures et coutumes des anciens Maoris des îles Marquises*. Papeete: Stepolde.

Roquefeuil, M. C. de
1823 *Voyage Round the World, Between the Years 1816–1819*. London: Richard Phillips and Co.

Rosenman, H. (editor and translator)
1992 *Two Voyages to the South Seas, Captain Jules S.-C. Dumont D'Urville*. Honolulu: University of Hawaii Press.

Routledge, K.
1919 *The Mystery of Easter Island*. Hazel, Watson, and Viney. London.

Sahlins, M. D.
1958 *Social Stratification in Polynesia*. Seattle: University of Washington Press.

Sahlins, M. D.
1992 Historical Ethnography. In *Anahulu: The Anthropology of History in the Kingsom of Hawaii*, Vol. 1, edited by of P. V. Kirch and M. Sahlins. Chicago: University of Chicago Press.

Sand, C., H. Valladas, H. Cachier, N. Tisnerat-Laborde, M. Arnold, J. Bole, and A. Ouetcho
2006 Oceanic Rock Art: First Direct Dating of Prehistoric Stencils and Paintings from New Caledonia. *Antiquity* 80(309):523–529.

Saquet, J.-L.
1990 *Le livre de Tahiti; te fenua*. Singapour: Editions Avant et Apres.

Shanks, M., C. Tilley
1987 *Social Theory and Archaeology*. Cambridge: Polity Press.

Shapiro, H. L., and R. C. Suggs
1959 New Dates for Polynesian Prehistory. *Man* 59:12–13.

Sharp, D. S., J. G. Kahn, C. M. Polito, and P. V. Kirch
2010 Rapid Evolution of Ritual Architecture in Central Polynesia Indicated by Precise 230Th/U Coral Dating. *Proceedings of the National Academies of Sciences*, 107(30):13234–13239 (available at www.pnas.org/cgi/doi/10.1073/pnas.1005063107).

Schütz, A. J.
1994 *The Voices of Eden: A History of Hawai'ian Language Studies*. Honolulu: University of Hawaii Press.

Silverthorne, H.
1936 *Society Islands Pounders*. Honolulu: Bernice P. Bishop Museum Occasional Papers. 11(17):1–17.

Simmons, D. R.
1977 *Regional Styles in Maori Woodcarving*. A Series of Lectures Presented to the Center for Continuing Education. Mimeographed. Auckland: Auckland Institute and Museum.

Sinoto, Y. H., and M. Kellum
1965 *Preliminary Report on Excavation in the Marquesas Islands, French Polynesia*. Mimeographed Report to the National Science Foundation. Bernice P. Bishop Museum.

Sinoto, Y. H.
1966 A Tentative Prehistoric Cultural Sequence in the Northern Marquesas Islands, French Polynesia. *Journal of the Polynesian Society* 75:287–303.

Sinoto, Y. H.
1967 Artifacts from Excavated Sites in the Hawai'ian, Marquesas, and Society Islands: A Comparative Study. In *Polynesian Culture History*, edited by G. A. Highland, R. W. Force, A. Howard, M. Kelly, and Y. H. Sinoto, pp. 341–61. Bernice P. Bishop Museum Special Publication 56. Honolulu.

Sinoto, Y. H.
1968 Position of the Marquesas Islands in East Polynesian Prehistory. In *Prehistoric Culture in Oceania*, edited by I. Yawata and Y. H. Sinoto, pp. 111–18. Honolulu: Bernice P. Bishop Museum Press.

Sinoto, Y. H.
1970 An Archaeological Based Assessment of the Marquesas as a Dispersal Center in East Polynesia. In *Studies in Oceanic Culture History*. R. C. Green and M. Kelley editors. Vol 1, pp. 105–132. Pacific Anthropological Records. Honolulu: Bernice P. Bishop Museum.

Schoon, T.
1947 New Zealand's Oldest Art Galleries. *New Zealand Listener* September 12:6–7.

Sinoto, Y. H.
1979 The Marquesas. In *The Prehistory of Polynesia*, edited by J. D. Jennings, pp. 110–134. Cambridge, MA: Harvard University Press.

Sinoto, Y. H.
1983 An Analysis of Polynesian Migrations Based on the Archaeological Assessments. *Journal de la Société des Océanistes* 39:57–67.

Sinoto, Y. H.
1996 Tracing Human Movement in East Polynesia: A Discussion of Selected Diagnostic Artifact Types. In *Mémoire de Pierre, Mémoire D'Homme; Tradition et Archéologie en Océanie. Hommage a José Garanger*, edited by M. Julien, M. Orliac, C. Orliac, B. Gérard, A. Lavondés, H. Lavondés, and C. Robineau, pp. 131–152. Paris: Publications de la Sorbonne.

Smith, C. S.
1965 The Burial Complex on the Island of Rapa Iti. In *Reports of the Norwegian Archaeological Expedition to Easter Island and the East Pacific*, Vol. 2, edited by T. Heyerdahl and E. N. Ferdon Jr., pp. 89–95. Stockholm, Sweden: Forum Publishing House.

Snow, P. A.
1950 Rock Carvings in Fiji. Transactions and Proceedings of the Fiji Society 4(3):71–82.

Söderström, J.
1937 Some Notes on Poi and Other Preserved Vegetables in the Pacific. *Ethnos* 4:235–242. Ethnographic Museum. Stockholm, Sweden.

Spriggs, M.
1981 Vegetable Kingdoms: Taro Irrigation and Pacific Prehistory. Unpublished Ph. D. dissertation, School of Archaeology and Anthropology, Australian National University. Canberra.

Spriggs, M.
1984 The Lapita Cultural Complex: Origins Distribution, Contemporaries, and Successors. *Journal of Pacific History* 19:202–23.

Spriggs, M.
1986 Landscape, Land Use, and Political Transformation in Southern Melanesia. In *Island Societies: Archaeological Approaches to Evolution and Transformation*, edited by P. V. Kirch, pp. 6–19. Cambridge: Cambridge University Press.

Spriggs, M., and A. Anderson
1993 Late Colonization of East Polynesia. *Antiquity* 67:200–217.

Stasack, E., R. I. Dorn, and G. Lee
1996 First Direct C14 Ages on Hawaiian Petroglyphs. *Asian Perspectives* 35(1):51–72.

Stasack, E. A., and D. S. Stasack
1997 *Petroglyph Recording Project Cave HV-76* (State site no. 50-10-62-20, 750). Keauhou, Puna-Ka'u Historic District Hawai'i Volcanoes National Park, Hawai'i.

Stasack, E. A., and D. S. Stasack
1999 *Lae'apuki/Panau Iki [Transect 2], Rock Art Recording Project Report.* Lae'apuki/Panau Iki Ahupua'a,

Puna District, Hawaii Volcanoes National Park.

Stasack, E. A., and D. S. Stasack
2005 Recent Rock Art Research on the Island of Hawai'i. Unpublished report.

Steadman, D. W.
1989 Extinction of Birds in Eastern Polynesia: A Review of the Record, and Comparisons with Other Pacific Island Groups. *Journal of Archaeological Science* 16:177–205.

Steadman, D. W., and B. Rolett
1986 A Chronostratigraphic Analysis of Landbird Extinction on Tahuata, Marquesas Islands. *Journal of Archaeological Science* 23:81–94.

Stewart, C. S.
1832 *A Visit to the South Seas in the U. S. Ship "Vincennes," During the Years 1829 and 1830*. London: Fisher, Son and Jackson, Newgate St.

Steinen, K. von den
1969 [1925] *Die Marquesaner und ihre Kunst. Tatauierung*. Berlin. Reprinted by Hacker Art Books, New York.

Steinen, K. von den.
1969(II) [1928] *Die Marquesaner und ihre Kunst. Plastik*. Berlin. Reprinted by Hacker Art Books, New York.

Steinen, K. von den
1969(III) [1928] Die *Marquesaner und ihre Kunst. Die Sammlungen*. Berlin. Reprinted by Hacker Art Books, New York.

Stokes, J. F. G.
1921 Field Report from Rapa Iti. Manuscript on file at the Bernice P. Bishop Museum, Honolulu.

Stolpe, H.
1890 Utvecklingsforeteelser i Naturfolkens Ornamentik; *Ethnografisk Undersokning*. Ymer II:193–250.

Stolpe, H.
1927 *Collected Essays in Ornamental Art*. Stockholm: Aftonbladets Tryckeri.

Storey, A. A., D. Quiroz, and E. A. Matisoo-Smith
2011 A Reappraisal of the Evidence for Pre-Columbian Introduction of Chickens to the Americas. In *Polynesians in America; Pre-Columbian Contacts with the New World*, edited by T. L. Jones, A. A. Storey, E. A. Matisoo-Smith, and J. M. Ramirez-Aliaga, pp 139–170. AltaMira Press. New York.

Stringer Rowe, G. (editor)
1982 *Fiji and the Fijians: The Islands and their Inhabitants*. Vol. 1. Suva: Fiji Museum.

Stuiver, M., Polach H. A.
1977 Discussion: reporting of ^{14}C data. *Radiocarbon* 19(3):355–363.

Stuiver, M., Polach H. A.
1993 High-Precision decadal calibration of the radiocarbon time scale. AD 1950–6000 BC. In Long, A. and R. S. Kra, editors, *Radiocarbon* 35(1):35–65.

Suggs, R. C.
1961 Archaeology of Nuku Hiva, Marquesas Islands, French Polynesia. *Anthropological Papers of the American Museum of Natural History*. New York.

Tautain, M.
1896a Sur le tatouage aux Iles Marquises. *L'Anthropologie* 7:247–288. Paris.

Tautain, M.
1896b Sur l'anthropophagie et les sacrifices humains aux îles Marquises. *L'Anthropologie* 7:443–452. Paris.

Tautain, M.
1896c Notes sur l'ethnographie des iles Marquises. *L'Anthropologie* 7:543–552. Paris.

Tautain, M.
1897 Notes sur les constructions et monuments des Marquises. L'Anthropologie 8 (1):538–558 et 667–678. Paris.

Tautain, M.
1898 Etude sur la dépopulation de l'Archipel des Marquises. L'Anthropologie 9(1):298–318, 418–436. Paris.

Te Ara. The Encycolpedia of New Zealand
Maori rock art – nga toi ana, pp 3. www.teara.gov.nz (accessed August 17, 2016).

Terrell, J.
1982 Joseph Kabris and His Notes on the Marquesas. *Journal of Pacific History* 17:101–112. Canteberra, Australia.

Terrell, J. (editor)
1988 *Von den Steinen's Marquesan Myths*. Canberra, Australia: Target Oceania/Journal of Pacific History. Canteberra: Australia.

Thomas, N.
1990 *Marquesan Societies: Inequality and Political Transformation in Eastern Polynesia*. Clarendon Press: Oxford.

Thomas, N.
1995 *Oceanic Art*. London: Thames and Hudson Ltd.

Thompson, P.
1989 *Maori Rock Art: The Ink That Will Stand Forever*. GP Books. Wellington.

Thornton, A.
1989 Some Reflections on Traditional Maori Carving. *Journal of the Polynesian Society* 98:147–166.

Tilley, C.
1991 *Material Culture and Text: The Art of Ambiguity*. New York: Routledge.

Torrey, W.
1848 *The Life and Adventures of William Torrey*. Boston: Press of A. J. Wright.

Trigger, B. G.
1967a The Determinants of Settlement Patterns. In *Settlement, Society, and Archaeology*, e, pp. 53–78. Palo Alto: National Press.

Trigger, B. G.
1967b Settlement Archaeology: Its Goals and Promise. *American Antiquity* 32:149–60.

Trigger, B. G.
1998 Reflections on Encounters with Archaeology. In *Bringing Back the Past: Historical Perspectives on Canadian Archaeology*, edited by P. J. Smith and D. Mitchell, pp. 77–92. Hull, Quebec: Canadian Museum of Civilization.

Tuggle, H. and P. Griffin (editors)
1973 Lapakahi, Hawaii: Archaeological Studies. *Asian and Pacific Archaeology Series 5*. Social Science Research Institute, University of Hawai'i.

Trotter, M. and B. McCulloch
1971 *Prehistoric Rock Art of New Zealand*. Wellington: Reed Books.

Valenziani, C.
1940 *Renaissance démographique en Océanie Française*. Rome: Tipografia Conzortio Nazionale.

Valeri, V.
1985 *Kingship and Sacrifice: Ritual and Society in Ancient Hawaii*. Chicago: The University of Chicago Press.

Van Tilburg, J. A.
1994 *Easter Island, Archaeology, Ecology, and Culture*. Washington, D.C.: Smithsonian Institution Press.

Vogan, A. J.
1937 Recent Archaeological Discoveries in the Western Pacific. *Journal of the Polynesian Society* 46(3):99–104.

Waller, S. J.
2011 *Rock Art Acoustics*. www. sites.google.com/site/rockartacoustics/.

Wallin. P.
1993 *Ceremonial Stone Structures: The Archaeology and Ethnohistory of the Marae Complex in the Society Islands, French Polynesia*. Uppsala, Sweden: Societas Archaeologica Upsaliensis.

Walsh, D. S. and B. Biggs
1966 *Proto-Polynesian Word List I*. Linguistic Society of New Zealand.

Weisler, M. I.
1998 Hard Evidence for Prehistoric Interaction in Polynesia. *Current Anthropology* 39:521–32.

Weisler, M. I.
2002 Centrality and the Collapse of Long-Distance Voyaging in East Polynesia. In *Geochemical Evidence for Trade and Exchange*, edited by M. D. Glascock, pp. 157–73. Westport CT: Bergin and Garvey.

Weisler, M. I., and R. Green
2011 Rethinking the Chronology of Colonization of Southeast Polynesia. In *Polynesians in America; Pre-Columbian Contacts with the New World*, edited by T. L. Jones, A. A. Storey, E. A. Matisoo-Smith, and J. M. Ramirez-Aliaga, pp. 223–246. New York: AltaMira Press.

Weisler, M. I., and P. V. Kirch
1985 The Structure of Settlement Space in a Polynesian Chiefdom: Kawela, Moloka'i, Hawaiian Islands. *New Zealand Journal of Archaeology* 7:129–158.

Weisler, M. I., E. Conte, and P. V. Kirch
2004 Material Culture and Geochemical Sourcing of Basalt Artifacts. In *Archaeological Investigations in the Mangareva Islands (Gambier Archipelago), French Polynesia*, edited by E. Conte and P. V. Kirch, pp. 128–148. Contribution Number 62, Archaeological Research Facility. University of California, Berkeley.

Wentworth, C. K.
1922 A Scale of Grade and Class Terms for Clastic Sediments. *Journal of Geology*. 30(5):377–392.

Whitley, D. S., and L. L. Loendorf (editors)
1994 *New Light on Old Art: Recent Advances in Hunter-Gatherer Rock Art Research.* Monograph 36. Los Angeles: Institute of Archaeology, University of California.

Willey, G. R.
1953 *Prehistoric Settlement Patterns in the Viru Valley, Peru.* Bulletin 155. Washington, D.C.: Bureau of American Ethnology.

Williams, G.
2008 North Otago, South Cantebury Maori rock art records history. www.odt.co.nz (accessed August 2016).

Wilson, J.
1997 [1797] *A Missionary Voyage in the Ship Duff.* Tahiti: Société des Etudes Océaniennes, Haere Po No Tahiti.

Wobst, H. M.
1977 Stylistic Behavior and Information Exchange. In *For the Director: Research Essays in Honor of James B. Griffin,* edited by C. E. Cleland, pp. 317–342. Ann Arbor: Museum of Anthropology Anthropological Paper 61. Museum of Anthropology, University of Michigan.

Wright, A. M.
2014 *Religion on the Rocks: Hohokam Rock Art, Ritual Practice, and Social Transformation.* Salt Lake City: University of Utah Press.

Yen, D. E.
1974 *The Sweet Potato and Oceania. An Essay in Ethnobotany.* Bernice P. Bishop Museum Bulletin 236. Honolulu: Bishop Museum Press.

Yen, D. E.
1975 Indigenous Food Processing in Oceania. In *Gastronomy; The Anthropology of Food and Food Habits,* edited by M. Arnott, pp 147–168. Chicago: Aldine.

Yen, D. E.
1991 Polynesian Cultigens and Cultivars: The Questions of Origin. In *Islands, Plants, and Polynesians: An Introduction to Polynesian Ethnobotany,* edited by P. A. Cox and S. A. Banack, pp. 67–95. Portland: Dioscorides Press.

About the Author

Sidsel Nørgaard Millerstrom, born in Norway, earned her BA at University of California, Berkeley and her MA at San Francisco State University, both with a major in Anthropology. Her Ph.D., also in Anthropology, was awarded in 2001 at University of California, Berkeley. Since 1984 she has focused her research on archaeological art and architecture in the Marquesas Islands, French Polynesia. At the center of her Ph.D. research was the interrelationship of anthropological sculptures and architecture, the spatial arrangement between petroglyphs, pictographs and architecture in the landscape, and the images' sociopolitical and religious cultural significance. While surveying a great numbers of valleys in the Marquesas Islands, her dissertation fieldwork took place in Hatiheu Valley, Nuku Hiva. She has conducted archaeology in various Polynesian islands, as well as rock art research in Fiji.

Her academic background has drawn her in numerous fascinating directions and her quest to see the world has taken her to spectacular archaeological sites. While working part-time for Pacific Legacy and teaching for several years at Diablo Valley College, she also taught anthropology at the University of Social Sciences in Ho Chi Minh City (2003–2009) and later teaching at Thu Da Mot University, Vietnam (2006). Twice (2004 and 2008) she taught anthropology at Semester at Sea, University of Virginia, while circling the world on *M/S Explorer*. Presently she is working for Far Horizons Archaeological and Cultural Trips as a lecturer and tour manager. She regularly present lecturers on *Aranui*, a ship that voyages between Tahiti and the Marquesas Islands. She has authored several scholarly articles. A version of this manuscript was published by the government in Tahiti, 2003.

Index

I

J

U

www.ingramcontent.com/pod-product-compliance
Ingram Content Group UK Ltd.
Pitfield, Milton Keynes, MK11 3LW, UK
UKHW062004041026
15591UKWH00016B/1159

9 780998 246062